Exploring Chemistry with Pyridine Derivatives

Edited by Satyanarayan Pal

Published in London, United Kingdom

Exploring Chemistry with Pyridine Derivatives
http://dx.doi.org/10.5772/intechopen.100792
Edited by Satyanarayan Pal

Contributors
Priyank Purohit, Gaurav Joshi, Meenu Aggrawal, Huseyin Istanbullu, Gulsah Bayraktar, Merve Saylam, Nikolay Borsch, Lilij Ageeva, Nikolay Kuvardin, Edayadulla Naushad, Shankar Thangaraj, Kaushal Kumar Joshi, Lozan Todorov, Irena Kostova, Sunil Junapudi, Yasodha Krishna Janapati, Medidi Srinivas, Sunithasree Cheweti, Bojjibabu Chidipi, Ana M. G. Silva, Andreia Leite, Carla Queirós, Marcus Bond, Adebimpe Adesina

First published in London, United Kingdom, 2023 by IntechOpen
IntechOpen is the global imprint of INTECHOPEN LIMITED, registered in England and Wales, registration number: 11086078, 5 Princes Gate Court, London, SW7 2QJ, United Kingdom

British Library Cataloguing-in-Publication Data
A catalogue record for this book is available from the British Library

Additional hard and PDF copies can be obtained from orders@intechopen.com

Exploring Chemistry with Pyridine Derivatives
Edited by Satyanarayan Pal
p. cm.
Print ISBN 978-1-80356-662-7
Online ISBN 978-1-80356-663-4
eBook (PDF) ISBN 978-1-80356-664-1

Meet the editor

Dr. Satyanarayan Pal is an associate professor in the Department of Chemistry, at Utkal University, India. He obtained his Ph.D. from the University of Hyderabad, India, in 2003. He completed post-doctoral fellowships with the prestigious Japan Society for Promotion of Science (JSPS) Fellowship Program, Nagoya University, Japan, and Brain Korea 21 Program, Seoul National University, South Korea. He has published thirty-four research papers in international journals and written one book chapter. Currently, Dr. Pal is developing Ir(III) and Pt(II) luminescent complexes and studying their bioactive properties.

Contents

Preface

Pyridine-based compounds play an important role in chemical and biological sciences. Proven as vital pharmacophores in drug design, they are currently part of numerous prescribed drugs to treat several critical human illnesses. The tremendous pharmacological activity and medicinal value of pyridine derivatives have attracted the attention of the scientific community.

This book presents recent developments in pyridine derivatives. It includes ten chapters in two sections on "Chemistry of Pyridine Derivatives" and "Applications of Pyridine Derivatives".

In the first section, Chapters 1 and 2 discuss the chemistry of pyridine derivatives and organic reactions of different categories. Chapter 3 describes a reaction involving pyridinium cations with copper halides in the formation of halocuprate complexes with diverse structural features.

In the second section, chapters examine the importance of pyridine-based compounds with documentation in notable fields of chemistry and biology. Applications are noted for both naturally available and synthetic analogues of pyridine compounds applied as drugs for treatment against a myriad of human diseases. Chapters 4–7 describe pyridine-based compounds applied against viral and bacterial infections, cancer, cardiac diseases, and other illnesses. Chapters 8–9 investigate the possibility of developing chemosensors based on pyridyl-based fluorophores and Schiff bases derived from formyl/amine-containing pyridine derivatives. An array of pyridine-functionalized fluorophores obtained from rhodamine, BODIPY, and 1,8-naphthalimide dyes have displayed excellent sensing capability towards various metal cations. Finally, Chapter 10 explores 2(4)-aminopyridines as chelators to platinum group metals such as iridium.

I am very thankful to all scientists who contributed their work to this book. I also wish to thank IntechOpen for giving me the opportunity to edit this volume.

Dr. Satyanarayan Pal
Associate Professor,
Department of Chemistry,
Utkal University,
Bhubaneswar, India

Section 1

Chemistry of Pyridine Derivatives

Chapter 1

Pyridine Nucleus as a Directing Group for Metal-Based C–H Bond Activation

Priyank Purohit, Gaurav Joshi and Meenu Aggarwal

Abstract

Carbon-hydrogen (C–H) bond activation involves a methodology for the construction of carbon-X (C–X) bonds where X can be carbon (C), oxygen (O), or the nitrogen (N), allowing the formation of C–C, C–O, or C–N bonds. Among them, the construction of the C–C bond within the aromatic moiety has remained a bottleneck because the abundance of C–H bonds in aromatic molecules possesses almost similar bond dissociation energies comparable to the C–C bond allowing leading to the poor reactivity and selectivity. Secondly, C–H bonds possess low polarity and thus confer them inertness. Considering this, directing group strategy came into existence, where the coordination ability of the heteroatoms such as O and N atoms within the ring was utilized for the direction of the reaction. The use of the heteroatom for the regioselective C–H bond activation is quite advantageous that could be explored immensely for their functionalization. In this chapter, we have congregated the information and put forth the evidence of C–H activation leading to the C–C bond formation in pyridine and pyridine-containing entities.

Keywords: C–H bond activation, meta directing C–H activation, regioselective C–H activation, pyridine template

1. Introduction

C–H activation or functionalization is a technique of activating and transforming the C–H bond into the C–X bond, allowing the C–C, C–N, or C–O bond construction [1]. Among these, the functionalization to form a C–C bond is widely used [2]. As the aromatic moieties consist of an array of C–C bonds with attached hydrogens (C–H bond), the selective activation of C–H bond is troublesome owing to similar bond dissociation energies to C–C bonds and low polarity of C–H bond [1, 3, 4]. The C–H is a saturated bond possessing only sigma bond, which must be preactivated. Traditionally, coupling or cross-coupling reactions (Suzuki, Heck, etc.) were immensely utilized to form these C–C bonds. However, these reactions confer additional steps to synthetic methodologies, including oxidative addition, reductive elimination, conversion to organic halides, triflates, along with boron or metal-based compounds. The available methods (coupling) that allow preactivation of the C–H bond to facilitate the construction of the C–C bonds (**Figure 1**) [4–6].

Owing to the drawbacks, direct C–H activation is seen as an alternative method possessing a cost-effective and eco-friendly system. The direct C–H activation allows the utilization of numerous transition metals as a catalyst with advantages over the traditional C–H bond activation pathway(s). Metal such as Ru, [7] Pd, [8], and Cu, [9] is frequently used for an efficient C–H activation leading to the C–C functionalization [3, 10]. The stability of the oxidation states of these transition metals remains one of the prerequisites peculiar features in catalyzing the C–C bond formation. Briefly, these transition metals primarily allow the C–H bond cleavage by forming an organometallic complex upon their reaction with the hydrocarbon. The C–H activation is preceded by an initial step that includes sigma and agostic interactions (**Figure 2a**) [11]. These interactions activate a C–H bond primarily by stabilizing metal intermediates possessing high energy and inducing the polarity in the C–H bond, thereby allowing the cleavage to occur. These interactions allow the transfer of electron density from the sigma orbital of C–H bonds to transition metals empty d-orbital. The sigma interaction proceeds via an intermolecular approach while the C–H bond interacts with the metal through is involved in the intramolecular approach in agostic interaction. The agostic complex forms the coordination sphere complex via the interaction of C–H with the metal-ligand. Further, the sigma interactions are considered weak, and the transition state complexes are usually not trapped or isolable [1, 11, 12].

Considering this interaction for preactivation of the C–H bond, the C–H activation proceeds via four effective mechanisms, which are determined by numerous factors, including the nature of metal (early or late transition metal) involved in catalysis , change in oxidation state during bond cleavage of metal, and the type of ligand involved [1, 2, 10]. These central mechanisms for C–H activation include the electrophilic substitution (ES) mechanism (**Figure 2b**) that usually occurs at an electropositive late transition metal complex leading to the formation of a four-membered centered transition state with no change in the oxidation state of the metal involved in the catalysis [13]. ES further does not need the involve lone pair involvement. The recent advancement of ES mechanism advanced mechanism under ES that has been identified includes processes such as ambiphilic metal-ligand activation (AMLA), concerted metallation deprotonation (CMD), electrophilic concerted metallation deprotonation (eCMD), and ligand-to-ligand hydrogen transfer (LLHT). The second mechanism includes oxidative addition (OA) [14]. OA involves the breaking of C–H bond (**Figure 2c**) by low-valent electron-rich metal complexes having neutral ligands (L-type) association. These associated ligands strongly donate the electron, thus

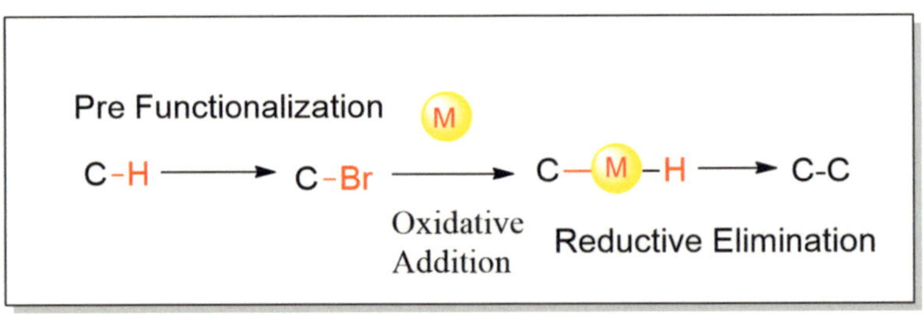

Figure 1.
Classical metal-based C–C bond forming reaction.

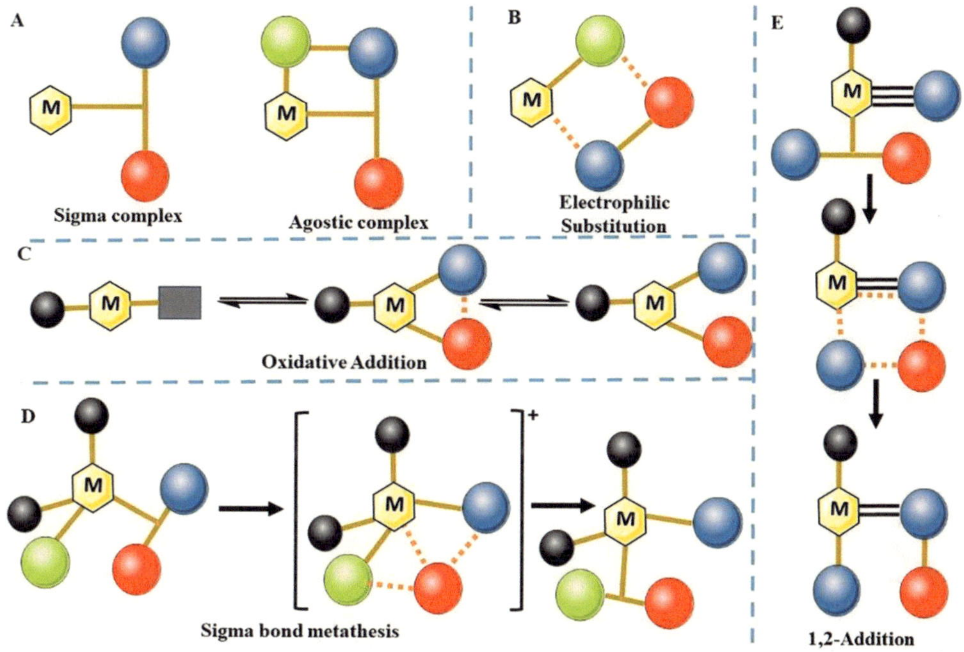

Figure 2.
The illustration depicts **A.** *the preactivation of the C–H bond via sigma and agnostic interaction;* **B.** *electrophilic substitution (ES) mechanism;* **C.** *oxidative addition (OA);* **D.** *sigma bond metathesis (SBM); and* **E.** *1,2-addition mechanism. The* **blue** *ball represents carbon;* **red** *represents hydrogen;* **green** *represents nitrogen or halide, whereas the hexagon represents the metal.*

creating the charge disparity between C–H bond and thereby inducing the enough polarity in the C–H bond for undergoing the activation. The breakage of the C–H bond is associated with an increase in metal formal oxidation state and coordination number by a factor of 2. The third mechanism associated with C–H bond direct activation is sigma bond metathesis (SBM) [15, 16]. This methodology (**Figure 2d**) is limited to metals in early transition series devoid of d-orbital electrons for oxidative addition. This proceeds via the formation of a four-centered transition complex where an H atom (C–H) is transferred to the metal-carbon bond (M-C). This allows the dissociation of the H-atom acceptor from the transition metal complex (M-C). The net change in oxidation state is usually restricted in this mechanism. The fourth mechanism is 1,2-addition [17]. This mechanism (**Figure 2e**) usually involves early transition metals but is associated with C–H activation across multiple bonds. The mechanism proceeds via the addition of H-atom from C–H fragment on a double or triple bond, allowing the reduction of atom or ligand bound to the metal, leading to a new M-C bond formation.

The transition metals in the C–H activation increase the atom economy by reducing the number of functional groups (FG) for making the required bonds. The other advantages include reducing reaction times, synthetic steps, and allowing more greener chemistry. However, the C–H activations offer various advantages, but at the same time, maintaining the regioselectivity due to uncontrolled and unspecific C–H bond activation is troublesome. This has now been omitted chiefly due to the use of the directing group strategy. Various functional-based (**Figure 3**) directing groups are used to activate the inert C–H bonds. Most functional groups have oxygen and

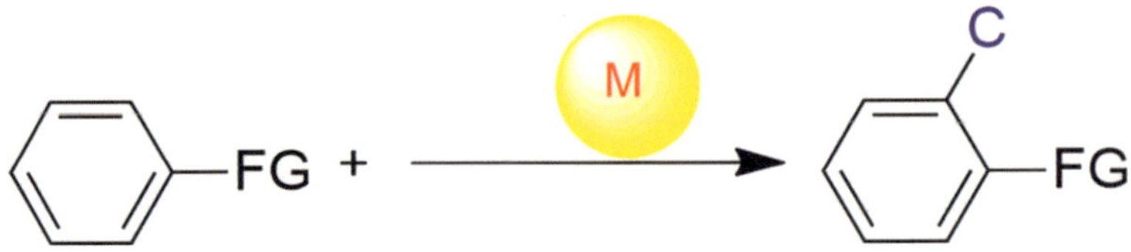

Figure 3.
Functional group-based C–H **bond** *activation.*

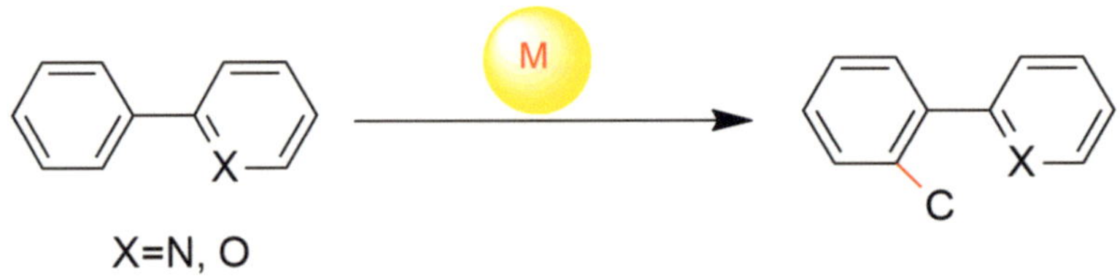

Figure 4.
Heteroatom-based C–H bond activation **with** *pyridine as directing group.*

nitrogen atoms within the core structure such as the amide, sulfonamide, phosphonamide, ester, acid, and other carbonyl-based groups [4]. The specific/coordinating functional group was a prerequisite in all those reported protocols, which was the demerit of those reaction design protocols. However, later the heterocycle-based aromatic ring was found suitable for the regioselective C–H activation. The heteroatoms such as N and, O inside the ring were used by various groups, with a detailed mechanistic investigation. It was utilized for the functionalization of the various medicinally important pharmacophores, such as indole, imidazole, pyridine, pyrimidine, etc., as depicted in **Figure 4** [18].

The chapter therefore is kept forth to discuss the mechanistic insight that includes the discussion on C–H activation in pyridine and pyridine-containing entities. The chapter will provide enough insights to the organic and medicinal chemists to further explore these privileged **heterocycles** ic for their use as pharmaceuticals or diagnostic agents.

2. Pyridine as a directing group

Pyridine, an aromatic compound, possesses uneven electronic distribution on the ring because of heteroatom, which results in the loss of the aromaticity. In comparison with the high aromatic benzene ring, it has less aromaticity because of the presence of the heteroatom, N. The nitrogen atom on the pyridine acts as a donor to bind with metal to form (pyridine)N-metal bond s many complexes, which is the critical factor of the ring to acts as directing group with the metal-based C–H activation. Pyridine provides regioselectivity (**Figure 5**) to the attached aryl group at ortho and meta positions. However, some of the reactions are reported where pyridine makes ortho selective metal complex on its own [19].

The metal and coordinating groups form a cyclic intermediate to get the space between the C–H bond and result in the C–C bond with desired regioselectivity. The pyridine nucleus was also used to synthesize chiral catalyst, using the coordinating capability of nitrogen to and metal with the appropriate direction [20].

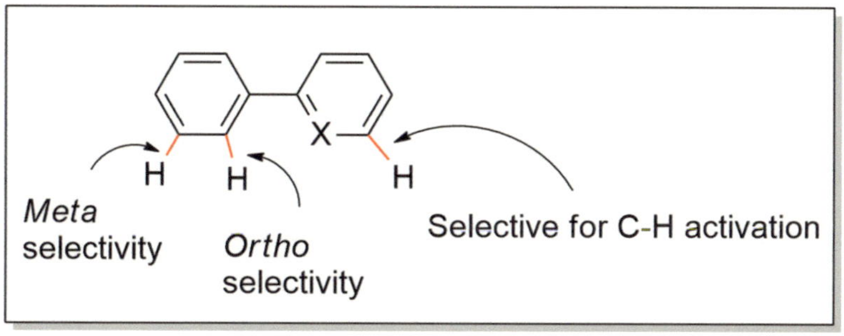

Figure 5.
Regioselectivity of pyridine nucleus.

2.1 *Ortho* C–H bond activation through pyridine directing group

Various other reactions are reported with the 2-aryl pyridine as a directing group for the ortho functionalization. In these reactions, many organometallic catalysts were used. The pyridine nucleus was were a directing group for the functionalization of the 2-aryl group with different functional groups through the metalacyclic system, where the reduction of the metal was the key to the newly constructed bond, as it is depicted the below in **Figure 6**.

Pyridine nucleus-based drugs are an essential class of the heterocycles that possess important medicinal values [21]. The hydrogen bonding capacity of nitrogen atoms because of their non-bonded electron makes them available to make a hydrogen bond with the target amino acids/protein/enzymes. US FDA has approved various pyridine-based nuclei with a very high success rate unlimited successful as the first pyridine-based drug was known as Omeprazole, a widely used drug since 1998 as proton pump inhibitor. Many drugs based on pyridine have been approved later as Netupitant (2014), Abemaciclib (2015), Lorlatinib (2018), Apalutamide (2018), and Ivosidenib (2019) [22–24].

Ortho arylation at the two positions with the metal gains momentum with the attachment of the sensitive functional group such as a halo, ester, cyano, etc. The ortho-substituted reaction protocol was extended with C–O, C–P, and C–S, which claims the directing group capability of the pyridine with various coupling partners. The scope of the pyridine directing group is depicted in the **Figure 7** with limited and important ed examples [25].

Pyridine undergoes substitution with allyl group under the influence of ruthenium catalyst (**Figure 8**) at the C_2 position of the pyridine ring via metal-based C–H activation. However, in the absence of catalyst, electrophilic aromatic substitution

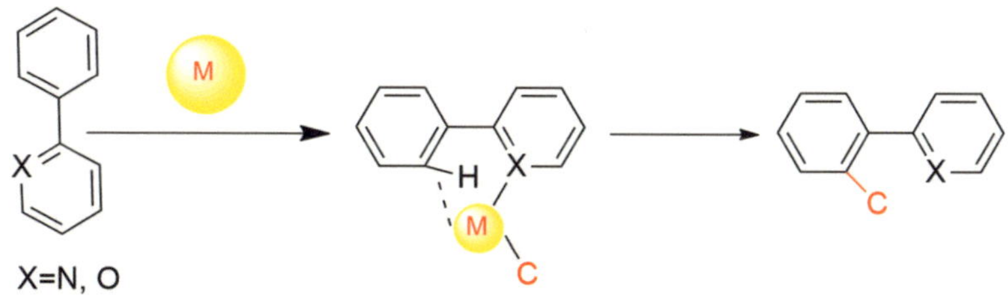

Figure 6.
Metal-based cyclic intermediate with 2-aryl pyridine.

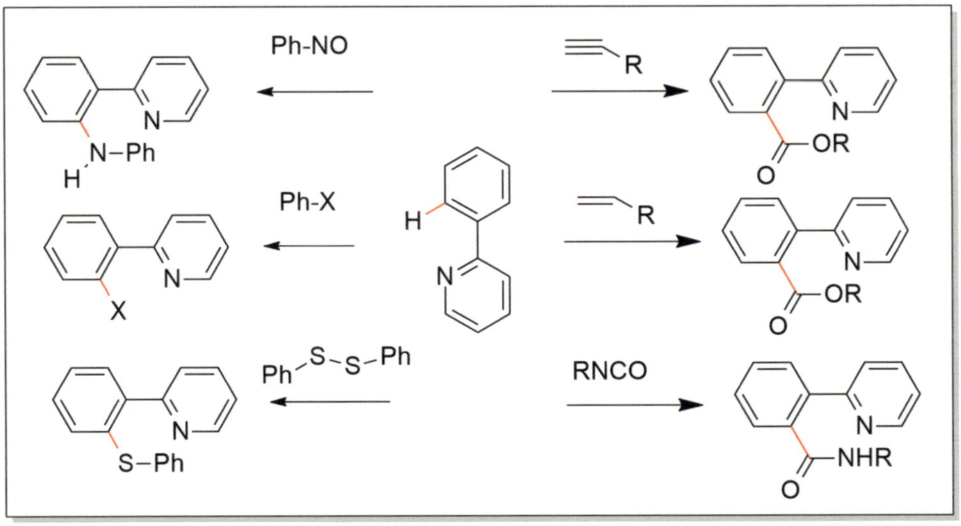

Figure 7.
Metal-based ortho substitution ation of 2-aryl pyridine.

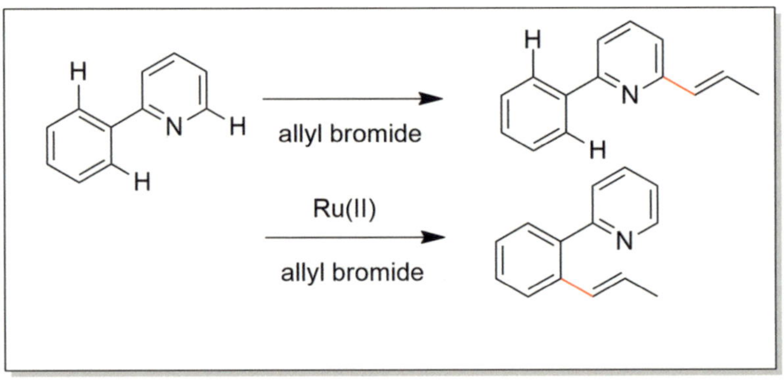

Figure 8.
C–H bond activation of pyridine or pyridine-linked ring.

was found to occur predominantly instead of C–H activation. The allylation chiefly take place at phenyl ring (C_2) rather than C_2 position of the pyridine ring in the absence of metal catalyst [9].

In pursuit of the ortho arylation with the chlorobenzene counterpart, which is considered the least reactive part because of the weak leaving property, the research group Crabtree and group developed a biomass-derived ligand that portrayed significantly improved catalytic activity (**Figure 9**) of ruthenium catalyst for *ortho* C–H bond arylation of 2-phenyl pyridine [10].

The 2-aryl-based scaffold was employed to substitute with azide to develop further a multi-nitrogen-bearing ring. The method of *ortho*-azidation was developed using copper catalyst (CuI), an oxidant, and benzotriazole sulfonyl azide as the azidating agent (**Figure 10**). The oxidant, $K_2S_2O_8$ was used to enhance the system's catalytic activity. The advantage of this protocol was claimed as a starting material for the many pharmaceutical products as apoptosis inducers and phosphate transport protein inhibitors [11].

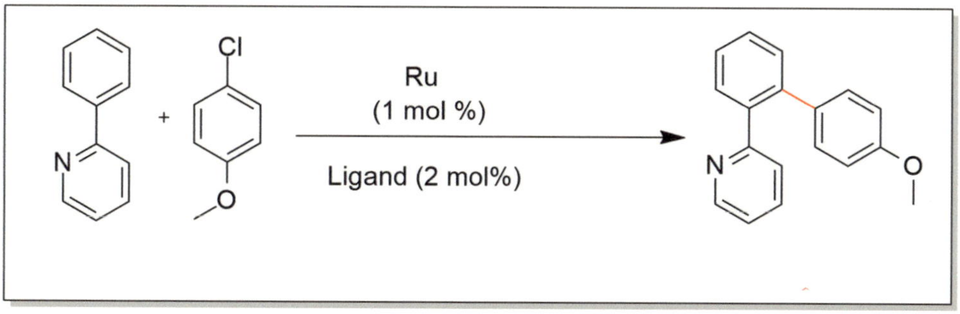

Figure 9.
Arylation of 2-phenyl pyridine through Ru biomass ligand direct C–H activation.

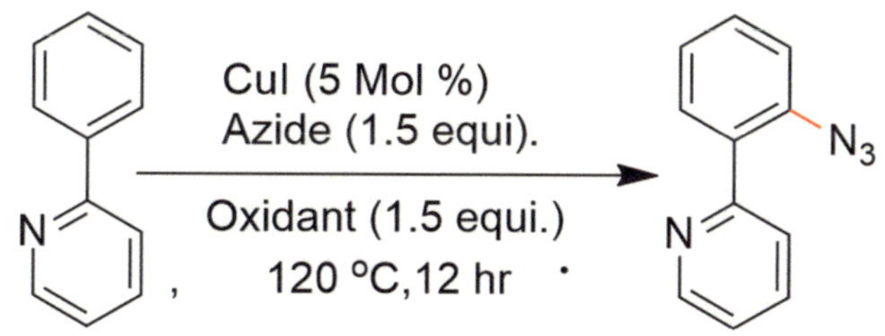

Figure 10.
Azidation of 2-phenyl pyridine.

In the ortho functionalization, the C–P bond was formed through the palladium-based cyclo-metallic system, wherein the nitrogen atom of pyridine was acting as a directing group to get the substitution on the 2-aryl pyridine (**Figure 11**) [12].

2.2 Meta C–H activation through pyridine directing group

Various reports for the meta-C–H activation were reported, with the help of the directing group assisting bridge, where the geometry played a pivotal role to activate the meta-C–H bonds. The assisting bridge was found suitable for the meta directing as depicted below in the **Figure 12** [13]. Some of them arise from the pyridine bases, as one of the important examples is the use of the direct ruthenium-catalyzed *meta*-bromination of arenes, which was utilized for the synthesis of Vismodegib. However, the mechanistic approach was found in their free radical mechanism [14].

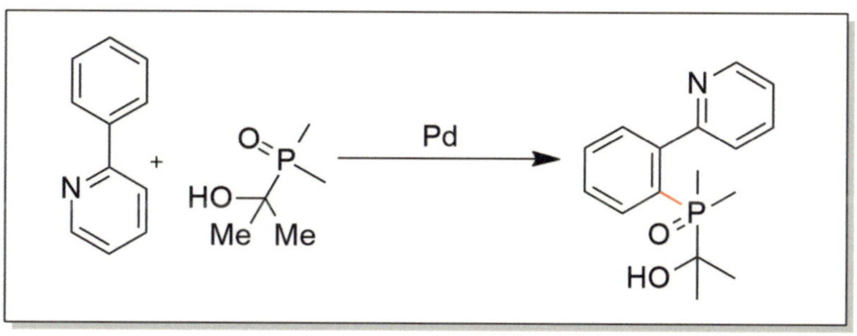

Figure 11.
Phosphonation of 2-phenyl pyridine.

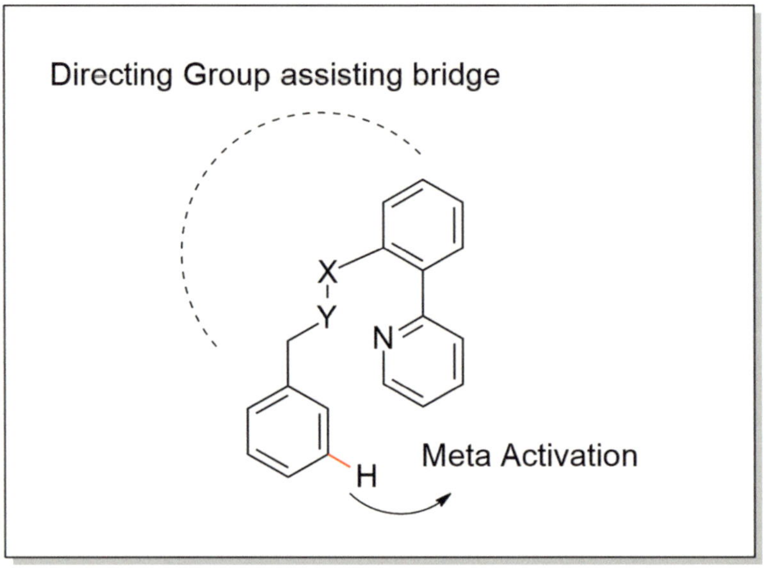

Figure 12.
Meta-directed C–H activation.

The palladium-catalyzed *meta*-selective C–H deuteration substrates with pyridine ring were used to develop a meta-directing protocol to functionalize a complex ring-based structure. The optimized protocol successfully activated (**Figure 13**) the pyridine-based template with acid and ester-based functional group. The ester linkage played a pivotal role in developing a bridge to activate the meta-C–H activation [14].

A scientific group reported using a pyridine template to get the *meta*-C–H activation of benzyl and phenyl ethyl alcohols through its stereo interference (**Figure 14**) on the metalacyclic intermediate. The claim over the Pd and its sigma coordination to the site of concern is proved with the help of the designed experiment and found success with this versatile catalytic system [26].

2.3 Pyridine vs. pyridine N-oxide as directing group

Pyridine *N*-oxides show reactivity toward nucleophile and an electrophile, while pyridine shows the most negligible reactivity for both of them. The oxidized form

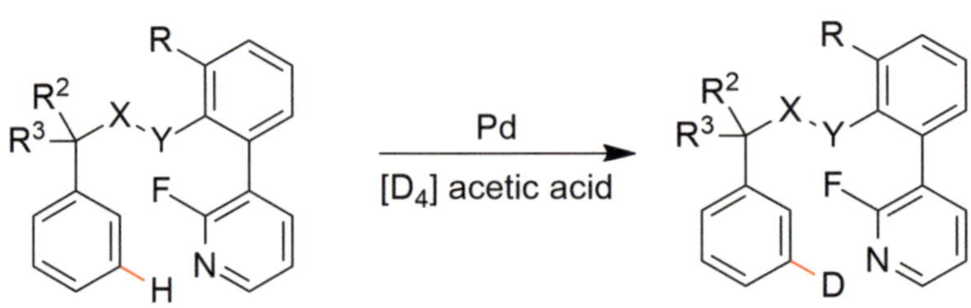

Figure 13.
Deuteration through pyridine template.

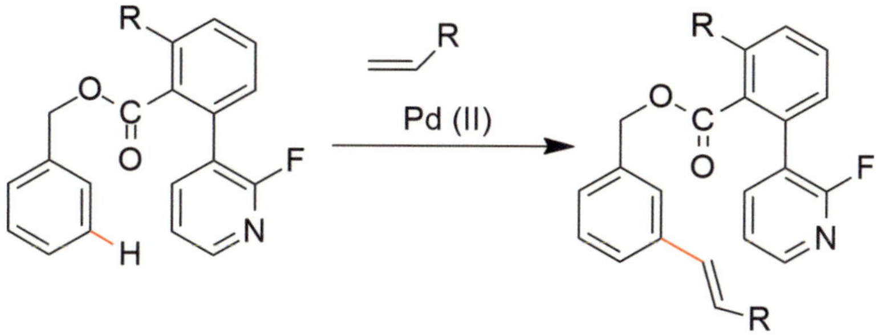

Figure 14.
Alkenylation through pyridine template.

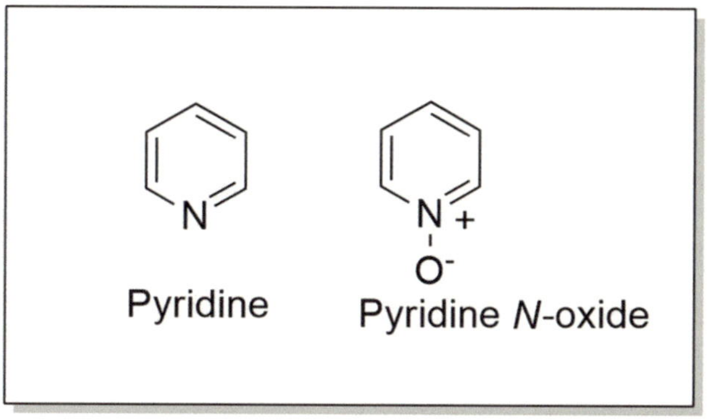

Figure 15.
Pyridine N-oxide and pyridine.

of pyridine is also considered a directing group for the metal-based C–H activation, with high regioselectivity [27]. The research group of Keith Fagnou used pyridine *N*-oxide extensively (**Figure 15**) for C–H activation-based methodology. Their methodologies were able to show the precise role of the Pyridine N-oxide as a regioselective directing group [28].

This directing group is to show the advantage of the pyridine nucleus as its oxidized form. Given the regioselectivity, one of the research groups claims different selectivity of the pyridine and its oxidized form (pyridine *N*-oxide) to the alkene counterpart. It also justifies that the shifting of regioselective functionalization is possible in the pyridine and its oxidized form [27].

3. Conclusion

The opening of the new C–H activation era has unlocked opened a wide range of options to develop a successful scaffold without disturbing the core structure and sensitive functional group. The ease and the minimal waste without using prefunctionalization of the C–H bond are the merits of this organometallic reaction. The importance of the reaction is that it can be utilized for the functionalization of the various heteroatoms–based scaffolds. The various scaffolds have been utilized for functionalization so far. Moreover, important and active moleculess are is also reported with good biological activity by various esteemed groups. Herein we summarized the functionalization of the pyridine nucleus with the help of organometals. The nitrogen of the pyridine was taken as a standard for directing the C–H activation, which resolved the issue of the regioselectivity. The problem of regionselectivity was also discussed here in the example of directing-group-based C–H activation. The reduction of the step and regioselectivity through the C–H activation protocol will have a significant impact on the chemistry and the pharmaceutical field through the reduction of cost. The reduction of the prefunctionalization step will also exert a beneficial action on the environment.

Acknowledgements

The authors are thankful to Graphic Era Hill University, Dehradun, India, for providing the required infrastructure.

Conflict of interest

The authors declare no conflict of interest.

Abbreviations

CuI	Copper (I) Iodide
Ortho	1 and 2 substituted aromatic compound.
Meta	1 and 3 substituted aromatic compound.
Para	1 and 4 substituted aromatic compound.
Metallocycle	A cyclic structure with metal.
$K_2S_2O_8$	Potassium persulfate
Halides	F, Cl, Br, I
N atom	Nitrogen
O atom	Oxygen
P atom	Phosphorus
S atom	Sulfur
Pd	Palladium
Ru	Ruthenium
FG	Functional group

DOI: http://dx.doi.org/10.5772/intechopen.105544

Author details

Priyank Purohit[1]*, Gaurav Joshi[1] and Meenu Aggarwal[2]

1 Graphic Era Hill University, Dehradun, India

2 Department of Chemistry, Aggarwal College Ballabgarh, Faridabad, Haryana, India

*Address all correspondence to: priyank.niper@gmail.com; prpurohit@gehu.ac.in

References

[1] Altus KM, Love JA. The continuum of carbon–hydrogen (C–H) activation mechanisms and terminology. Communications Chemistry. 2021;**4**(1):1-11

[2] Davies HM, Morton D. Recent advances in C–H functionalization. The Journal of Organic Chemistry. 2016;**81**(2):343-350

[3] Dalton T, Faber T, Glorius F. C–H activation: Toward sustainability and applications. ACS Central Science. 2021;7(2):245-261

[4] Periana RA, Bhalla G, Tenn WJ, Young KJH, Liu XY, Mironov O, et al. Perspectives on some challenges and approaches for developing the next generation of selective, low temperature, oxidation catalysts for alkane hydroxylation based on the CH activation reaction. Journal of Molecular Catalysis A: Chemical. 2004;**220**(1):7-25

[5] Bolm C. Cross-coupling reactions. The Journal of Organic Chemistry. 2012;77(12):5221-5223

[6] Yin J. Carbon–carbon coupling reactions catalyzed by heterogeneous palladium catalysts. Chemical Reviews. 2007;**107**(1):133-173

[7] Arockiam PB, Bruneau C, Dixneuf PH. Ruthenium(II)-catalyzed C–H bond activation and functionalization. Chemical Reviews. 2012;**112**(11):5879-5918

[8] Chen X, Engle KM, Wang D-H, Yu J-Q. Palladium(II)-catalyzed C–H Activation/C–C cross-coupling reactions: Versatility and practicality. Angewandte Chemie International Edition. 2009;**48**(28):5094-5115

[9] Guo X-X, Gu D-W, Wu Z, Zhang W. Copper-catalyzed C–H functionalization reactions: Efficient synthesis of heterocycles. Chemical Reviews. 2015;**115**(3):1622-1651

[10] Crabtree RH, Lei A. Introduction: CH activation. Chemical Reviews. 2017;**117**(13):8481-8482

[11] Etienne M, Weller AS. Intramolecular C–C agostic complexes: C–C sigma interactions by another name. Chemical Society Reviews. 2014;**43**(1):242-259

[12] Harvey BG, Ernst RD. Transition-metal complexes with (C–C)→ M agostic interactions. European Journal of Inorganic Chemistry. 2017;**2017**(9):1205-1226

[13] Ess DH, Goddard WA, Periana RA. Electrophilic, ambiphilic, and nucleophilic C– H bond activation: Understanding the electronic continuum of C– H bond activation through transition-state and reaction pathway interaction energy decompositions. Organometallics. 2010;**29**(23):6459-6472

[14] Wang DY, Choliy Y, Haibach MC, Hartwig JF, Krogh-Jespersen K, Goldman AS. Assessment of the electronic factors determining the thermodynamics of "oxidative addition" of C–H and N–H bonds to Ir (I) complexes. Journal of the American Chemical Society. 2016;**138**(1):149-163

[15] Thompson ME, Baxter SM, Bulls AR, Burger BJ, Nolan MC, Santarsiero BD, et al. Sigma.-Bond metathesis for carbon-hydrogen bonds of hydrocarbons and S c–R (R= H, alkyl, aryl) bonds of permethylscandocene derivatives. Evidence for noninvolvement of the. pi. system in electrophilic activation of

aromatic and vinylic CH bonds. Journal of the American Chemical Society. 1987;**109**(1):203-219

[16] Labinger JA, Bercaw JE. Understanding and exploiting C–H bond activation. Nature. 2002;**417**(6888):507-514

[17] Webb JR, Burgess SA, Cundari TR, Gunnoe TB. Activation of carbon–hydrogen bonds and dihydrogen by 1, 2-CH-addition across metal–heteroatom bonds. Dalton Transactions. 2013;**42**(48):16646-16665

[18] Bolm C. Cross-coupling reactions. Organic Letters. 2012;**14**:2925-2928

[19] Heck RF. Palladium reagents. Organic Syntheses. 1990. London: Academic Press 1990. ISBN-10: 0123361419. ISBN-13: 978-0123361417

[20] Yeung CS, Dong VM. Catalytic dehydrogenative cross-coupling: Forming carbon– carbon bonds by oxidizing two carbon– hydrogen bonds. Chemical Reviews. 2011;**111**(3):1215-1292

[21] Ling Y, Hao Z-Y, Liang D, Zhang C-L, Liu Y-F, Wang Y. The expanding role of pyridine and dihydropyridine scaffolds in drug design. Drug Design and Developmental Theraphy. 2021;**15**:4289-4338

[22] Altaf AA, Shahzad A, Gul Z, Rasool N, Badshah A, Lal B, et al. A review on the medicinal importance of pyridine derivatives. Journal of Drug Design and Medical Chemistry. 2015;**1**(1):1-11

[23] Alizadeh SR, Ebrahimzadeh MA. Antiviral activities of pyridine fused and pyridine containing heterocycles, a review (from 2000 to 2020). Mini Reviews in Medicinal Chemistry. 2021;**21**(17):2584-2611

[24] Khan E. Pyridine derivatives as biologically active precursors: Organics and selected coordination complexes. ChemistrySelect. 2021;**6**(13):3041-3064

[25] Ackermann L, Vicente R, Kapdi AR. Transition-metal-catalyzed direct arylation of (hetero) arenes by C–H bond cleavage. Angewandte Chemie International Edition. 2009;**48**(52):9792-9826

[26] Chu L, Shang M, Tanaka K, Chen Q, Pissarnitski N, Streckfuss E, et al. Remote meta-C–H activation using a pyridine-based template: Achieving site-selectivity via the recognition of distance and geometry. ACS Central Science. 2015;**1**(7):394-399

[27] Kanyiva KS, Nakao Y, Hiyama T. Nickel-catalyzed addition of pyridine-N-oxides across alkynes. Angewandte Chemie. 2007;**119**(46):9028-9030

[28] Campeau LC, Fagnou K. Synthesis of 2-Aryl pyridines by palladium-catalyzed direct arylation of pyridine N-oxides. Organic Syntheses. 2003;**88**:22-32

Chapter 2

The Chemistry of Benzo and Carbocyclic Derivatives of Pyridine

Adebimpe D. Adesina

Abstract

The chemistry of pyridine and its derivatives is of considerable importance in the synthesis of intermediates leading to biologically active compounds and novel materials. Generally, derivatives of pyridine are stable and relatively unreactive but can be attacked by electrophiles at ring nitrogen and certain carbon atoms. Pyridines undergo radical substitution reactions preferentially at the 2-position. Simple pyridines and their benzo derivatives are weak bases that form salts with strong acids. Various Lewis acids form complexes with pyridine and its benzo derivatives. The quaternization of pyridine and its benzo derivatives using alkyl and acyl halides have been used as versatile synthetic intermediates to biologically active compounds as final products. Precursors to cyanine dyes have been prepared by means of the 1,4-addition of pyridines and quinolines to acrylamide. *N*-oxides, obtained by the oxidation of pyridine and its benzo analogues, are versatile intermediates in organic synthesis.

Keywords: benzo derivatives, pyridine, quinoline, isoquinoline, synthetic intermediates, electrophilic substitution, nucleophilic substitution

1. Introduction

Pyridine was first isolated in a pure state from bone oil by Anderson [1] who had earlier obtained picoline from coal tar. He established the molecular formula of pyridine and showed it to be a tertiary base, capable of forming quaternary salts. A Kekule-type structure was proposed for pyridine **1** by Korner (**Figure 1**) [2]. The proposed structure was confirmed by the reduction of pyridine to piperidine, by the reverse oxidation and by the synthesis of piperidine.

In addition to being attacked by electrophiles, strong nucleophiles can also react, at the α- or γ- ring carbon atoms of the pyridine ring [3, 4].

Quinoline **2** and isoquinoline **3** are the two possible structures in which a benzene ring is annelated to a pyridine ring. The effect that the benzene ring has on the reactivity of the pyridine ring, and *vice versa* should be considered. Electrophilic substitution favors the benzenoid ring, rather than the pyridine ring with preferred substitution at the 5- (**Figure 2**) and 8- positions.

The electron-deficiency of the carbons in pyridines, particularly the α- and γ-carbons, and the ability of the heteroatom to accommodate negative charge in the

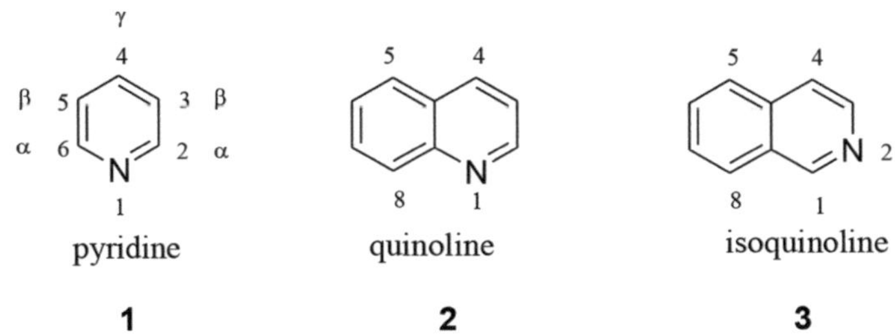

Figure 1.
Structure of pyridine and its benzo-fused analogues.

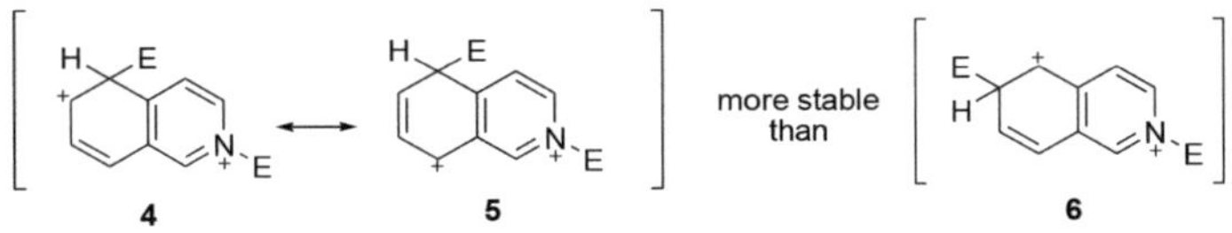

Figure 2.
Electrophilic substitution of isoquinoline.

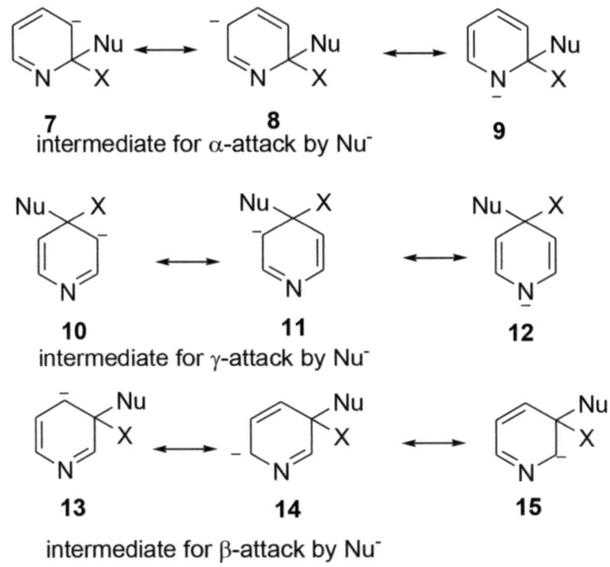

Figure 3.
Selectivity of nucleophilic attack on halopyridines.

intermediate thus produced, makes nucleophilic addition and, especially nucleophilic displacement of halide (and other good leaving groups), a very important feature of pyridine chemistry (**Figure 3**) [5]. Quinoline and isoquinoline are reactive to nucleophiles in the pyridine ring, especially at the positions α and γ to the nitrogen and, further, are more reactive in this sense than pyridines.

2. Synthesis

The synthesis of a pyridine ring can be achieved in many ways. Some of these will be described and exemplified.

2.1 Condensation reactions

One of the methods for constructing the pyridine nucleus is by way of condensation reactions. This is done by the combination of an amino group with two carbonyl groups followed by the loss of two or more equivalents of water. A final oxidation step was often necessary to obtain the aromatic ring system. Most condensations leading to a pyridine derivative **17** proceed through an intermediate which can be related to a 1,5-dicarbonyl compound **16** (**Figure 4**).

The Chichibabin pyridine synthesis is an example of the condensation method for synthesizing pyridine rings. The reaction involves the condensation of aldehydes, ketones, α, β-unsaturated carbonyl compounds, or any combination of these, with ammonia.

Frank and Seven [6] have reported the modified synthesis of pyridine by heating the carbonyl compounds or derivatives with aqueous ammonia and catalytic amounts of ammonium acetate to produce good yields of single products. But-2-enal was reacted with ammonia to form 5-ethyl2-methylpyridine (**Figure 5**). However, the use of a steel autoclave at high temperatures and pressures was a drawback in this process.

An improved Chichibabin synthesis was also investigated by Weiss [7] and a mechanism was proposed for the formation of the pyridine ring. The mechanism of the reaction of benzaldehyde **20** with acetophenone **21** involved an aldol condensation to form **22**, followed by a Michael-type reaction to give a 1,5-dicarbonyl **23,** which then condenses with ammonia to form a dihydropyridine **24**, which, in turn, is dehydrogenated to a pyridine **25** (**Figure 6**).

2.2 Cycloaddition reactions

Some 6π cycloadditions have been used to form pyridines. The first to be reported was the addition of a dienophile **28** to an oxazole **27** [8, 9]. When acrylonitrile was used, hydrogen cyanide was lost to aromatise and the oxazole oxygen retained to give 3-hydroxypyridines, while with the use of acrylic acid, the oxygen was lost as water (**Figure 7**).

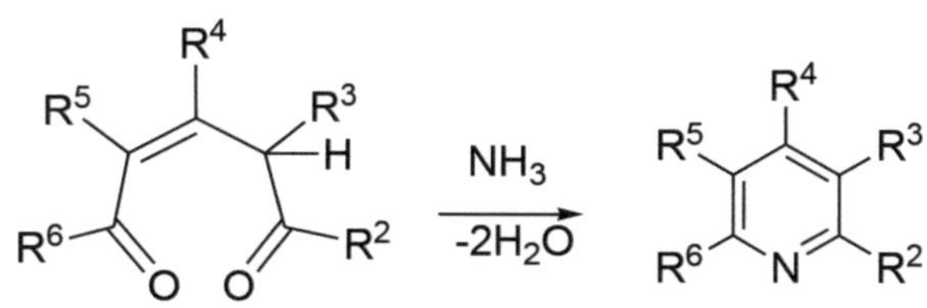

Figure 4.
Typical pyridine ring synthesis.

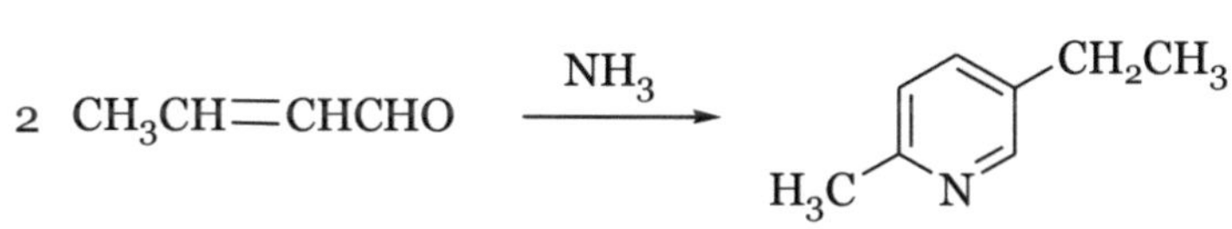

Figure 5.
An example of the Chichibabin synthesis.

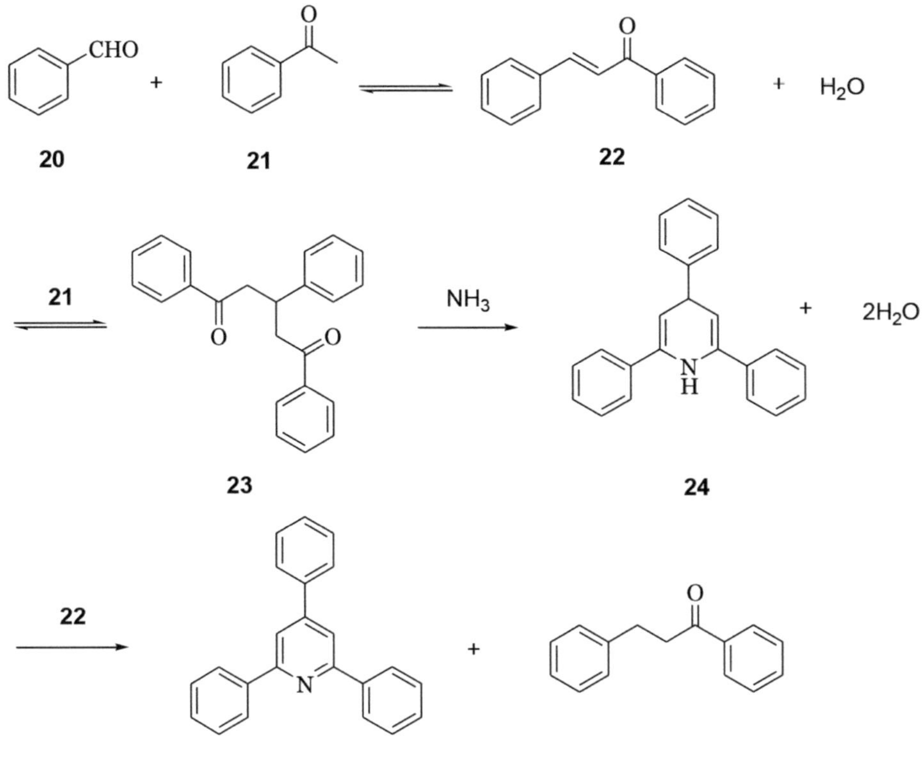

Figure 6.
An improved Chichibabin synthesis of pyridine.

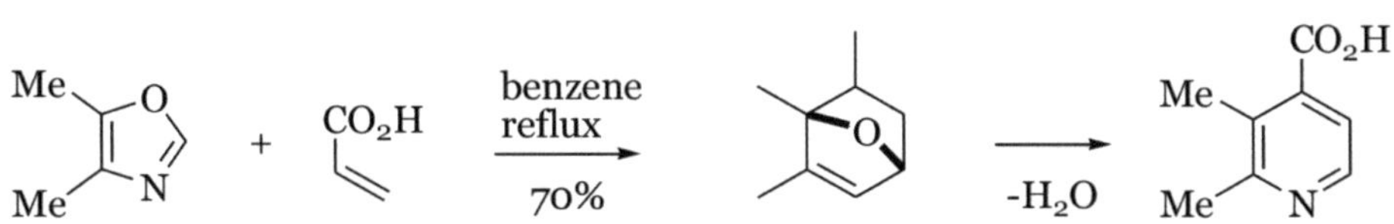

Figure 7.
Synthesis of pyridines via cycloadditions.

The interaction of propargylamine **32** with a cyclic ketone **31**, produced an enamine **33**, followed by a ring closure which when effected with a gold catalyst, gave a carbocyclic pyridine derivative **34** (**Figure 8**) [10].

2.3 Cyclization reactions

Pyridines can be formed by the cyclization of nitriles at either carbon or nitrogen. Cyclizations at nitrogen were more common and incorporated the nitrogen into the pyridine ring.

Methyl-substituted pyridine derivatives have been synthesized from the cyclization of cyclic precursors **36** which were prepared from the treatment of β-ketoesters

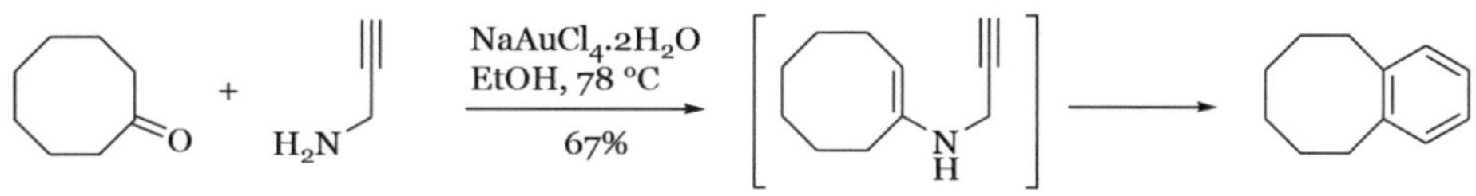

Figure 8.
Synthesis of a carbocyclic pyridine derivative.

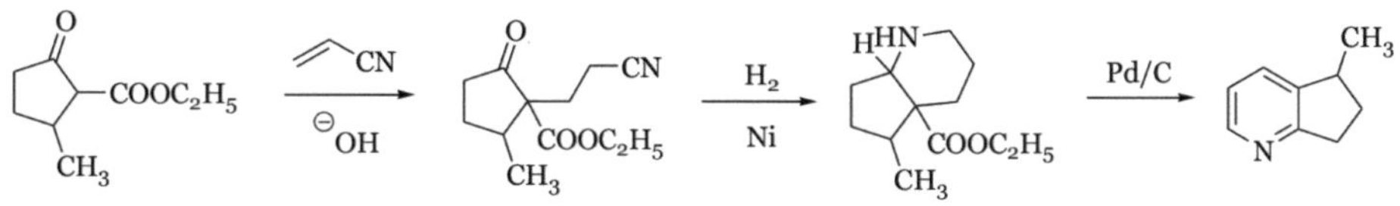

Figure 9.
Synthesis of pyridine from nitrile cyclization.

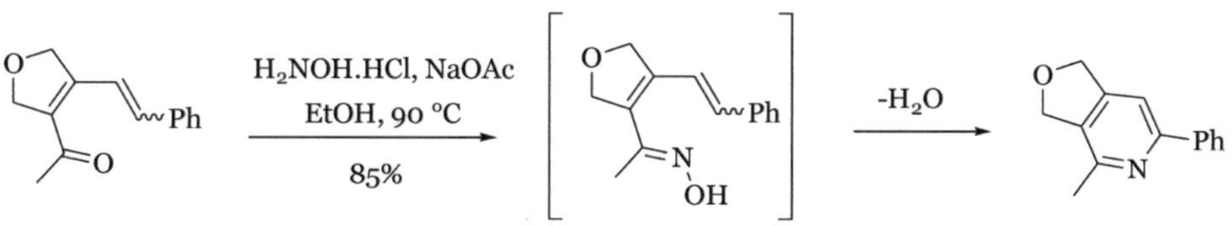

Figure 10.
Azatriene cyclizations to form pyridines.

35 with acrylonitrile (**Figure 9**) [11]. The dehydrogenation of the piperidine ring in the final step also resulted in the loss of the ester group.

The fusion of pyridines to other ring systems has been investigated via thermal electrocyclization [12]. The pyridines were formed from the oxidation of dihydropyridines which were generated from the electrocyclization of aza-1,3,5-trienes. However, the use of an oxime or hydrazine derivative, followed by the elimination of water or an amine *in situ* gave the pyridine directly (**Figure 10**).

3. Reaction with electrophilic reagents

3.1 Addition to nitrogen

3.1.1 Protonation and salt formation

Pyridines behave like tertiary aliphatic or aromatic amines in reactions that involves bond formation using the lone pair of electrons on the ring nitrogen. Simple pyridines and their benzo derivatives are weak bases that form crystalline, frequently hygroscopic, salts with most protic acids [3, 4].

Chromium salts of pyridine have become important reagents in organic synthesis because of their mild oxidizing capability. Pyridinium chlorochromate (Corey's reagent), pyridinium dichromate, and $(Py)_2CrO_3$ (Collins' reagent) are the most widely used.

3.1.2 Alkylation

Alkyl halides and sulfates react readily with pyridine and its benzo derivatives at room temperature, giving quartenary *N*-substituted pyridinium salts, which have been used as versatile synthetic intermediates to biologically active compounds or as final products [13–15]. Quaternization of pyridine with alkyl halides or related compounds is an example of Menschutkin reaction (**Figure 11**).

A review on quartenary salts of pyridines and related compounds describing their synthesis, physicochemical properties, possible applications, and their biological activities has been published [16].

3.1.3 Acylation

Acylation of pyridines can be achieved at temperatures as low as −78°C. Acid halides react readily with pyridines to generate *N*-acylpyridinium salts in solution, and in some cases, as crystalline, non-hygroscopic solids (**Figure 12**) [17]. *N*-Acylpyridinium salts have been found to be more reactive than their *N*-alkyl counterparts and are susceptible to attack by nucleophiles.

3.1.4 Halogenation

Pyridines and their benzo derivatives react with halogens to give *N*-halogenopyridinium salts. The complexes of pyridine with chlorine have been well studied [18]. Pyridine iodo compounds can be prepared by treating $TiI_3[AsF_6]$ with pyridines, from which the pyridinium salt $[C_5H_5NI]^+[AsF_6]^-$ has been isolated and characterized [19]. Several syntheses of *N*-fluoropyridinium salts have been reported.

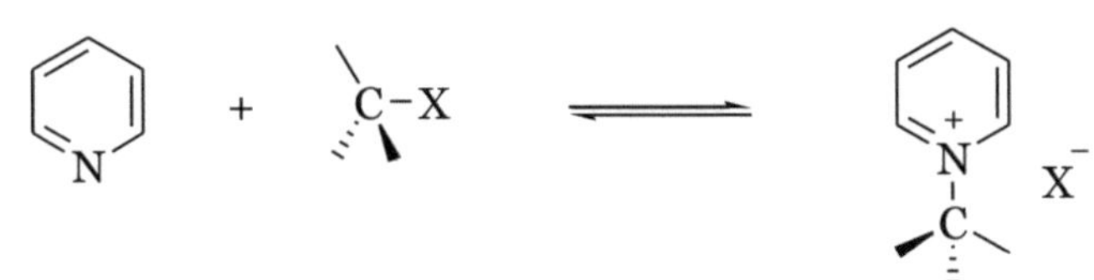

1 **42** **43**

Figure 11.
Alkylation of pyridine.

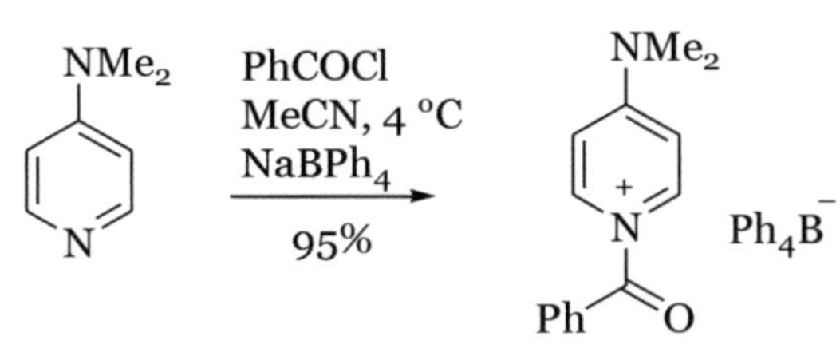

44 **45**

Figure 12.
Acylation at nitrogen of 4-dimethylaminopyridine (DMAP).

These compounds have received growing interest because of their use as fluorinating agents [20].

N′, *N′*-Difluoro-2,2′-bipyridinium bis(tetrafluoroborate) **47,** prepared in one pot by introducing BF3 gas into 2,2′-bipyridine **46** at 0°C followed by fluorine gas diluted with nitrogen, has been shown to be a highly reactive electrophilic fluorinating agent (**Figure 13**) [21].

3.1.5 N-*oxidation*

N-Oxides, obtained from the oxidation of pyridine and its benzo analogues, are versatile intermediates in organic synthesis [22–24]. Reagents used for the *N*-oxide formation include peracids, [3] H2O2/AcOH, dioxiranes, [25] organic hydrotrioxides, [26] Caro's acid, oxaziridines [27] and oxygen with ruthenium trichloride as catalyst [28].

Similarly, there are many ways to deoxygenate pyridine *N*-oxides: samarium iodide, chromous chloride, stannous chloride with low-valent titanium, ammonium formate with palladium and catalytic hydrogenation at room temperature can be used [29–33]. The most frequently used methods have involved oxygen transfer to trivalent phosphorus [34] or divalent sulfur [35] (**Figure 14**).

3.2 Electrophilic attack at carbon

In most cases, electrophilic substitution of pyridines occurs very much less readily than for the correspondingly substituted benzene. This is because the electrophilic reagent, or a proton in the reaction medium, adds first to the pyridine nitrogen, generating a pyridinium cation, which is naturally very resistant to attack by an electrophile.

The electron-withdrawing effect of nitrogen in pyridine is profound at the 2- and 4-positions and diminished at the 3-position. When electrophilic attack does occur, it is generally at the 3-position.

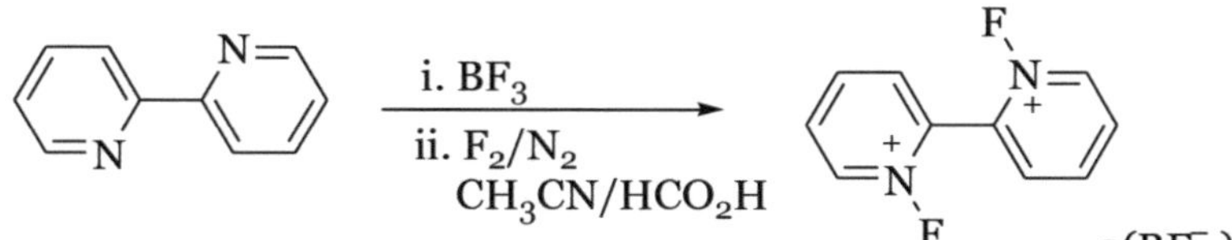

46 **47**

Figure 13.
Fluorination of pyridine compounds at nitrogen.

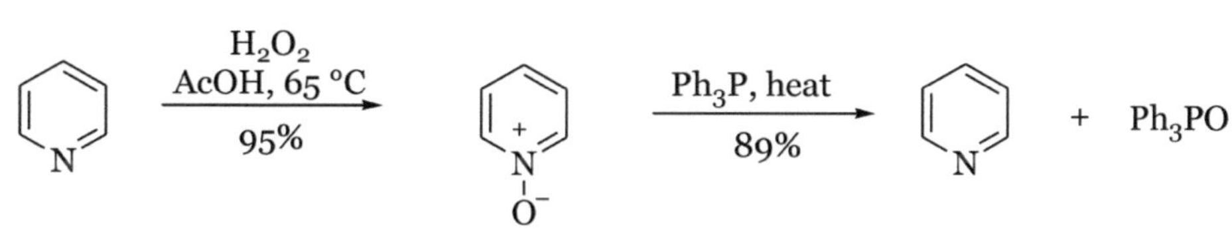

1 **48**

Figure 14.
Oxidation of pyridine at nitrogen.

3.2.1 Nitration

The electron-deficient nature of pyridine makes its direct nitration difficult even under rigorous conditions, whereas pyridine *N*-oxide, pyridines and pyridinamines can be nitrated more easily [36].

Initial reaction of pyridines with dinitrogen pentoxide in sulfur dioxide proceeds by addition at 2-position forming a 1,2-dihydropyridine intermediate. Transfer of the nitro group to a β-position, via a [1,5]-sigmatropic migration, is then followed by elimination of the nucleophile, regenerating the aromatic system to give 3-nitropyridines **49** (**Figure 15**) [37].

3.2.2 Halogenation

The halogenation of pyridines can be achieved using a variety of reagents which are not always mild and compatible with other functionalities in the molecule. Due to the electron-deficiency of the pyridine ring, electrophilic halogenations are mostly difficult.

The reaction of bromine with pyridine in oleum has produced 3-bromopyridine **51** in good yield [38]. The reactive species in the process involves pyridinium-1-sulfonate. Similarly, 3-chloropyridine **50** has been produced by chlorination at 200°C, or at 100°C in the presence of aluminum chloride, although in low yield (**Figure 16**) [39].

3.2.3 Sulfonation

The reaction of pyridine with concentrated sulfuric acid only gave low yields of 3-sulfonic acid after prolonged reaction time at 320°C. However, a higher yield was achieved with the addition of mercuric sulfate in catalytic quantities at a somewhat lower temperature (**Figure 17**) [40].

The sulfonation of quinoline has been achieved under conditions of 30% oleum at 90°C, occurring at the 8-position to give **53** in good yield, whereas isoquinoline gave the 5-acid. At higher temperatures, under thermodynamic control, other isomers are produced, for example quinoline-8-sulfonic acid is isomerised to the 6-acid **54** (**Figure 18**) [41, 42].

3.2.4 Oxidation

Pyridines require vigorous conditions to be oxidized as they are generally resistant to oxidizing agents. Pyridines have been converted into 2-pyridones **55** using copper sulfate (**Figure 19**) [43]. A similar conversion using zinc sulfate heptahydrate or

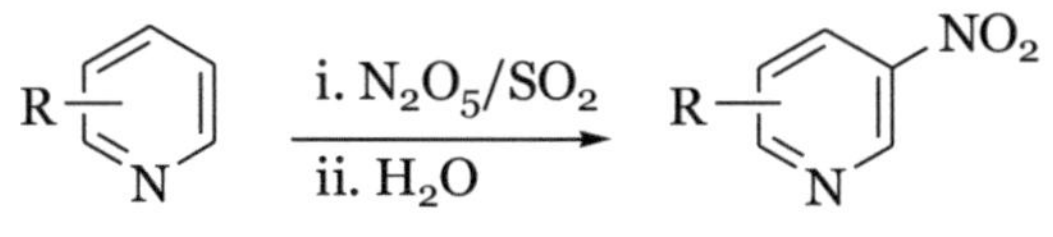

49

R = H, 2-Me, 3-Me, 4-Me, 4-Ph, 3-Ac, 4-Ac, 3-Cl, 4-CN, Quinoline, Isoquinoline

Figure 15.
Nitration of pyridine and substituted pyridine.

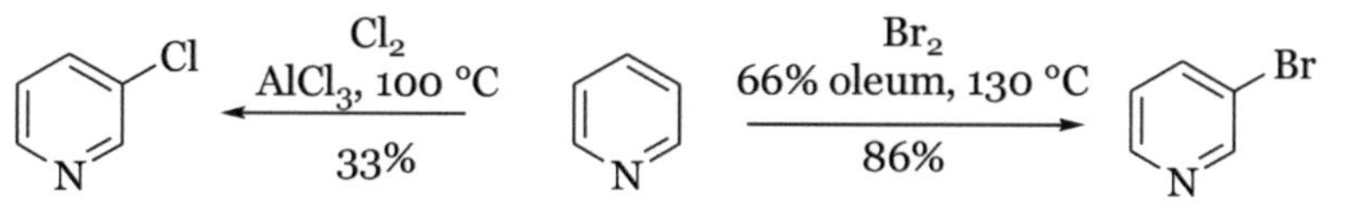

50 **51**

Figure 16.
Chlorination and bromination of pyridine.

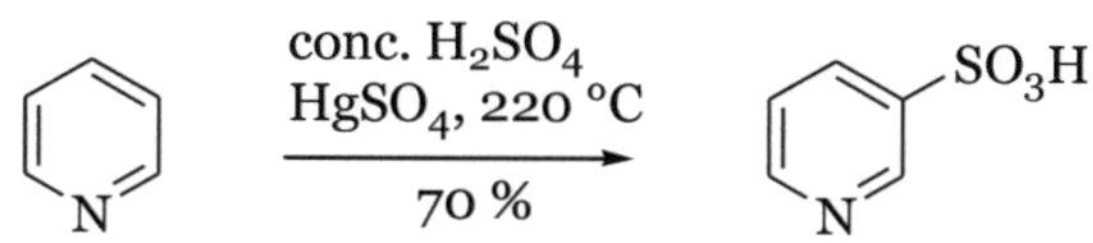

52

Figure 17.
Sulfonation of pyridine.

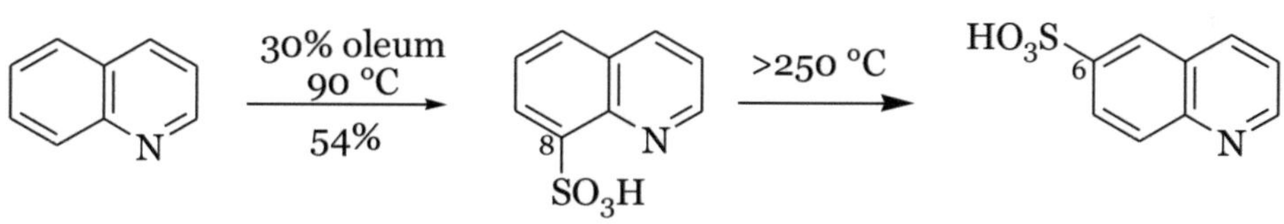

53 **54**

Figure 18.
Sulfonation of quinoline.

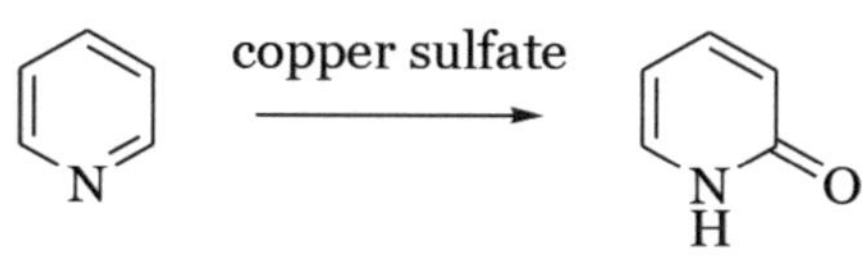

55

Figure 19.
Oxidation of pyridine.

tricadmium sulfate octahydrate and oxygen has also been reported, although with low yield [44].

When quinoline was oxidized under ozonolysis conditions, it gave pyridine-2,3-biscarboxaldehyde. The oxidation of quinoline or isoquinoline with permanganate can occur in either the benzene or pyridine ring, depending on the conditions. Electron-withdrawing or donating groups can direct the oxidation to either the benzene or pyridine ring. The oxidation of 5-aminoisoquinoline occurred in the benzene ring; however, 5-nitroquinoline gave the product of pyridine ring oxidation [4].

4. Reaction with nucleophilic reagents

Nucleophilic substitution reactions are characteristic of pyridines just as electrophilic substitution reactions are characteristic of benzene and electron-rich heteroaromatic compounds such as pyrrole and furan. The nucleophilic substitution of hydrogen usually involves a hydride transfer in the last step [5].

4.1 Nucleophilic attack at carbon

Although many nucleophiles react with halogenated pyridines effecting the displacement of halogen, only strong nucleophiles react with simple pyridine. However, pyridine *N*-oxide and certain pyridines readily undergo nucleophilic substitution [4].

Nitro group has been introduced into the position 1 of isoquinoline using a mixture of potassium nitrite, dimethylsulfoxide and acetic anhydride [45]. The mechanism is shown in the quaternisation reaction of a complex of dimethylsulfoxide and the anhydride at nitrogen followed by the key step, the nucleophilic addition of nitrite to the heterocycle (**Figure 20**).

4.1.1 Alkylation and arylation

Reaction with alkyl- or aryl-lithiums proceeds in two discrete steps: addition to give a dihydro-pyridine *N*-lithio-salt which can then be converted into the substituted aromatic pyridine by oxidation, disproportionation or elimination of lithium hydride (**Figure 21**) [46]. The *N*-lithio salts can be observed spectroscopically and, in some cases, isolated as solids [47].

4.1.2 Amination

Amination of pyridines and related heterocycles, generally at a position α to the nitrogen, is called the Chichibabin reaction, [48–50] the pyridine reacting with sodamide in toluene, xylene or dimethylaniline with the evolution of hydrogen. The 'hydride' transfer and production of hydrogen probably involve interaction of amino-pyridine product, acting as an acid, with the anionic intermediate. Vicarious

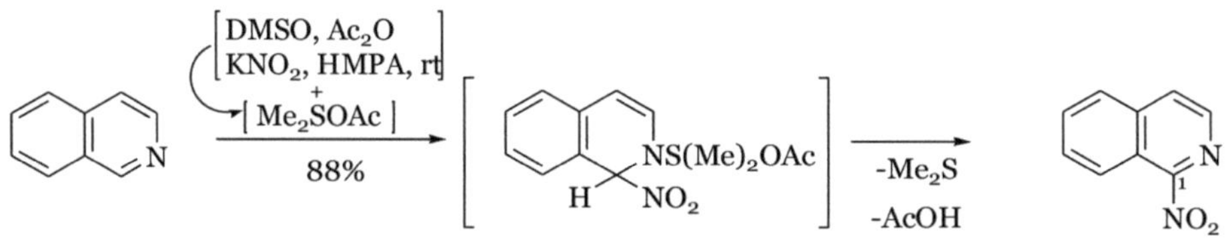

56

57

Figure 20.
An example of nucleophilic attack at carbon of isoquinoline.

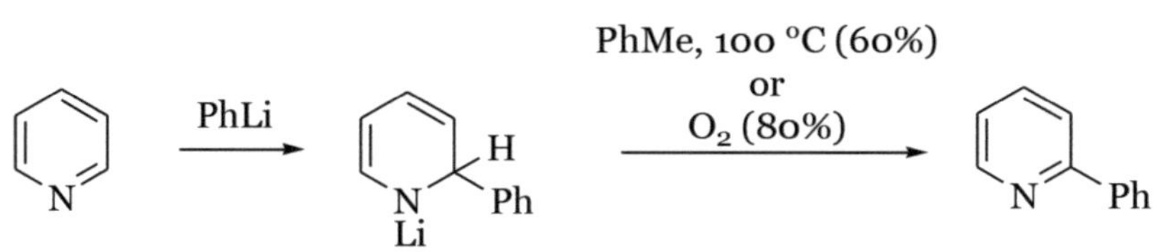

58

59

Figure 21.
Arylation of pyridine.

nucleophilic substitution permits the introduction of amino groups *para* (or *ortho* if *para* blocked) to nitro groups by reaction with 1-amino-1,2,4-triazole **61** (**Figure 22**).

The amination of quinoline with potassium amide in liquid ammonia can, depending on conditions, give 2- or 4-aminoquinoline. The quinoline-2-aduct rearranges to the more stable 4-aminated adduct at higher temperatures (**Figure 23**) [51]. Isoquinoline, however, reacts with potassium amide in liquid ammonia at room temperature to give 1-aminoisoquinoline [52, 53].

4.1.3 Silylation

The reaction of pyridine with trimethylsiliconide anion has afforded 4-trimethylsilylpyridine efficiently. This process probably proceeds via a 1,4-dihydro-adduct (which can be trapped as its *N*-CO2Et derivative by addition of ethyl chloroformate), to give the fully aromatic product via hydride shift to silicon (**Figure 24**) [54, 55].

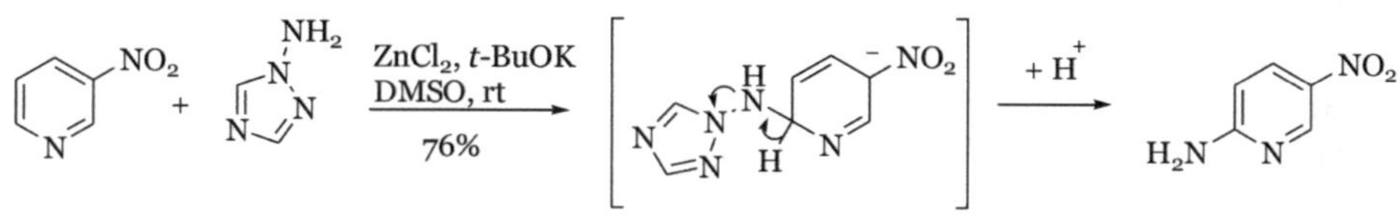

60 **61** **62** **63**

Figure 22.
Amination of pyridine.

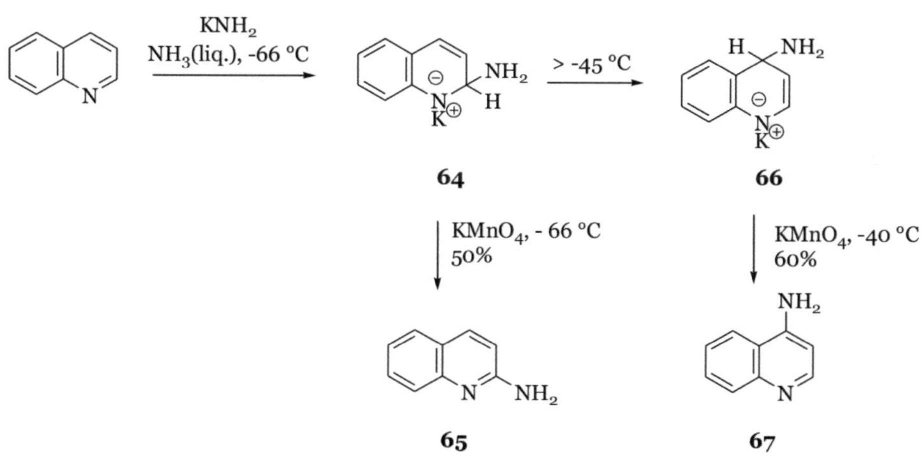

64 **66** **65** **67**

Figure 23.
Amination of quinoline.

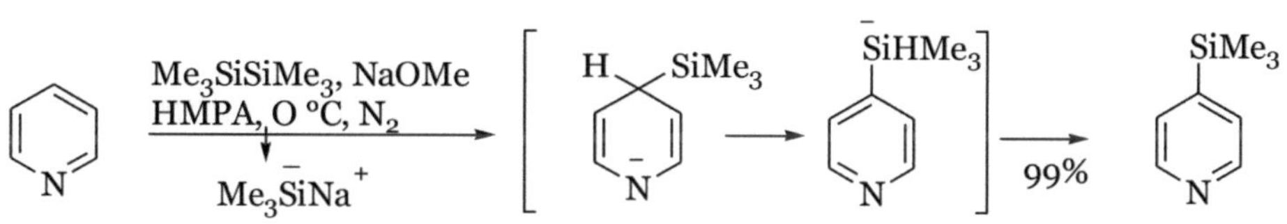

68 **69** **70**

Figure 24.
Silylation of pyridine.

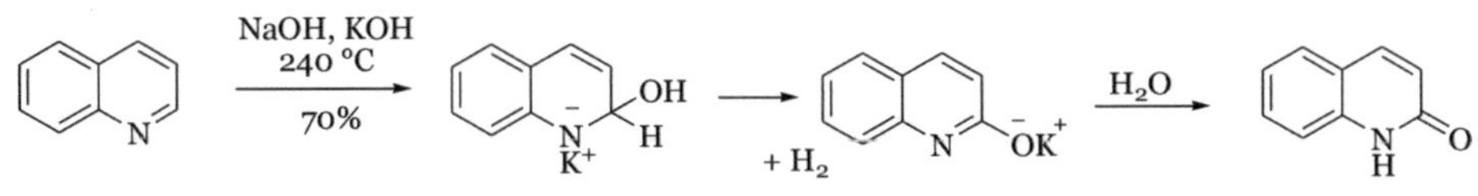

Figure 25.
Hydroxylation of quinoline.

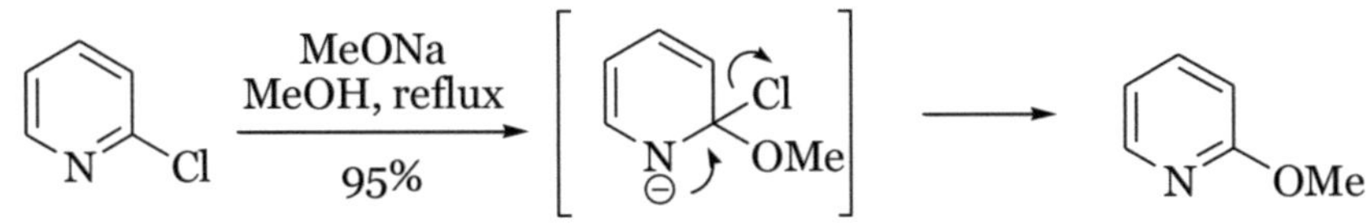

Figure 26.
Nucleophilic substitution of pyridine.

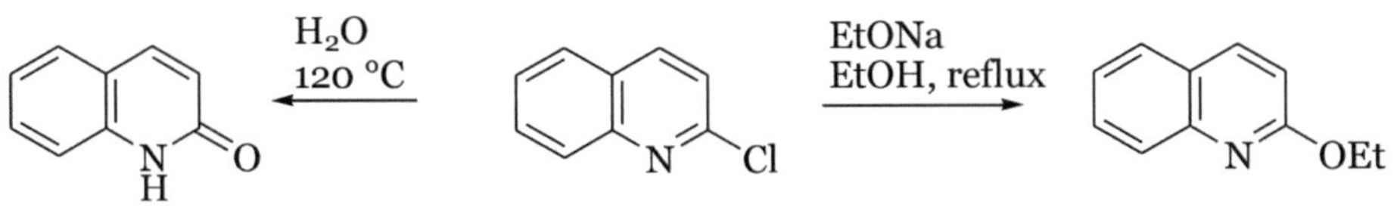

Figure 27.
Nucleophilic substitution of quinoline.

4.1.4 Hydroxylation

Hydroxide ion attacks pyridine only at very high temperatures to produce 2-pyridone in low yield. This can be usefully contrasted with the much more efficient reaction of hydroxide with quinoline and isoquinoline and with pyridinium salts [56].

Quinoline and isoquinoline can be directly hydroxylated with potassium hydroxide at high temperature with the evolution of hydrogen to give 2-Quinolone and 1-isoquinolone as the isolated products (**Figure 25**).

4.2 Nucleophilic substitution with displacement of good leaving groups

Halogen, and some other good leaving groups such as nitro, alkoxysulfonyloxy and methoxy at α- or γ- positions of the pyridine ring are easily displaced by nucleophiles via an addition-elimination mechanism. The nucleophilic substitution of halopyridine and haloquinoline are shown in the **Figures 26** and **27** respectively.

5. Metallation and reactions of *C*-Metallated pyridines, quinolines and isoquinolines

5.1 Direct ring C-H metalation

The heating of pyridine in MeONa-MeOD at 165°C causes an H-D exchange at all positions via small concentrations of deprotonated species. An example of the use of

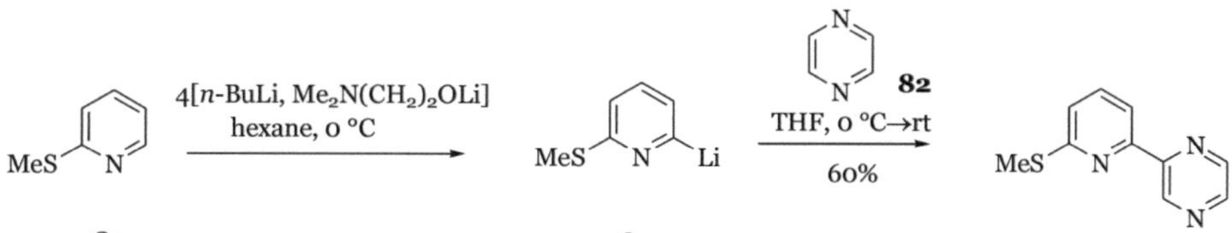

Figure 28.
Lithiation of pyridines.

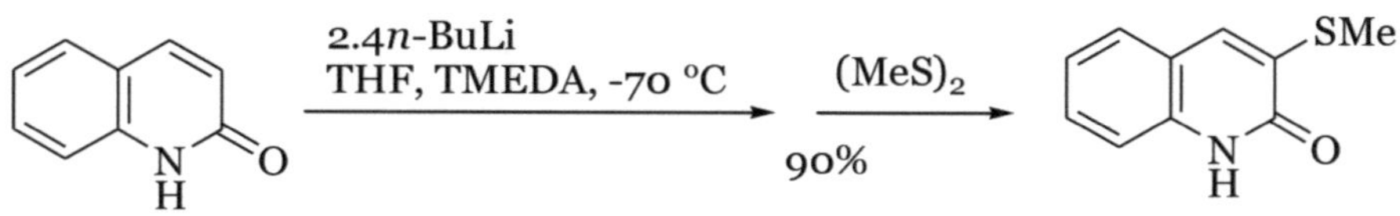

Figure 29.
Lithiation of quinolone.

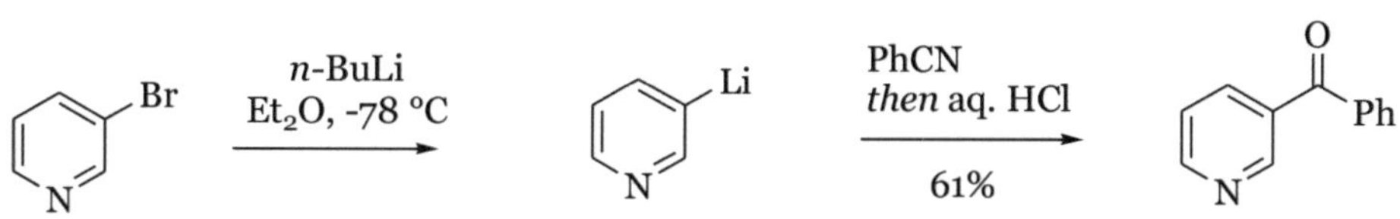

Figure 30.
Metal-halogen exchange of pyridine.

lithiated pyridines, is their nucleophilic addition to azines **82**, to produce bihetaryls **83** on oxidation during work-up (**Figure 28**) [57].

2-Lithiation of 1-substituted 4-quinolones and 3-lithiation of 2-quinolone provides derivatives with the usual nucleophilic propensity (**Figure 29**) [5].

5.2 Metal-halogen exchange

Lithio-pyridines behave as typical organometallic nucleophiles, as in the reaction of 3-bromopyridine with n-butyllithium in ether at −78°C (**Figure 30**) [5].

Nucleophilic addition is a competing reaction in the preparation of lithio-quinolines and isoquinolines via metal-halogen exchange, however the use of low temperatures allow metal-halogen exchange at both pyridine [58] and benzene ring positions [59] in quinolines, and the isoquinoline-1-[60] and 4-positions, [61] subsequent reaction with electrophiles generating *C*-substituted products (**Figure 31**).

6. Photochemical reactions

The ultraviolet irradiation of pyridines can produce highly strained species that can lead to isomerised pyridines or can be trapped. When *N*-methyl-2-pyridone **92** was

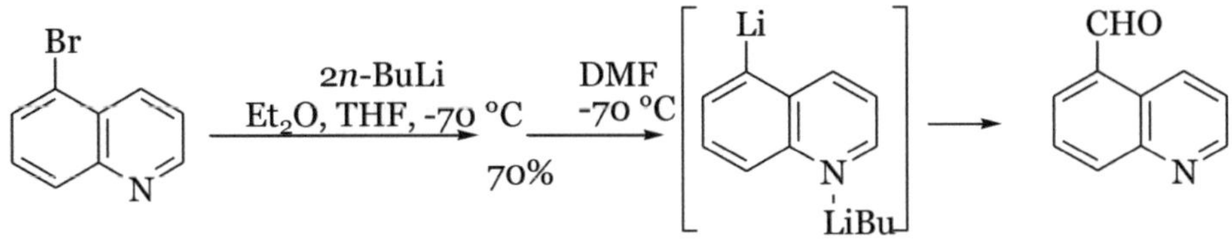

89 **90** **91**

Figure 31.
Metal-halogen exchange of quinoline.

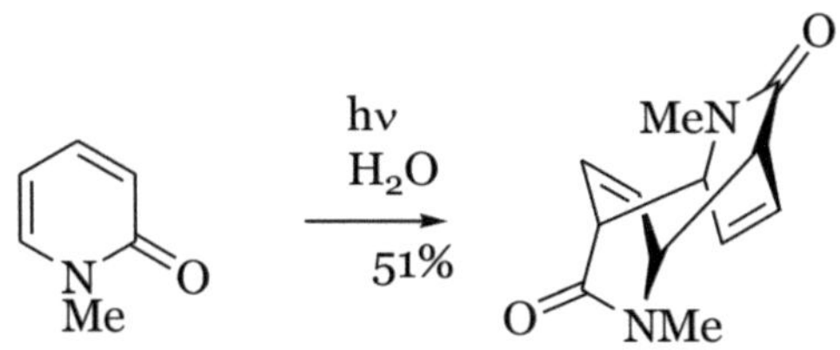

92 **93**

Figure 32.
Ultraviolet irradiation of pyridone.

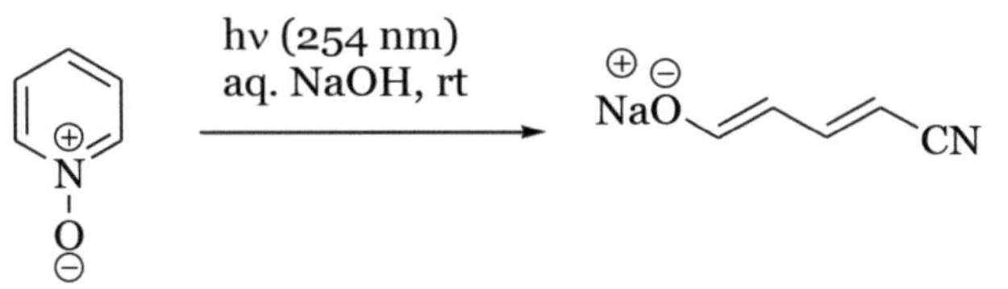

94 **95**

Figure 33.
Photolysis of pyridine-N-oxide.

irradiated in aqueous solution, a mixture of regio- and stereoisomeric 4π plus 4π photo-dimers **93** were produced (**Figure 32**).

The photolysis of pyridine *N*-oxides in alkaline solution induced ring opening to cyano-dienolates (**Figure 33**) [62].

2-Quinolones undergo 2 + 2 photo dimerization involving the C-3-C-4 double bond [63].

7. Conclusion

The synthesis and reactions of pyridine and its benzo derivatives have been extensively discussed. The Chichibabin synthesis is a notable example of the condensation method of preparing pyridines. Electrophilic substitution reactions occur less readily than the nucleophilic reactions. These reactions have been used for the preparation of versatile intermediates and precursors for biologically active compounds.

Author details

Adebimpe D. Adesina
Federal University of Agriculture, Abeokuta, Ogun State, Nigeria

*Address all correspondence to: adesinaad@funaab.edu.ng

References

[1] Anderson T. Transactions of the Royal Society Edinburgh. 1849;**16**(123): 463

[2] Korner GG. Science National Economic Palermo. 1869:5. reprinted in Calm, A. and Buchka, K. v. Die Chemie des Pyridins und seiner Derivate, Braunschweig, 1889-1891, which contains an excellent account of the early history of these bases

[3] Scriven EFV. In: Katritzky AR, Rees CW, editors. Comprehensive Heterocyclic Chemistry I. Oxford: Pergamon; 1984. 2, 165, 216, 220

[4] Comins DL, Joseph SP. In: Katritsky AR, Rees CW, Scriven EFV, editors. Comprehensive Heterocyclic Chemistry II. Elsevier, Oxford; 1996. 5, 37, 41, 68, 76, 78, 80, 82, 98

[5] Joule JA, Mills K. Heterocyclic Chemistry. 5th ed. West Sussex: Blackwell; 2010

[6] Frank RL, Seven RP. Pyridines. IV. A study of the Chichibabin synthesis. Journal of the American Chemical Society. 1949;**71**:2629-2635

[7] Weiss M. Acetic acid-ammonium acetate reactions. An improved Chichibabin pyridine synthesis. Journal of the American Chemical Society. 1952; **74**:200-202

[8] Naito T, Yoshikawa T, Ishikawa F, Isoda S, Omura Y, Takamura I. Synthesis of 3-pyridinols. I. Reaction of 5-unsubstituted oxazoles with acrylonitrile. Chemical & Pharmaceutical Bulletin. 1965;**13**:869

[9] Ya Kondrat'eva G, Huan C-H. Doklady Akademii Nauk SSSR. 1965;**164**: 816. (Chem. Abstr., 1966, 64, 2079)

[10] Abbiati G, Arcadi A, Bianchi G, Giuseppe SD, Marinelli F, Rossi E. Sequential amination/annulation/aromatization reaction of carbonyl compounds and propargylamine: A new one-pot approach to functionalized pyridines. The Journal of Organic Chemistry. 2003;**68**:6959

[11] Lochte HL, Pittman AG. Notes - The nitrogen compounds of petroleum distillates. XXIX. Identification of 5-methyl-6,7-dihydro-1,5-pyridine. The Journal of Organic Chemistry. 1960;**25**: 1462

[12] Trost BM, Gutierrez AC. Ruthenium-catalyzed cycloisomerization-6pi-cyclization: A novel route to pyridines. Organic Letters. 2007;**9**:1473

[13] Donohoe TJ, Johnson DJ, Mace LH, Thomas RE, Chiu JYK, Rodrigues JS, et al. The ammonia-free partial reduction of substituted pyridinium salts. Organic & Biomolecular Chemistry. 2006;**4**:1071

[14] Baudoux J, Judestein P, Cahard D, Plaquevent J-C. Design and synthesis of novel ionic liquid/liquid crystals (IL2Cs) with axial chirality. Tetrahedron Letters. 2005;**46**:1137

[15] Borisov AB, Belsky VK, Goncharova TV, Borisova GN, Osmanov VK, Matsulevich ZV, et al. Sulfenyl halides in the synthesis of heterocycles. 2. Cyclization in reactions of hetarenesulfenyl chlorides with 3,3-dimethyl-1-butene. Chemistry of Heterocyclic Compounds. (English Translation). 2005;**41**:771

[16] Sliwa W. Quaternary salts of pyridines and related compounds. Current Organic Chemistry. 2003;7:995

[17] King JA, Bryant GL. Preparation and characterization of crystalline N-acylammonium salts. The Journal of Organic Chemistry. 1992;**57**:5136

[18] Breslow R, Brandl M, Hunger J, Turro N, Cassidy K, Krogh-Jespersen K, et al. Pyridine complexes of chlorine atoms. Journal of the American Chemical Society. 1987;**109**:7204

[19] Tornieporth-Oetting I, Klapoetke T, Passmore J, Anorg Z. The reactivity of the I3+ cations to ammonia, nitriles and pyridine. Allgemeine Chemistry. 1990; **586**:93 (Chem. Abstr., 1991, 114, 16 578)

[20] Silvester MJ. Fluoroheterocyclic compounds: Synthesis, reactions, and commercial applications. Aldrichimica Acta. 1991;**24**:31

[21] Adachi K, Ohira Y, Tomizawa G, Ishihara S, Oishi S. Electrophilic fluorination with N,N'-difluoro-2,2'-bipyridinium salt and elemental fluorine. Journal of Fluorine Chemistry. 2003;**120**:173

[22] Marchais-Oberwinkler S, Nowicki B, Pike VW, Halldin C, Sandell J, Chou Y-H, et al. N-Oxide analogs of WAY-100635: New high affinity 5-HT1A receptor antagonists. Bioorganic & Medicinal Chemistry. 2005;**13**:883

[23] Krebs FC. Functionalisation of the hinge region in receptor molecules for explosive detection. Tetrahedron Letters. 2003;**44**:6643

[24] Razus AC, Birzan L, Nae S, Cristian L, Chiraleu F, Cimpeanu V. Azulene-1-azopyridine 1'-oxides. Dyes and Pigments. 2003;**57**:223

[25] Murray RW, Singh M, Jeyaraman R. Dioxiranes. 20. Preparation and properties of some new dioxiranes. Journal of the American Chemical Society. 1992;**114**:1346

[26] Shereshovets VV, Bachanova LA, Komissarov VD, Vostrikov NS, Tolstikov GA. Izvestiya Akademi Nauk SSSr, Seriya Khimicheskaya. 1982:1922. (Chem. Abstr., 1983, 98, 4465)

[27] Youssif S. Recent trends in the chemistry of pyridine N-oxides. ARKIVOC. 2001;**2001**:242

[28] Jain SL, Sain B. Ruthenium catalyzed oxidation of tertiary nitrogen compounds with molecular oxygen: An easy access to N-oxides under mild conditions. Chemical Communications. 2002;(10):1040

[29] Zhang Y, Lin R. Some deoxygenation and reduction reactions with samarium diiodide. Synthetic Communications. 1987;**17**:329

[30] Akita Y, Misu K, Watanabe T, Ohto A. Deoxygenation of heterocyclic N-oxides by chromium (II) chloride. Chemical & Pharmaceutical Bulletin. 1976;**24**:1839

[31] Malinowski M, Kaczmarek L. Titanium (0) reagents; 2. A selective and efficient deoxygenation of halogen containing heteroaromatic N-oxides. Synthesis. 1987;(11):1013

[32] Balicki R, Kaczmarek L, Malinowski M. Selective reduction of the N–O bond in heteroaromatic N-oxides by TiCl4/SnCl2. Synthetic Communications. 1989;**19**:897

[33] Balicki R. Efficient deoxygenation of heteroaromatic N-oxides with ammonium formate as a catalytic hydrogen transfer agent. Synthesis. 1989;(8):645

[34] Katritzky AR, Lam JN. Heterocyclic N-oxides and N-imides. Heterocycles. 1992;**33**:1011

[35] Olah GA, Arvanaghi M, Vankar YD. Synthetic methods and reactions. 87. Deoxygenation of pyridine N-oxides with sodium iodide-trimethyl(ethyl) amine-/sulfur dioxide complexes. Synthesis. 1980;(8):660

[36] Katritzky AR, Fan W-Q. Mechanisms and rates of the electrophilic substitution reactions of heterocycles. Heterocycles. 1992;**34**:2179

[37] Bakke JM. Nitropyridines, their synthesis and reactions. Journal of Heterocyclic Chemistry. 2005;**42**:463

[38] den Hertog HJ, den Does LV, Laandheer CA. Recueil des Travaux Chimiques des Pays-Bas. 1962;**91**:864

[39] Pearson DE, Hargreave WW, Chow JKT, Suthers BR. The swamping catalyst effect. III. The halogenation of pyridine and picolines. The Journal of Organic Chemistry. 1961;**26**:789

[40] McElvain SM, Goese MA. The sulfonation of pyridine and the picolines. Journal of the American Chemical Society. 1943;**65**:2233

[41] Beisler JA. A short synthesis of several gambir alkaloids. Tetrahedron. 1970;**26**: 1961

[42] McCasland GE. The preparation of 8-quinolinesulfonic acid. The Journal of Organic Chemistry. 1946;**11**:277

[43] Tomasik P, Woszczyk A, Abramovitch RA. Oxidation of pyridines with copper sulfate. Journal of Heterocyclic Chemistry. 1979;**16**:1283

[44] Gillard RD, Hall DPJ. Simple oxidations of pyridines: Zinc sulphates or natural sand as remarkably specific catalysts. Journal of the Chemical Society, Chemical Communications. 1988;(17):1163

[45] Baik W, Yun S, Rhee JU, Russell GA. DMSO-Ac2O promoted nitration of isoquinolines. One step synthesis of 1-nitroisoquinolines under mild conditions. Journal of Chemical Society, Perkin Transactions. 1996;(15):1777

[46] Evans JCW, Allen CFH. Organic Synthesis, Collection, II. 1943:517

[47] Illuminati G, Stegel F. The formation of anionic σ-adducts from heteroaromatic compounds: Structures, rates and equilibria. Advances in Heterocyclic Chemistry. 1983;**34**:305

[48] Leffler MT. Organic Reactions. 1942; **1**:91

[49] McGill CK, Rappa A. Advances in the Chichibabin reaction. Advances in Heterocyclic Chemistry. 1988;**44**:2

[50] Vorbruggen H. Advances in amination of nitrogen heterocycles. Advances in Heterocyclic Chemistry. 1990;**49**:117

[51] Zoltewicz J, Helmick LS, Oestreich TM, King RW, Kandetzki PE. Addition of amide ion to isoquinoline and quinoline in liquid ammonia. Nuclear magnetic resonance spectra of anionic sigma complexes. The Journal of Organic Chemistry. 1973;**38**:1947

[52] Berstrom FW. Justus Liebigs Annalen der Chemie. 1935;**515**:34

[53] Ewing GW, Steck EA. Absorption spectra of heterocyclic compounds. I. Quinolinols and isoquinolinols. Journal of the American Chemical Society. 1946;**68**: 2181

[54] Postigo A, Rossi RA. A novel type of nucleophilic substitution reactions on nonactivated aromatic compounds and benzene itself with trimethylsiliconide anions. Organic Letters. 2001;**3**:1197

[55] Postigo A, Vaillard SE, Rossi RA. Reactions of trimethylstannide and trimethylsiliconide anions with aromatic and heteroaromatic substrates. Journal of Physical Organic Chemistry. 2002;**15**:889

[56] Chichibabin AE. Chemische Berichte. 1879;**1923**:56

[57] Gros P, Fort Y. Direct synthesis of unsymmetrical bis-heterocycles from 2-heterosubstituted 6-lithiopyridines. Journal of Chemical Society, Perkin Transactions. 1998;**1**:3515

[58] Gilman H, Soddy TS. Carbonation of lithium derivatives of certain quinolines and isoquinolines. The Journal of Organic Chemistry. 1957;**22**:565

[59] Wommack JB, Barbee TG, Thoennes DJ, McDonald MA, Pearson DE. Synthesis of quinoline- and isoquinolinecarboxaldehydes. Journal of Heterocyclic Chemistry. 1969; **6**:243

[60] Fernandez M, de la Cuesta E, Avendano C. Metallation of 2 (1H)-quinoline: Synthesis of 3-substituted compounds. Synthesis. 1995;(11):1362

[61] Baradarani MM, Dalton L, Heatley F, Joule JA. Cyclising nucleophilic addition to azinium systems. Part 1. Reaction of 3-indol-2-ylpyridine, 3-indol-2-ylquinoline, 4-indol-2-ylisoquinoline and pyrido [3, 4-a]carbazoles with acetic anhydride. Journal of Chemical Society, Perkin Transactions. 1985;**1**:1503

[62] Buchardt O, Christensen JJ, Nielsen PE, Koganty RR, Finsen L, Lohse C, et al. Photochemical studies. XXII. Photochemical ring-opening of pyridine N-Oxide to 5-Oxo-2-pentenenitrile and/or 5-Oxo-3-pentenenitrile. A reassignment of structure. Acta Chemica Scandinavica, Series B. 1980;**34**:31

[63] Woodward RB, Doering WE. The total synthesis of quinine. Journal of the American Chemical Society. 1945;**67**:860

Chapter 3

Structural Diversity in Substituted Pyridinium Halocuprates(II)

Marcus R. Bond

Abstract

The flexible coordination sphere of the Jahn-Teller active Cu(II) ion provides access to a full spectrum of coordination geometries from 4-coordinate (tetrahedral or square planar) to 6-coordinate elongated octahedral. This is further enhanced in anionic halide complexes by the ability of the halide ligand to bridge between Cu(II) centers to generate extended oligomeric or polymeric complexes. Coordination geometry and extended structure of the anionic complex is very sensitive to the nature of the organic counterion. This is especially true for planar substituted pyridinium cations in which minor changes in the nature or position of the substituted group can generate completely different halocuprate(II) structures. Early work focused on reducing ligand-ligand repulsion through strong hydrogen bonding with the organic cation in order to manipulate the Cu(II) coordination sphere. However, many unique structures have been found in which quaternary pyridinium cations were employed-including the remarkable thermochromic compound (1,2,6-trimethylpyridinium)$_2$C uCl$_4$- in which strong hydrogen bonding is absent. More recently aminopyridinium cations, which further increase structural diversity not only through the possibility of having mono- or di-protonated cations but also the ability of monoprotonated cations to coordinate to the Cu(II) center through the amino group, have been investigated.

Keywords: substituted pyridinium compounds, structural chemistry, copper(II) complexes, Jahn-Teller effect

1. Introduction

The d^9 Cu^{2+} ion is, perhaps, the best known example of a Jahn-Teller active ion with an extremely flexible coordination sphere—to the extent that it has been referred to as "a chameleon of coordination chemistry" [1]. To summarize standard arguments [2]: in octahedral coordination the degenerate electronic ground state of d^9 Cu^{2+} is further stabilized by distortion (typically by elongation of one octahedral axis) to yield a non-degenerate electronic ground state (**Figure 1**). (In tetrahedral coordination a flattening distortion toward the square planar limit serves a similar purpose).

The elongated octahedral geometry can be described as 4 + 2 coordinate with four short coordinate covalent bonds (typical Cu-Cl bond lengths in the 2.2–2.4 Å range) and two longer semicoordinate bonds (typical Cu···Cl bond lengths ranging from 2.7

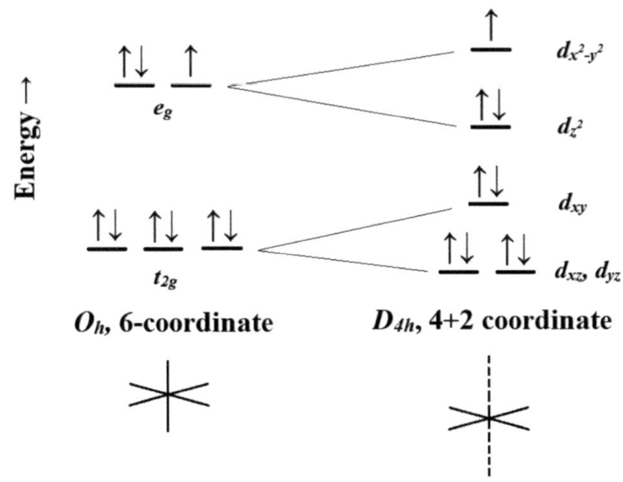

Figure 1.
Schematic diagram of the d-orbital splitting of an octahedral $CuCl_6^{4-}$ complex undergoing an elongated axis Jahn-Teller distortion.

to well over 3 Å). The two semicoordinate bonds can be of different lengths leading to 4 + 1 + 1′ coordination. Further elongation of the longer bond eventually leads (conceptually) to removal of that ligand and results in 4 + 1 coordination. In some situations the semicoordinate bond of a 4 + 1 complex is short enough (Cu-Cl distance of 2.6 Å or less) to become a coordinate bond and yielding a five coordinate geometry that is usually found somewhere on the continuum between trigonal bipyramidal and square pyramidal due to a second order Jahn-Teller distortion [3]. Removal of the other semicoordinate ligand yields a 4-coordinate complex that is usually found in a flattened tetrahedral geometry with *trans* Cl-Cu-Cl angles between 130 and 140°. However, these angles are found with a range of values, including 180° in the square planar limit. Square planar $CuCl_4^{2-}$ complexes are rare, and square planar $CuBr_4^{2-}$ complexes are almost completely unknown—a fact attributed to the stronger ligand-ligand repulsion between the larger bromide ions. Thus a wide range of coordination numbers and geometries is available to a copper(II) complex, as depicted in **Figure 2**.

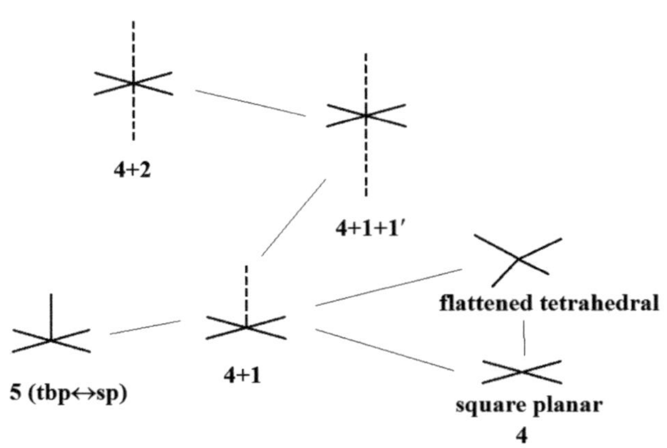

Figure 2.
Coordination numbers and geometries available to a Cu^{2+} complex.

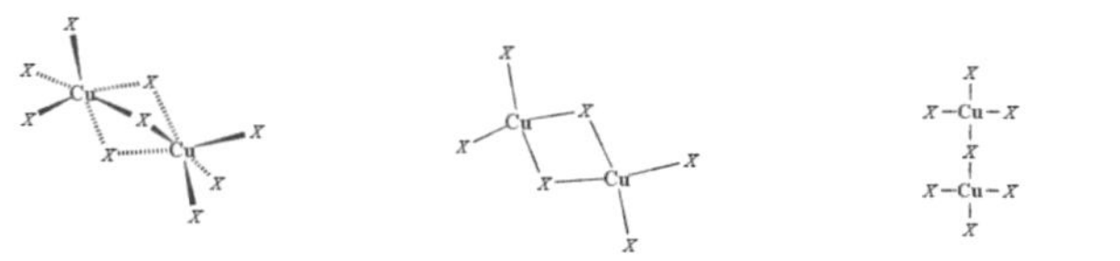

Figure 3.
Bridging modes available to halocuprate(II) complexes.

The focus of this chapter is on copper(II) halide complexes—ergo the bond length and angle examples previously given. Chloride complexes have been more thoroughly investigated than bromide complexes [4]. This may be due to the wide variety of colors exhibited by chloride complexes of differing geometries and coordination numbers: ranging from reds to orange to yellow to greens. The author has found in some situations crystals of three or four different colors growing in the same beaker as identifiably distinct compounds. In contrast, the visible spectra of bromide complexes is dominated by ligand-to-metal charge transfer to give an intensely dark purple color with little variation across compounds [5, 6]. In both cases, however, the chloride and bromide ligands can bridge between copper(II) centers, as shown in **Figure 3**, to increase structural complexity by forming oligomeric or polymeric species.

2. The utility of pyridinium cations

2.1 Anionic halocupreates(II)

Halocuprate(II) complexes, whether isolated monocopper(II) or oligomeric or polymeric, are anionic and in crystalline solids are always accompanied by a cationic species. Earliest studied compounds used highly symmetric alkali metal or ammonium cations [7–10] but a wide variety of conterions have been used, including cationic inorganic or organometallic complexes. A broad array of organic cations, often readily and commercially available, have been most frequently employed [2].

2.2 Structures with organoammonium cations

Organoammonium cations can quickly become bulky with larger groups and higher degrees of substitution which prevents formation of polymeric complexes. Consider, for example, the $(Et)_x(Me)_{4\text{-}x}$ series of chlorocuprates with an approximate 1:1 ratio of organic cation to $CuCl_2$. For tetramethylammonium (x = 0) a tribridged chain of face sharing $CuCl_6$ octahedra is found in $(Me_4N)CuCl_3$ [11]. For x = 1, in $EtMe_3N)_4Cu_5Cl_{14}$ a linear chain is also found, but with a mix of bi- and tribridging that "stretches" the chain in order to accommodate the bulkier organic cation [12]. For x = 2, a $(Cu_4Cl_{11}{}^{3-})_n$ with even more frequent bibridging is found [13]. Organic cations with x = 3 and 4 are so bulky that a continuous chain is no longer possible, and isolated $Cu_3Cl_9{}^{3-}$ [2, 14] and $Cu_4Cl_{12}{}^{4-}$ [15] oligomers are found. Primary alkylammonium cations favor formation of layer perovskite A_2CuX_4 (A = monopositive cation and X = Cl, Br) compounds in which layers of corner sharing CuX_6 octahedra are separated by bilayers of organic cations with $_{-}NH_3$ head groups directed toward the inorganic layer to form multiple N-H...X hydrogen bonds [16].

2.3 Structures with anilinium versus structures with pyridinium cations

Substituted planar aromatic cations, i.e. anilinium or pyridinium, provide a wealth of counterion possibilities while avoiding the bulkiness found with organoammonium ions. With a protonated $_NH_3^+$ head group, an anilinium cation can function structurally as a primary ammonium cation. Indeed, (anilinium)$_2$CuCl$_4$ exists as a layer perovskite system [17]. With pyridinium cations, however, the ring nitrogen acts as a single hydrogen bond donor (when protonated) that generally forms a single direct or a bifurcated hydrogen bond to halide(s) on a neighboring complex. The ring nitrogen can also be readily quaternized to examine halocuprate(II) structures in the absence of N-H hydrogen bonding. Pyridinium compounds have been more heavily studied than anilinium compounds: a Cambridge Structural Database (CSD) substructure search [18] on the anilinium core versus the pyridinium core with tetrachlorocuprate(II) complexes yields 24 compounds (14 of which are layer perovskites) and 120 compounds, respectively. This difference might be due to the tendency for anilinium cations to decompose in the presence of Cu(II) (presumably acting as a one-electron oxidation catalyst). In the author's experience, crystal growth under ambient conditions of anilinium chlorocuprate(II) compounds often yields brown or black residues. Indeed, there are no reported structures of ring substituted methyl or dimethylanilinium chlrocuprate(II) compounds in the CSD, whereas there are a handful of chlorozincate(II) compounds (where Zn(II) with a d^{10} configuration does not have access to a + 1 oxidation state) and numerous ring substituted methyl or dimethylpyridinium chlorocuprate(II) compounds.

3. A_2CuX_4 compounds containing isolated CuX_4^{2-} complexes

3.1 General properties

Compounds containing isolated CuX_4^{2-} complexes are readily prepared by slow evaporation of a solution, e.g. hydrohalic acid or alcoholic, containing a stoichiometric amount of organic cation halide and copper(II) halide, and examples are regularly reported. These most commonly contain flattened tetrahedral complexes with *trans* *X*-Cu-*X* angles in the range 130–140°. Strong hydrogen bonding between the organic cation and the halide ions of the inorganic complex is thought to reduce ligand-ligand repulsion and allow for larger *trans* *X*-Cu-*X* angles. Examples of these complexes are more rare, especially those with larger *trans* angles, Crystals containing chloro complexes with the commonly found *trans* angle are yellow/orange in color and become progressively more green in coloras the *trans* angle increases to reach an intensely dark green color at the square planar limit (180°) [19].

3.2 Square planar complexes

A recent example of square planar $CuCl_4^{2-}$ is in the isonicotinamidium (H-INAc) salt where strong bifurcated hydrogen bonds from the protonated ring nitrogen stabilize the *sp* geometry (**Figure 4**). This particular compound is also interesting since there is a companion compound in which neutral isonicotinamide molecules coordinate to the copper(II) center as terminal ligands in di-μ_2-chloro polymeric chains. Exposure of these chains to moist HCl vapor protonates the pyridine and generates the *sp* complex in a reversible process [20]. In some cases green *sp* complexes undergo

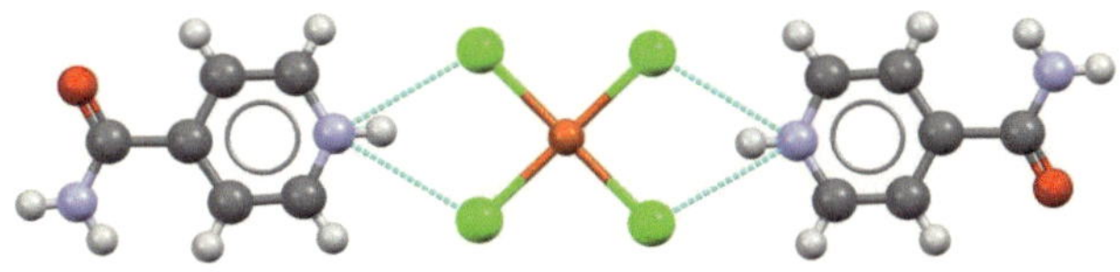

Figure 4.
Ball-and-stick model of the formula unit of bis(isonicotinadium) tetrachlorocuprate(II) showing the strong, bifurcated hydrogen bonds that stabilize the sp geometry.

abrupt (green to yellow) thermochromic phase transitions to *tet* complexes on heating as increased thermal motion weakens the hydrogen bonding that stabilizes the *sp* geometry [19]. While no such transition is reported for the H-INAc salt, it is possible that heating might result in deprotonation of the pyridine before a transition occurs.

3.3 Polymorphism and supramolecular interactions

Polymorphic crystalline forms of these systems can be obtained and studied with different polymorphs possible upon crystallization from different solvents or using different methods. An example occurs with 2,6-dimethylpyridinium in which a monoclinic structure (*C2/c*) is obtained from acidic aqueous solution [21] and an orthorhombic (*Pbcn*) polymorph is obtained from ethanol [22]. Supramolecular interactions were examined in both polymorphs, which illustrates a common application of A_2CuX_4 systems. Since they are readily prepared, it is convenient to use them to study supramolecular interactions with variations in pyridinium ring substitution, e.g. a recent study of halogen bonding in (chloromethyl)pyridinium salts [23].

3.4 Catalytic ring substitution

Willett et al. provided a classic series of papers detailing Cu(II) as a catalytically active species in serendipitous ring substitution reactions of pyridines. For example, recrystallization of 2-amino, 3-methylpyridine with $CuBr_2$ in a slightly acidic solution yielded partial bromination of the pyridine to give (2-amino, 5-bromo, 3-methylpyridinium)(2-amino, 3-methylpyridinium) tetrabromocuprate(II) [24]. Likewise, recrystallization of 2,6-diaminopyridine with $CuCl_2{\cdot}2H_2O$ in slightly acidic solution yields (2,6-diamino, 3,5-dichlropyridinium) tetrachlorocuprate(II) [25] (the bromo analog more recently reported [26]).

3.5 Structural complexity

A_2CuX_4 systems can also provide examples of structural complexity, e.g. through complex packing arrangements or symmetrically inequivalent CuX_4^{2-} complexes with different degrees of flattening. Well known older examples, the incommensurate phase of $[(CH_3)_4N]CuCl_4$ [27] and the thermochromic compound $[CH_3CH_2NH_3]_2CuCl_4$ (which contains three distinctly different $CuCl_4^{2-}$ complexes with one unit cell axis length ~ 45 Å) [28], do not contain pyridinium cations but there are more recent examples that do. The compound [bis(pyridinium-3-ylmethyl) ammonium]$_4(CuCl_4)_5Cl_2$ contains four distinct $CuCl_4^{2-}$ complexes with *trans* Cl-Cu-Cl angles ranging from 128 to 155° [29]. The high symmetry compound (1,3,4-trimethylpyridinium)$_2CuCl_4$ crystallizes in orthorhombic *Fdd*2 with complex anions found between layers of organic cations. The diamond glide symmetry generates a four organic cation layer repeat sequence and leads to a ~ 35 Å *b*-axis length. The corresponding bromide compound is in lower symmetry monoclinic $P2_1/c$ with symmetrically inequivalent organic cations that are segregated into separate layers, as another form of structural complexity [30].

This *Fdd*2 structure is found across a range of (1,3,4-trimethylpyridinium)$_2MCl_4$ compounds (*M* = Co, Ni, Zn [31–33]) but larger metal ions (Mn, Cd [34, 35]) crystallize in monoclinic *C*2/*c*. A CSD search shows that *C*2/*c* is the second most commonly reported space group for pyridinium A_2CuCl_4 compounds (~40 structures) and slightly exceeds the number of compounds reported in the most commonly reported space group for all compounds, monoclinic $P2_1/c$. (The most commonly reported space group from pyridinium A_2CuCl_4 structures is triclinic $P\overline{1}$ with ~60 structures.) Since in the *C*2/*c* structure both organic cations are symmetrically equivalent, this suggests a strategy in pursuing structurally complex compounds by mixing different organic cations to give $A'ACuCl_4$ structures. A few examples are known, such as the 2-amino-3-methylpyridinium example cited above in which organic cations are similar, or the (dimethylammonium)(3,5-dimethylpyridinium)$CuCl_4$ structure [36] where the organic cations are quite different. A systematic study could be conducted by redissolving existing stocks of A_2CuCl_4 compounds in a 1:1 molar ratio and recrystallizing.

4. Quasi-planar oligomers

4.1 Overview

Halocuprate(II) complexes can form linear multicopper complexes through edge sharing of neighboring CuX_4 complexes. At the simplest level this is a dicopper(II) complex which, with a monopositive organic cation, has the typical formulation $A_2Cu_2X_6$ for a 1:1 organic cation:Cu(II) stoichiometry More copper rich stoichiometries are needed for longer $Cu_nX_{2n+2}{}^{2-}$ oligomers. Crystallization of a particular type of oligomer is not predictable, unlike the 2:1 stoichiometry A_2CuX_4 compounds which are readily formed. Thus it is common when exploring the halocuprate(II) structural landscape to prepare solutions of different stoichiometries, e.g. 2:1, 1:1, and 1:2. The stoichiometry of the crystals obtained from solution is not necessarily the same as starting stoichiometry. These compounds require crystallographic analysis to establish their identities as dissolution destroys the compounds.

4.2 Stacking of quasiplanar oligomers

With bulky organic cations the oligomers are isolated and are formed from edge-sharing CuX_4 flattened tetrahedra. For planar or less bulky cations, the oligomers are now quasi planar, formed from edge sharing of CuX_4 square planes, and are no longer isolated with halide ions from one oligomer form semicoordinate bonds with Cu(II) centers on neighboring oligomers to aggregate into stacks [4]. Neighboring oligomers are offset from each other by a half-integral multiple of a CuX_4 edge length with the pattern simply communicated by a bracketed pair. For example, 2[1/2,1/2] indicates that neighboring dicopper oligomers are offset from one another by ½ an edge length parallel to the long axis of the oligomer and ½ an edge length perpendicular (the 2 in front of the bracket identifies these as dicopper(II) complexes). These stacking

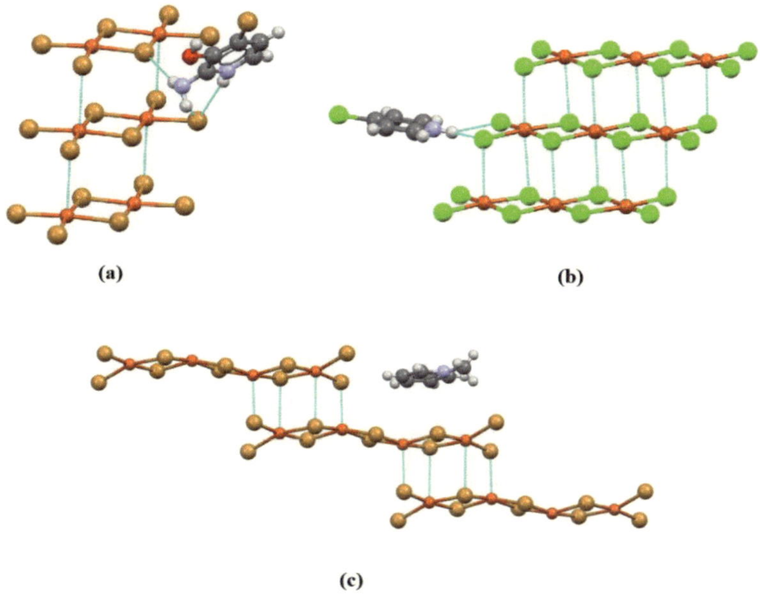

Figure 5.
(a) 2[1/2,1/2] stacking in (2-amino-4-bromo-3-hydroxopyridinium)$_2$Cu$_2$Br$_6$·2H$_2$O (water molecules omitted for clarity) [38], (b) 3[1/2,1/2] stacking in (4-chloropyridinium)$_2$Cu$_3$Cl$_8$ [39], and (c) 4[5/2,1/2] stacking in (1-meth ylpyridinium)$_2$Cu$_4$Br$_{10}$ [40].

patterns vary by compound, and can become more complicated with different oligomers in the stack having different stacking environments [37]. A selection of different oligomer structures with associated stacking patterns are shown in **Figure 5**.

4.3 Pyridinium cation interactions in oligomer stacking

While many different kinds of cations generate oligomer stacks, distinctions can be observed for pyridinium cations. Where hydrogen bonding is present, the oligomer is often terminated with a bifurcated hydrogen bond which mimics the bibridged structure within the oligomer, as shown in **Figure 6**.

Oligomer stacks can often be visualized as sections of a layer from the CuX_2 parent structure. This is particularly true for situations where hydrogen bonding is not possible and the structures can be described as CuX_2 layers in which organic cation pairs replace $(Cu_nX_{2n-2})^{2+}$ fragments, as illustrated in **Figure 7** for (1-methylp yridiniuim)$_2$Cu$_4$Br$_{10}$ [40]. In this case the shape of the organic cation may have more to do with templating the inorganic structure rather than directed intermolecular interactions.

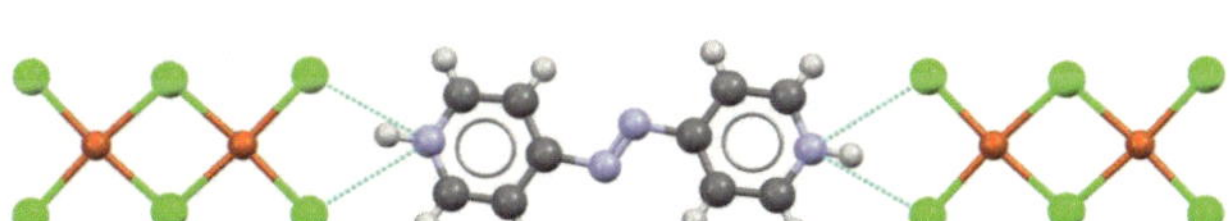

Figure 6.
Hydrogen bonding scheme in (4,4'-diazenediyldipyridinium) Cu$_2$Cl$_6$ [41].

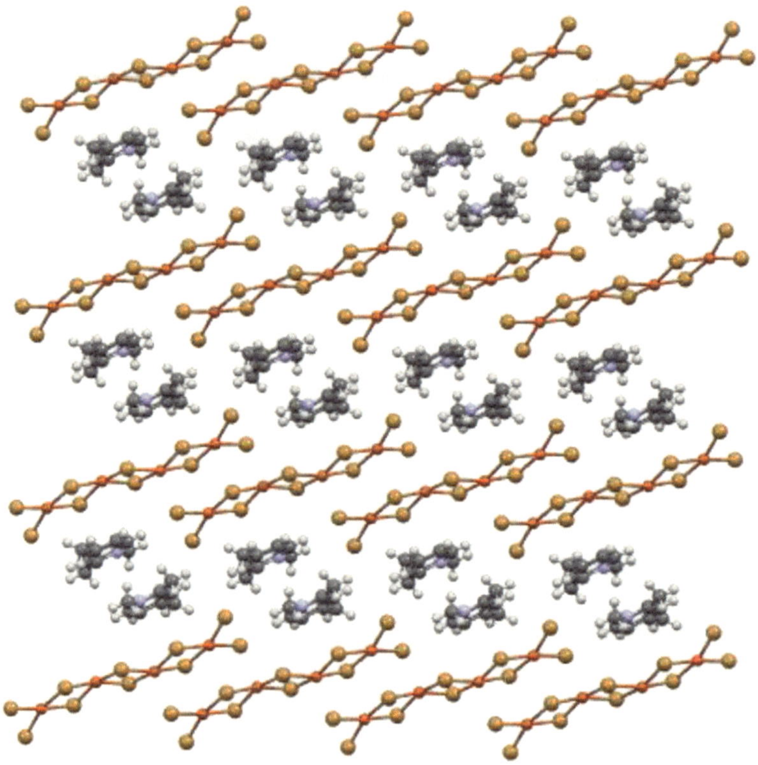

Figure 7.
Layer structure of (1-methylpyridinium)$_2$Cu$_4$Br$_{10}$.

4.4 High-nuclearity oligomers

Oligomers containing more than four Cu(II) centers are rare. Only one example of a $Cu_5X_{12}^{2-}$ oligomer is known (in 2-chloro-1-methylpyridinium)$_2$$Cu_5Br_{10}$ [40]) in which oligomers are still found in isolated stacks (the neutral pentacopper oligomer Cu_5Cl_{10}(n-$CH_3CH_2CH_2OH$)$_2$ has long been known [42]). 1,2 dimethylpyridinium crystallizes with a hexacopper oligomer ($Cu_6Cl_{14}^{2-}$) and a heptacopper oligomer ($Cu_7Br_{16}^{2-}$) [43]. For these longer oligomers the stacks are no longer isolated, but overlap one another to form layers with the organic cations sandwiched between layers, as illustrated in **Figure 8**.

2-Chloro-1-methylpyridinium also crystallizes with a $Cu_7Br_{16}^{2-}$ oligomer in a structure that is almost identical to that of 1,2-dimethylpyridinium. The chloro and methyl groups are disordered, indicating no directed intermolecular interaction with the oligomer and templating of the oligomer on the cation shape may be more important [40].

The longest reported oligomer is found in (3,5-dibromopyridinium)$Cu_{10}Br_{22}$. The authors rationalize formation of the decacopper(II) oligiomer in terms of halogen bonding contacts with the organic cation [44]. This laboratory has obtained also a $Cu_{10}Br_{22}^{2-}$ for 2,6-dimethylpyridinium shown in **Figure 9** [45]. In spite of similarity in shape of the cation and similar unit cell parameters, the two structures are not superimposable. Longer oligomers are certainly possible, but these high copper(II) stoichiometries are rarely investigated so that further discoveries are likely to be serendipitous.

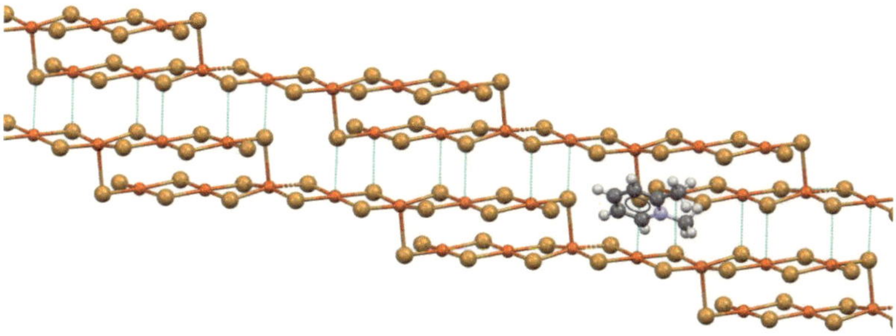

Figure 8.
Overlapping of oligomer stacks to form a layer with one organic cation shown for the compound (1,2-dimethylpyr dinium)$_2$Cu$_7$Br$_{16}$.

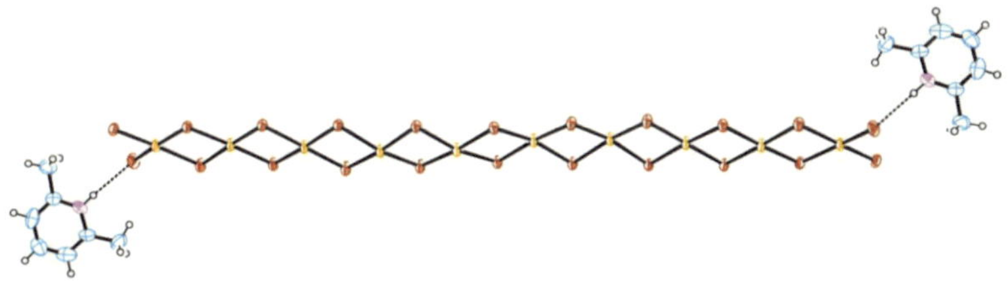

Figure 9.
Thermal ellipsoid plot (at the 50% level) of the formula unit of (2,6-dimethylpyridinium)$_2$Cu$_{10}$Br$_{22}$ at 295 K. H-atoms are drawn as circles of arbitrary radii.

4.5 Asymmetrically bridged dicopper(II) oligomers

An exception to the symmetrically bridged oligomers discussed above are situations where two CuX_4 square planes stack offset from each other to form long semicoordinate bonds between halide ligands of one complex and the Cu(II) center of the other. This leads to two asymmetric bridges with one short Cu-X bond and one long Cu···X bond, as shown in **Figure 10**. Stacks of CuX_4 square planes are not known, but such stacking would lend itself to formation of linear chain structures—which are known.

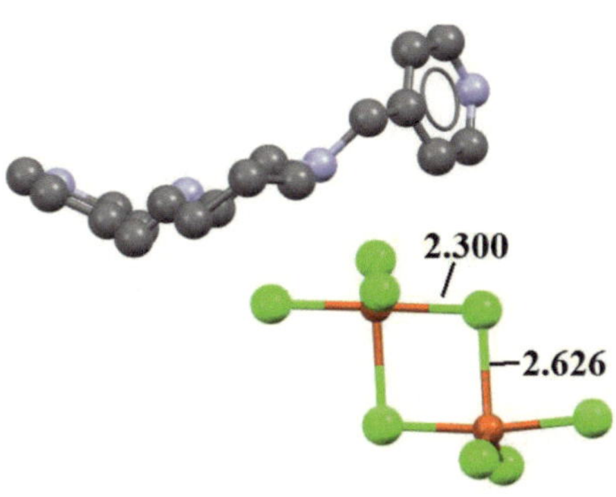

Figure 10.
Asymmetric dicopper(II) oligomer in 4,4'-((cyclohexane-1,2-diylbis(ammoniumdiyl))bis(methylene)) bis(pyridin-1-ium Cu$_2$Cl$_8$·H$_2$O [46]. Hydrogen atoms and water molecule are omitted for clarity. Bond distances in the dimer bridge are shown in units of Å.

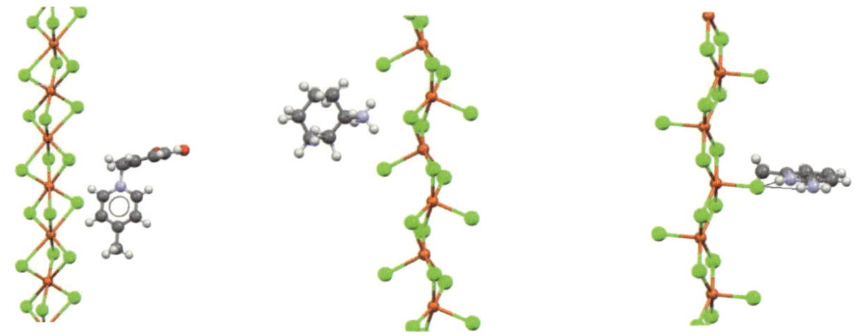

Figure 11.
(a) The tribridged chain in 1-(4′-nitrobenzyl)-4-methylpyridinium $CuCl_3$, (b) the bibridged chain in cyclohexylammonium $CuCl_3$ showing the ammonium head group in postion to hydrogen bond to canted apical chloride ligands, and (c) the bibridged chain in 2-amino-6-methylpyridinium $CuCl_3$ showing hydrogen bonds from the pyridine and amino N atoms that stabilize the apical chloride ligand as non-bridging.

5. $ACuX_3$ linear chains

The $CsNiCl_3$ structure consisting of chains of face-sharing $NiCl_6$ octahedra separated by monopositive cations is the parent structure for $ACuX_3$ linear chains. In the parent structure each Ni(II) ion is linked to its neighbor by three symmetric bridges. However, due to the axial Jahn-Teller distortion of the $CuCl_6$ octahedron, in $ACuX_3$ chains each Cu(II) is linked to its neighbor by two asymmetric bridges and only one symmetric bridge. In the absence of hydrogen bonding interactions from the counterion, in the case of Cs, $(CH_3)_4N^+$, or the quaternary4-methyl-1-(4′-nitrobenzyl)-4-methylpyridinium (shown in **Figure 11a** [47]) cations, the tribridged chain is observed. Hydrogen bonding from the cation to the halides of the chain provides charge compensation to the halides that permits lengthening of the Jahn-Teller elongated Cu-Cl bond. For strong enough hydrogen bonding the semicoordinate bond is broken and the chain is converted into a symmetrically bibridged chain of $CuCl_5$ square pyramids. This is illustrated by the (cyclohexylammonium)$CuCl_3$ structure where hydrogen bonding from the ammonium head group leads to elongation of the Jahn-Teller axial bond to a distance of 3.48 Å (as compared to elongated distances of 2.76 and 2.96 Å in the chain shown in 11(a)). At this distance the chloride ligand is, at best, weakly interacting with the neighboring Cu(II) center and a nascent $CuCl_5$ square pyramid has formed with two symmetric bridges now connecting Cu(II) centers with the apical Cu-Cl bond still canted at an acute angle relative to the chain axis (see **Figure 11b** [48]). In 2-amino-6-methylpyridinium $CuCl_3$ both the pyridine nitrogen and amino group serve as hydrogen bond donors forming multiple hydrogen bonds to the apical chloride and providing sufficient charge compensation that the apical Cu-Cl bond is perpendicular to the chain axis and no longer involved in bridging (see **Figure 11c** [49]).

6. Quaternary pyridinium cations

6.1 Overview

While hydrogen bonding has traditionally been seen as a means to control halocuprate(II) geometry, quaternary pyridinium cations provide a means of examining the effect of a lack of N-H hydrogen bonding on structure. Common methods used (in this laboratory) for preparation of quaternary pyridinium cations are (1) reaction of a substituted pyridine with an excess of iodomethane, then anion exchange with an excess of the appropriate silver halide in H_2O or (2) direct combination of condensed chloro- or bromomethane (in excess) with chilled substituted pyridine in a pressure vessel that is then sealed and allowed to warm to room temperature for 24 hr. Examples of structures containing quaternary pyridinium cations have been mentioned in passing already. Here two particularly interesting systems are discussed.

6.2 "Knobby" chains in (1,4-dimethylpyridinium)$_4$$Cu_5Cl_{14}$

The quaternary 1,4-dimethylpyridinium cation might be expected to crystallize with a tribridged $(CuCl_3)_n$ chain due to the lack of hydrogen bonding—just as the quaternary cation example cited in the previous section and as is, notably, the case for (1,4-dimethylyrdinium) $PbBr_3$ [50]. Instead it templates a highly unusual $(Cu_5Cl_{14})_n$ "knobby" chains in which $CuCl_4$ flattened tetrahedral "knobs" edge-share so as to bridge adjacent Cu(II) ions in the central chain [12]. The chain structure is distinctive since it exhibits the three major coordination numbers of Cu(II). Besides the flattened tetrahedral "knobs" on the outside of the chain, the central Cu(II) ion of the Cu_5Cl_{14} repeat unit has the elongated octahedral 4 + 2 coordination. This bibridges on either side to 5-coordinated Cu(II) complexes with the intermediate *sqp*/*tbp* geometry. These 5-coordinate complexes then bibridge to 5-coordinate complexes on neighboring Cu_5Cl_{14} units to complete the chain (**Figure 12**). With the lack of directed intermolecular interactions, the inorganic structure must template on the cation shape, although it is difficult to discern specifically the driving force behind formation of the structure. The knobs on the chain are found between stacks of organic cations with bridging chlorides close to the pyridinium N atoms as the most prominent point of interaction. There has been no reported attempt to prepare the bromide analog.

6.3 *sp* to *tet* phase transitions in (1,2,6-trimethylpyridinium)$_2$CuX_4

The second interesting case is the (1,2,6-trimethylpyridinium)$_2$CuX_4 system [51]. Both the chloride and the bromide salts contain square planar CuX_4^{2-} in a low temperature phase (below 60°C for the chloride and below −48°C for the bromide). Both compounds undergo a solid-solid phase transition on heating to a high temperature phase in which CuX_4^{2-} is flattened tetrahedral, resulting in a thermochromic transition for the chloride salt (from dark green to yellow). As previously described, these

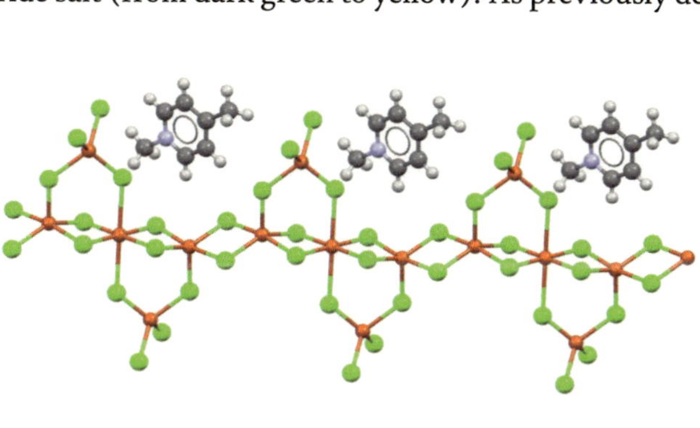

Figure 12.
A section of the "knobby" chains found for (1,4-dimethylpyridinium)Cu_5Cl_{15}.

sp to *tet* transitions are thought to occur due to a weakening of hydrogen bonding. Furthermore *sp* $CuBr_4^{2-}$ is not expected, even with strong hydrogen bonding, due to the greater ligand-ligand repulsion of the larger bromide. So the occurrence of *sp* complexes and *sp* to *tet* transitions in systems without strong hydrogen bonding is highly unusual, if not unprecedented. (It is also worth mentioning that the transitions go from higher symmetry (monoclinic *C*2/m) to lower symmetry (triclinic $P\bar{1}$), which is also quite unusual.)

The quaternary ammonium cations in the low temperature structures form zipper-like ribbons with *sp* $CuCl_4^{2-}$ complexes between the ribbons, as shown in **Figure 13**, that act to template the *sp* geometry. Crystallographic mirror planes are perpendicular to this layer and bisect both the organic cation and the $CuCl_4^{2-}$ complex. The three dimensional structure is built up by stacking these layers so that organic cations of one layer sit above or below $CuCl_4^{2-}$ complexes in another so that the complexes are truly isolated. The structural transformation that occurs on heating disrupts this ribbon structure and results in two symmetrically inequivalent organic cations with aromatic planes tilted with respect to each other.

The 1,2,3-trimethylpyridinium cation has a similar shape as 1,2,6-trimethyl-pyridinium, and also crystallizes as zipper-like ribbons with chlorocuprate(II) complexes between the ribbons to from layers. However the complexes formed are not isolated $CuCl_4^{2-}$ but asymmetrically bridged $[CuCl_3(H_2O)]_2$ dicopper complexes [52]. **Figure 14** illustrates the similar layer structure, right down to similar symmetry: monoclinic *C*2/m. As before, the mirror plane is perpendicular to the layer and bisects the organic cation. In this case, however, the N-atom lies off the mirror plane resulting in two-fold positional disorder that is not present in the 1,2,6-tri-methylpyridinium analog. Does positional disorder stabilize a different structure?

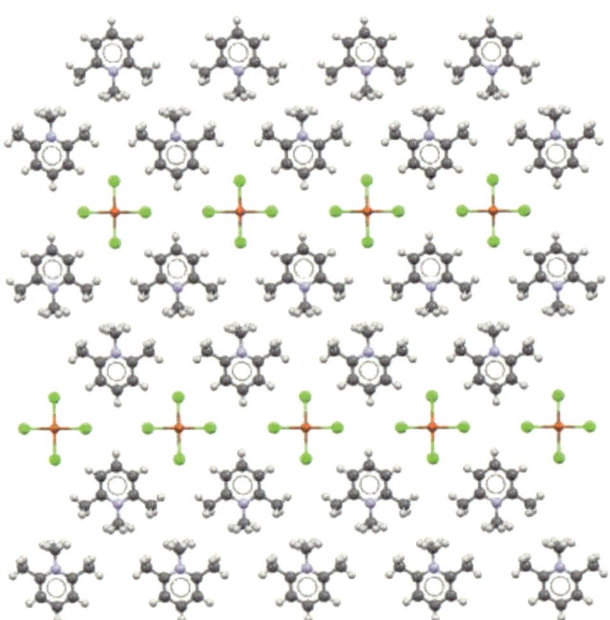

Figure 13.
Layer structure of (1,2,6-trimethylpyridinium)$_2$CuCl$_4$ showing the zipper-like ribbons of organic cations with methyl groups directed toward the center of the ribbon and with isolated sp $CuCl_4^{2-}$ between the ribbons. Mirror plane symmetry is perpendicular to the layer and bisects the organic cations and $CuCl_4^{2-}$ complexes.

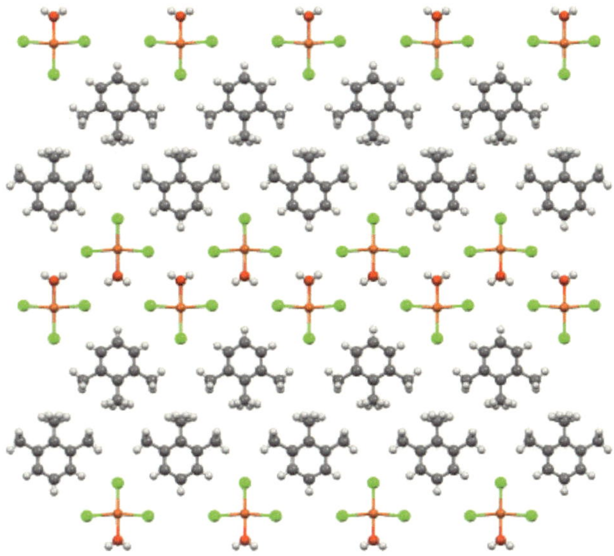

Figure 14.
Layer structure in (1,2,3-trimethylpyridinium) $CuCl_3(H_2O)$ showing the zipper-like ribbons of organic cations separating inorganic complexes. Another layer stacks with offset inorganic complexes to form asymmetrically bridged dimers. Mirror plane symmetry is perpendicular to the layer and bisects the organic cations and the inorganic complex to produce two-fold positional disorder of the organic cation.

That is a difficult question to answer. Nevertheless, the *sp* CuX_4 complex appears to be inaccessible with the 1,2,3-trimethylpyridinium cation. While (1,2,3-trimethylpyridinium)$_2CuBr_4$ is known, it is a conventional flattened tetrahedral complex that is isostructural to (1,2,3-trimethylpyridinium)$_2CoCl_4$ (both in monoclinic *C*2/*c*). Preliminary work from this laboratory indicates that mixed crystals of (1,2,6-trimethylpyridinium)$_x$(1,2,3-trimethylpyridinium)$_{2-x}CuCl_4$ do contain *sp* complexes in a situation where positional disorder is reduced [53]. In any case, these two examples indicate how minor changes in cation can produce major differences in halocuprate(II) structure. It would be interesting to study structures produced by the shape-similar 2,3,4- and 3,4,5-trimethylpyridinium cations which would now also introduce N-H hydrogen bonding interactions. Since these pyridines are not commercially available, a collaboration with a synthetic organic chemist is underway to prepare them.

7. Systems with 3-aminopyridines

Aminopyridines of various types have been prominent in the preparation of halocuprate(II) compounds, with some examples cited already. 3-Aminopyridines, in contrast to 2- or 4-aminooyridines, are capable of protonating both the pyridine and amino N-atoms. Typically the pyridine N-atom protonates first, and if a monoprotonated cation is desired care must be taken to crystallize compounds from solutions that are weakly acidic to avoid diprotonation. At the same time, the monoprotonated cation is capable of coordinating Cu(II) through the amino N-atom—which enables even further structural diversity. Willett et al. reported the earliest compounds with 3-aminopyridine and copper(II) halides. The 3-ammoniumpyridinium cation is found in CuX_4 layer perovskite structures for both the chloride and bromide by virtue

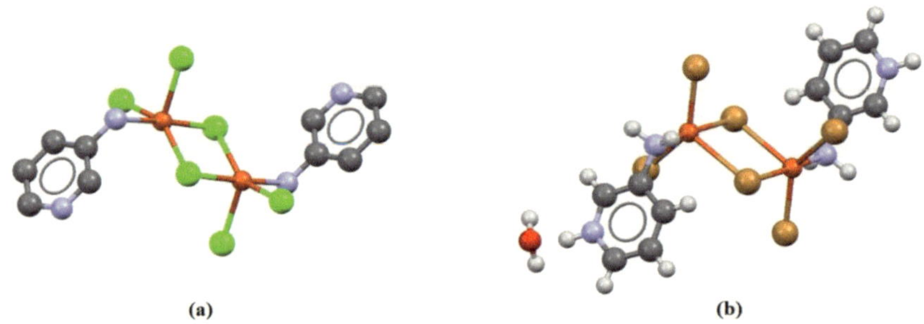

Figure 15.
(a) The symmetrically bridged [$CuCl_3$(3-aminopyridinium]$_2$ dicopper complex and, (b) the asymmetrically bridged [$CuBr_3$(3-aminopyridinium)]$_2$ dicopper(II) complex as a monohydrate.

of the $-NH_3$ head group [54]. Other reported compounds have coordinated 3-aminopyridinium ligands: a symmetrically and asymmetrically bridged dicopper(II) complex for the chloride and the bromide (as a monohydrate in the latter case), respectively, as shown in **Figure 15** [55].

A structure for (3-aminopyridinium)$_2CuCl_4$ has been reported as a typical compound containing flattened tetrahedral $CuCl_4^{2-}$. This reported structure, however, is almost identical to that of (2-aminopyridinium)$_2CuCl_4$, in both unit cell constants and atom positions, and is, in all likelihood, misreported [56]. In order to check this structure, this laboratory undertook crystal growth from acidic aqueous solution and managed to obtain crystals of (3-aminopyridinium)$_2CuCl_4$ by means that can only be described as serendipitous. These crystals gave completely different unit cell constants than the, likely, misreported structure.

In an effort to rationally synthesize crystals of this compound, crystals were grown from various organic solvents (1-proponal, acetonitrile, and tetrahydrofuran) by a thermal gradient technique in sealed, screwcap test tubes placed in a heater block with wells maintained at 40°C. Green crystals of (3-ammoniumpyridinium)$CuCl_4$ were loaded into individual test tubes containing each organic solvent and crystal growth commenced. Another portion of these green crystals were ground together with a stoichiometric amount of 3-aminopyridine and a red-orange solid obtained. Portions of this solid were similarly loaded for crystal growth.

Two new compounds (**Figure 16**) have been obtained in crystal growth from 1-proponal: (1) green crystals of an asymmetrically bridged dicopper complex isomeric to the symmetrically bridged dimer reported by Willett el al.; and (2) red

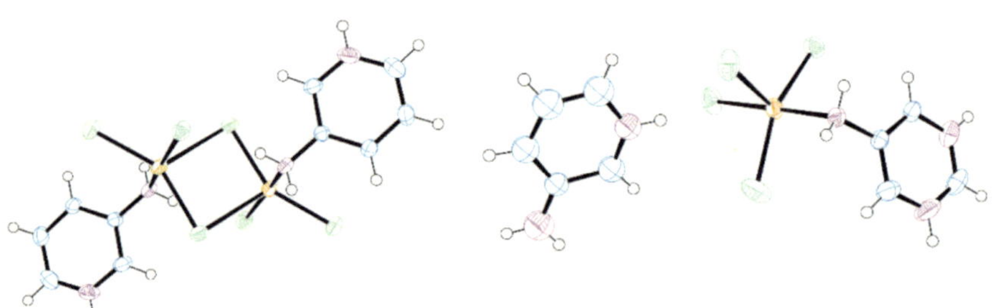

Figure 16.
Thermal ellipsoid plots of (a) the asymmetrically bridged dicopper(II) complex [$CuCl_3$(3-aminopyridine]$_2$ and (b) the formula unit of (3-aminopyridinium) [$CuCl_4$(3-aminopyridinium)].

crystals of a monocopper complex with a coordinated 3-aminopyridinium ligand and a 3-aminopyridinium lattice cation. The latter compound, [3-aminopyridinium] [(3-aminopyridinium)tetrachlorocuprate(II)], is identical in formulation to (3-aminopyridinium)$_2$CuCl$_4$ but with one organic cation moved to the inner coordination sphere. (Crystal growth from acetonitrile yields the known compounds (3-ammoniumpyridinium)CuCl$_4$ and (3-aminopyridinium)$_2$CuCl$_4$) [57]. So far five different compounds have been obtained from the 3-aminopyridine:CuCl$_2$ system just by varying crystal growth conditions. Studies are underway to investigate compounds of the corresponding bromides and of substituted 3-aminopyridines such as 3-amino-2-methylpyridine.

8. Conclusion

The use of pyridinium cations as counterions for halocuprate(II) complexes has provided a wealth of unusual structures due, in part, to the thin profile of the cation, the variety of possible substituent groups, and the easy ability to form a quaternary cation. Previous work has relied heavily on pyridines that are commercially available, but future advances may greatly benefit from targeted pursuit of synthetically prepared pyridines. Mixed cation structures, particularly of A_2CuX_4 systems, have been rarely studied and offer the potential for discovery of new compounds with structural complexity. Different crystallization conditions and solvents have been used in the past to prepare different polymorphs, but now find use in preparation of diverse 3-aminopyridinium halocuprate(II) compounds. While pyridinium halocuprate(II) compounds have been widely studied and dispayed an amazing range of structural diversity, recent discoveries show that they still have the capacity to surprise.

Notes

Structure graphics software used

Ball-and-stick diagrams were plotted using *Mercury 4.0* [58]. Thermal ellipsoid plots were drawn using *ORTEP-3 for Windows* [59].

Crystal data for (2,6-dimethylpyridinium)$_2$Cu$_{10}$Br$_{22}$

Triclinic, $P\bar{1}$, 295 K, a = 9.4862(5) Å, b = 10.0507(5) Å, c = 13.0217(5) Å, α = 104.108(3)°, β = 90.442(3)°, γ = 92.708(3)°, V = 1202.5(1) Å^3, Z = 2. Reflections total/observed = 10,393/3691. θ(max) = 35.139°. Number of least squares parameters = 218. $R_{ovserved}$ = 0.0926, $wR_{observed}$ = 0.2117, goodness of fit = 1.001, $\Delta\rho$(max/min) = 2.181/−2.585 e$^-$/Å^3.

Author details

Marcus R. Bond
Department of Chemistry and Physics, Southeast Missouri State University, Cape Girardeau, MO, USA

*Address all correspondence to: mbond@semo.edu

References

[1] Reinen D. Cu^{2+}, a chameleon in Coodination chemistry. Comments on Inorganic Chemistry. 1983;**2**:227-246. DOI: 10.1080/02603598308078120

[2] Willett RD. Multiple stereochemistry in copper(II) halides. Coordination Chemistry Reviews. 1991;**109**:181-205. DOI: 10.1016/0010-8545(91)80005-X

[3] Reinen D, Friebel C. Cu^{2+} in 5-coordination: A case of a second-order Jahn-teller effect. 2. $CuCl_5^{3-}$ and other $Cu^{II}L_5$ complexes: Trigonal bipyramid or square pyramid. Inorganic Chemistry. 1984;7:791-798. DOI: 10.1021/ic00175a001

[4] Smith DW. Chlorocuprates(II). Coordination Chemistry Reviews. 1976;**21**:93-158. DOI: 10.1016/S0010-8545(00)80445-2

[5] Barnes JC, Hume DN. Copper(II) bromide complexes. I. A spectrophotometric study. Inorganic Chemistry. 1963;**2**:444-448. DOI: 10.1021/ic50007a004

[6] Braterman PS, Copper(II) Bromide Complexes. II. A discussion of the Tetrabromocuprate(II) Spectrum. Inorganic Chemistry. 1963;**2**:448-452. DOI: 10.1021/ic50007a005

[7] McGinnety JA. Cesium tetrachlorocuprate. Structure, crystal forces, and charge distribution. Journal of the American Chemical Society. 1972;**94**:8406-8413. DOI: 10.1021/ja00779a020

[8] Willett RD, Dwiggins C, Kruh RF, Rundle RE. The crystal structures of $KCuCl_3$ and NH_4CuCl_3. The Journal of Chemical Physics. 1963;**38**:2429-2436. DOI: 10.1063/1.1733520

[9] Wells AF. The crystal structure of $CsCuCl_3$ and the crystal chemistry of complex halides ABX_3. Journal of the Chemical Society. 1947:1662-1670. DOI: 10.1039/JR9470001662

[10] Willett RD. Crystal structure of $(NH_4)_2CuCl_4$. The Journal of Chemical Physics. 1964;**41**:2243-2244. DOI: 10.1063/1.1726253

[11] Willett RD, Bond MR, Haije WG, Soonieus OPM, Maaskant WJ. Crystal structures of three phases of tetramethylammonium trichlorocuprate(II) (TMCuC). Inorganic Chemistry. 1988;**27**:614-620. DOI: 10.1021/ic00277a010

[12] Bond MR, Willett RD, Rubenacker GV. Crystal structures and magnetic behavior of two novel copper(II) halide chains. $(C_7H_{10}N)_4Cu_5Cl_{14}$ and $(C_5H_{14}N)_4Cu_5Cl_{14}$: Multiple copper(II) halide geometries. Inorganic Chemistry. 1990;**29**:2713-2720. DOI: 10.1021/ic00340a004

[13] Fujii Y, Wang Z, Willett RD, Zhang W, Landee CP. Crystal structure of $(Et_2Me_2N)_3Cu_4Cu_{11}$: An antiferromagnetic chain of Ferromagnetically coupled tetramers. Inorganic Chemistry. 1995;**34**:2870-2874. DOI: 10.1021/ic00115a013

[14] Akkina S, Bond M. Preliminary crystal structure for tris(triethylmethylmmonium) tetra(μ2-chlorido)pentachloridotricupraate(II). ChemRxiv. Cambridge: Cambridge Open Engage; 2022. This content is a preprint and has not been peer-reviewed. DOI: 10.26434/chemrxiv-2022-zw3wk

[15] Willett RD, Geiser U. Crystal structure of tetrakis(tetraethylammonium) dodecachlorotetracuprate(II), a new

structural type of a tetranuclear copper(II) halide complex. Inorganic Chemistry. 1986;**25**:4558-4561. DOI: 10.1021/ic00245a021

[16] Long GS, Wei M, Willett RD. Crystal structures and magnetic properties of a novel layer perovskite system: 3-PicoliniumylammoniumCuX_4 (X = Cl, Br). Inorganic Chemistry. 1997;**36**:3102-3107. DOI: 10.1021/ic960849+

[17] Larsen KP. The crystal structure of Anilinium Tetrachlorocuprate(II). Acta Chemica Scandinavica. 1974;**28a**:194-200. DOI: 10.3891/acta.chem.scand.28a-0194

[18] Groom CR, Bruno IJ, Lightfoot MP, Ward SC. The Cambridge structural database. Acta Crystallographica. 2016;**B72**:171-179. DOI: 10.1107/S2052520616003954

[19] Willet RD, Haugen JA, Lebsack J, Morrey J. Thermochromism in Copper(II) chlorides. Coordination geometry changes in $CuCl_4^{2-}$ anions. Inorganic Chemistry. 1974;**13**:2510-2513. DOI: 10.1021/ic50140a040

[20] Fellows SM, Prior TJ. Polymorphism and solid-gas-solid reactions of isonicotinic acid, isonicotinamide, and nicotinable copper chloride compounds. Crystal Growth and Design. 2017;**17**:106-116. DOI: 10.1021/acs.cgd.6b01295

[21] Ali BF, Al-Far R, Haddad SF. Hydrogen bonded, π···π stacked and X···π framework structures in Bis(2,6-Lutidinium) Tetrahalocuprate(II) Complexes. Journal of Chemical Crystallography. 2010;**40**:696-701. DOI: 10.1007/s10870-010-9724-8

[22] Awwadi FF, Haddas SF. Polymorphismin2,6-dimethylpyridinium tetrachlorocuprate(II): Theoretical and crystallographic studies. Journal of Molecular Structure. 2012;**1020**:28-32. DOI: 10.1016/j.molstruc.2012.04.015

[23] Protsenko AN, Shakirova OG, Protsenko AG, Kuratieva NV, Fowles SM, Turnbull MM. Effect of isomeric cations of 3(2)-(chloromethyl) pyridine on the the structure and properties of copper(II) and cobalt(II) complexes. Journal of Molecular Structure. 2021;**1240**:130561. DOI: 10.1016/j.molstruc.2021.130561

[24] Place H, Willett RD. Structure of catalytically related species involving copper(II) halides. III. 2-Amino-5-bromo-3-methylpyridinium 2-amino-3-methylpyridinium tetrabromocuprate(II). Acta Crystallographica. 1987;**C43**:1497-1500. DOI: 10.1107/S0108270187091327

[25] Willett RD, West DX. Structures of catalytically related species involving copper(II) halides. IV. Bis(2,6-diamino-3,5-dichloropyridinium) tetrachlorocuprate(II). Acta Crystallographica. 1987;**C43**:2300-2303. DOI: 10.1107/S0108270187087985

[26] Willett RD, Haddad SF, Twamley B. Bis(2,6-diamino-3,5-dibromo) tetrabromocuprate(II). Acta Crystallographica. 2000;**C56**:e437. DOI: 10.1107/S0108270100012105

[27] Sugiyama J, Wada M, Sawada A, Ishibashi Y. Successive phase transitions in $\{N(CH_3)_4\}CuCl_4$. Journal of the Physical Society of Japan. 1980;**49**:1405. DOI: 10.1143/JPSJ.49.1413

[28] BloomquistDR, PressprichMR. Willett RD Thermochromism in Copper(II) halide salts. 4. $[(C_2H_5)_2NH_2]_2CuCl_4$, structure of the high temperature phase and physical characterization of its two phases. Journal of the American Chemical Society. 1988;**110**:7391-7398. DOI: 10.1021/ja00230a020

[29] Wu J-Y, Zhong M-S, Chiang M-H, Bhattacharaya D, Lee Y-W, Lai L-L. Anion-directed Copper(II) Metallocages, coordination chain, and complex double salt: Structures, magnetic properties, EPR spectra, and density functional study. Chemistry-A European Jounral. 2016;**22**:7238. DOI: 10.1002/chem.201505215

[30] BondMR.Bis(1,3,4-trimethylpyridinium) tetrachloridocuprates(II) and bis(1,3,4-trimethylpyridinium) tetrabromidocuprate(II): An examination of the A_2CuX_4 *Fdd*2 structure type. Acta Crystallographica. 2009;**C65**:m279-m283. DOI: 10.1107/S0108270109023968

[31] Bond M. Bis(1,3,4-trimethylpyridin-1-ium) tetrachloro-nickel(II). CSD Communication. 2019:CCDC 1937961. DOI: 10.5517/ccdc.ced.cc231lwn

[32] Bond M. Bis(1,3,4-trimethylpyridin-1-ium) tetrachloro-cobalt(II). CSD Communication. 2019:CCDC 1545929. DOI: 10.5517/ccdc.ced.cc239wx7

[33] Bond M. Bis(1,3,4-trimethylpyridin-1-ium) tetrachloro-zinc(II). CSD Communication. 2019:CCDC 1945500. DOI: 10.5517/ccdc.ced.cc239g2z

[34] Bond M. Bis(1,3,4-trimethylpyridin-1-ium) tetrachloro-manganese(II). CSD Communication. 2019:CCDC 1941035. DOI: 10.5517/ccdc.ced.cc234t13

[35] Bond M. Bis(1,3,4-trimethylpyridin-1-ium) tetrachloro-cadmium(II). CSD Communication. 2019:CCDC 1937858. DOI: 10.5517/ccdc.ced.cc231hk7

[36] Awwadi F, Willett RD, Twamley B, Schneider R, Landee CP. Strong Rail Spin 1/2 Antiferromagnetic Ladder Systems: (Dimethylammonium)(3,5-Dimethylpyridinium)CuX_4, X = Cl, Br. Inorganic Chemistry. 2008;**47**:9327-9332. DOI: 10.1021/ic800905e

[37] Bond MR, Willett RD. An update of pseudoplanar, bibridged $Cu_nX_{2n+2}{}^{2-}$, $Cu_nX_{2n+1}L^-$, and $Cu_nX_{2n}L_2$ oligomer stacking patterns: Structure of 1,2-dimethylpyridinium bis(μ-chloro) trichloroaquadicuprate(II). Inorganic Chemistry. 1989;**28**:3267-3269. DOI: 10.1021/ic00315a038

[38] Willet RD. Structure of catalytically related species involving copper(II) halides. V. $C_5H_6BrN_2O^+.Br^-$ and $2C_5H_6BrN_2O^+.Cu_2Br_4{}^{2-}.2H_2O$. Acta Crystallographica. 1988;**C44**:450-453. DOI: 10.1107/S0108270187010928

[39] Zordan F, Espallargas GM, Brammer L. Unexpected structural homologies involving hydrogen-bonded and halogen-bonded networks in halopyridinium halometallate salts. CrystEngComm. 2006;**8**:425-432. DOI: 10.1039/B602518H

[40] Kelley A, Akkina S, Deverapally GK, Nalla S, Pasam D, Madhabushi S, et al. Nine compounds containing high-nuclearity $[Cu_nX_{2n+2}]^{2-}$ (n = 4, 6, or 7; X = Cl, Br) quasi-planar oligomers. Acta Crystallographica. 2011;**C67**:m22-m34. DOI: 10.1107/S0108270110049024

[41] Klein A. 4,4′-Diazenediyl dipyridinium bis(μ-chloro)-tetrachloro-dicopper(II). CSD Communication. 2019:CCDC 1951255. DOI: 10.5517/ccdc.csd.cc23hfqs

[42] Pon G, Willett RD. Redetermination of $[Cu_5Cl_{10}(n\text{-}C_3H_7OH)_2]$. Acta Crystallographica. 1996;**C52**:1122-1123. DOI: 10.1107/S0108270195012418

[43] Bond MR, Place H, Wang Z, Willett RD, Liu Y, Grigereit TE, et al. Structures and magnetic susceptibility studies of four new high-Nuclearity Copper(II) halide oligomers. Inorganic Chemistry. 1995;**34**:3134-3141. DOI: 10.1021/ic00116a003

[44] Haddad S, Awwadi F, Willett RD. A planar Bibridged $Cu_{10}Br_{22}^{2-}$ oligomer: Dimensional reduction and recombination of the $CuBr_2$ lattice via the N-H···Br and the C-Br···Br synthons. Crystal Growth and Design. 2003;**3**:501-505. DOI: 10.1021/cg030009n

[45] Nalla S, Bond MR. Bis (2,6-dimethylpyridinium) icosidyobromidodecacuprate(ii). CCDC 2206136: Experimental Crystal Structure Determination; 2022. DOI: 10.5517/ccdc.csd.cc2d1npt

[46] Li H, Famulari A, Xin L, Zhou H, Zhang P, Guo F. Stoichiometry mechanosynthesis and interconversion of metal salts containing $[CuCl_3(H_2O)]^-$ and $[Cu_2Cl_8]^{4-}$. CrystEngComm. 2019;**21**:7017-7024. DOI: 10.1029/C9CE00911F

[47] Han S, Liu X-Y, Cai Z-F, Wu Z-P, Yin W-T, Xie X-D, et al. An unexpected coordination polymer containing one-dimensional chlorine bridged copper(II) magnetic chain induced by organic cation: Synthesis, crystal structure, and magnetic properties. Inorganic Chemistry Communications. 2012;**24**:91-94. DOI: 10.1016/j.inoche.2012.08.017

[48] Groenendijk HA, HWJ B, van Duyneveldt AJ, Gaura RM, Landee CP, Willett RD. Crystal structure and magnetic properties of cyclohexylammonium trichlorocuprate(II): A quasi 1d Heisenberg S = 1/2 ferromagnet. Physica B+C. 1981;**106**:47-58. DOI: 10.1016/0378-4363(81)90011-5

[49] Geiser U, Gaura RM, Willett RD, West DX. Structure and magnetism in $ACuCl_3$ salts containing bibridged chains with square-pyramidal coordination geometry. Inorganic Chemistry. 1986;**25**:4203-4212. DOI: 10.1021/ic00243a029

[50] Raptopoulou CP, Terzis A, Mousdis GA, Papavassiliou GC. Preparation, structure, and optical properties of $[CH3SC(NH_2)_2]_3SnI_5$, $[CH_3SC(NH_2)_2][HSC(NH_2)_2]SnBr_4$, $(CH_3C_4H_4NCH_3)PbBr_3$, and $[C_6H_5CH_2SC(NH_2)_2]_4Pb_3I_{10}$. Zeitschrift f u r Naturforschung B. 2002;**57**:645-650. DOI: 10.1515/znb-2002-0609

[51] Kelley A, Nalla S, Bond MR. The square-planar to flattened-tetrahedral CuX_4^{2-} (X = Cl, Br) structural phase transition in 1,2,6-trimethylpyridinium salts. Acta Crystallographica. 2015;**B71**:48-60. DOI: 10.1107/S205252061402664X

[52] Nalla S, Bond MR. $[CuCl_3(H_2O)]^-$ complexes aggregated to form hydrate columns in methyl substituted pyridinium or piperidinium salts. Acta Crystallographica. 2011;**C67**:m185-m194. DOI: 10.1107/S0108270111017306

[53] Al-Mashala H, MacAinsh B, Bond M. Order Versus Disorder in 1,2,3- and 1,2,6- Trimethylpyridinium Chlorocuprate(II) Salts. ChemRxiv. Cambridge: Cambridge Open Engage; 2022. This content is a preprint and has not been peer-reviewed. DOI: 10.26434/chemrxiv-2022-0sqnf

[54] Willett R, Place H, Middelton M. Crystal structures of three new copper(II) halide layered perovskites: Structural, crystallographic, and magnetic correlations. Journal of the American Chemical Society. 1988;**110**:8639-8650. DOI: 10.1021/ja00234a010

[55] Blancette JT, Willett RD. Magnetic and structural correlations in bis(aminopyridinium) hexachlorodicuprate and hexabromodicuprate dehydrate. Inorganic Chemistry. 1988;**27**:843-849. DOI: 10.1021/ic00278a019

[56] Kumar DK, Ballabh A, Jose DA, Dastidar P, Das A. How robust is the N-H···Cl_2-Cu synthon? Crystal structures of some Perchlorocuprates. Crystal Growth & Design. 2005;**5**:651-660. DOI: 10.1021/cg0497086

[57] Bond MR, Balkarnshingh A. The Many Moods of the 3-Aminopyrdinium Clhorocuprate(II) system. In Abstracts of the 2021 American crystallographic association meeting. Acta Crystallographica. 2021;**A77**:a73. DOI: 10.1107/S010876732109927X

[58] Macrae CF, Sovago I, Cottrell SJ, Galek PTA, McCabe P, Pidcock E, et al. Mercury 4.0: From visualization to analysis, design and prediction. Journal of Applied Crystallography. 2020;**53**:226-235. DOI: 10.1107/S1600576719014092

[59] Farrugia LJ. WinGX and ORTEP for windows: An update. Journal of Applied Crystallography. 2012;**45**:849-854. DOI: 10.1107/S0021889812029111

Section 2

Applications of Pyridine Derivatives

Chapter 4

Naturally Isolated Pyridine Compounds Having Pharmaceutical Applications

Edayadulla Naushad and Shankar Thangaraj

Abstract

Heterocyclic moieties form important constituents of biologically active natural products and synthetic compounds of medicinal interest. Nitrogen heterocycles constitute important pharmacophores in drug design, especially pyridine derivatives, which are among the most frequently cited heterocyclic compounds. The isolated as well as synthesized pyridine compounds exhibited various pharmacological properties due to their diverse physiochemical properties like water solubility, weak basicity, chemical stability, hydrogen bond-forming ability, protein-binding capacity, cell permeability, and size of the molecules attracted the attention of medicinal chemists for the past few years. Their interesting molecular architecture seeks attention to isolate derivatives of medicinal interest from natural source. In this chapter, we plan to describe the isolated natural products having pyridine moiety and their pharmacological importance.

Keywords: pyridine, naturally isolated, nitrogen heterocyclic compounds, pharmaceutical applications

1. Introduction

Heterocyclic moieties form important constituents of biologically active natural products and synthetic compounds of medicinal interest. Thus, it is not surprising that the chemistry of heterocyclic compounds continue to receive special attention in drug discovery efforts. For more than decades, heterocycles have established one of the largest areas of exploration in organic chemistry. They contributed to the expansion of humanity from biological and industrial point of view as well as to the understanding of bioprocesses and to the efforts to advance the excellence of life [1]. Due to their diverse physiological potential, pharmacists have recently become pinched toward scaffolds with the intention of synthesizing an extensive range of novel bioactive molecules particularly natural product compounds.

Pyridine (C6H5N), an isostere of benzene, was initially isolated from the picoline by Anderson in 1846. Later, the structure of pyridine was elucidated by Wilhelm Korner (1869) and James Dewar (1871). Pyridine is one of the nuclear reactants of more than 7000 existing drug molecules of pharmaceutical importance. Pyridine-based natural products consist of a variety of interesting compounds with diverse

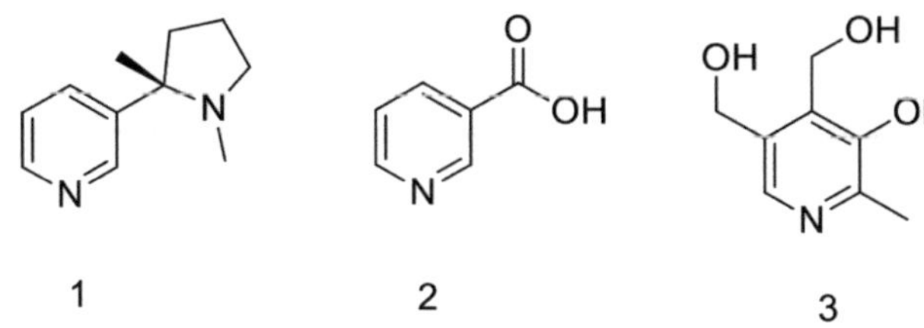

Figure 1.
Nicotine, niacin, and pyridoxine.

structures that originate from the five kingdoms of life. Nicotine, niacin (vitamin B_3 or nicotinic acid), and pyridoxine (vitamin B_6) are extreme recognized compounds with an aromatic π electron pyridine moiety (**Figure 1**). The structures having other oxidation states of pyridine, such as tetrahydropyridine, dihydropyridine, piperidine, or pyridone moieties, are fewer existed than the pyridine-based natural products [2].

2. Characteristic features of pyridine

In plants, pyridine compounds are mostly originated as alkaloids. In biological systems, a redox reaction of nicotinamide adenine dinucleotide (NAD) reduces

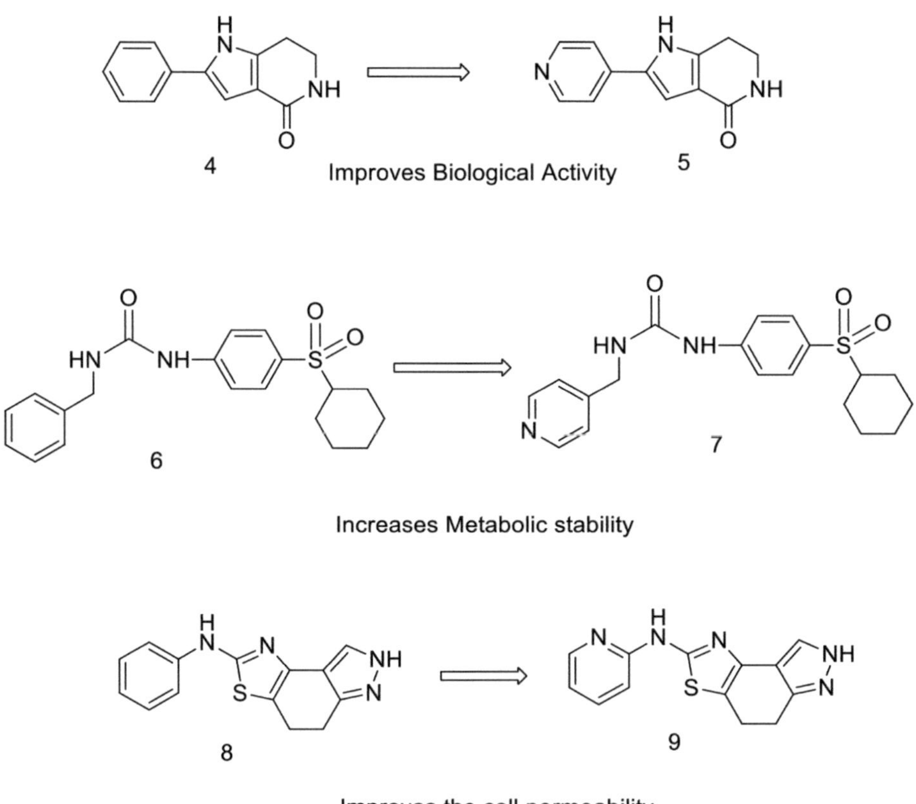

Figure 2.
Effect of pyridine on physiochemical parameters.

its pyridine moiety into dihydropyridine compounds, rendering NADH. Related redox reactions also exist in anabolic reactions involving NAD phosphate (NADP+/NADPH) interconversion. According to the Food and Drug Administration of the United States (FDA), pyridine-and dihydropyridine-containing drugs constitute nearly 14% and 4% of all Nitrogen containing heterocyclic drugs approved by the agency [3]. Among the 18%, the most important therapeutic areas of attention are communicable infections, swelling, the nervous system, and cancer treatment.

In pharmaceuticals, a pyridine-based synthesized compound enhances its biological potency, enhances penetrability and metabolic solidity, and fixes protein-binding issues [4]. The incorporation of pyridine ring is an important strategy in the drug discovery. Vanotti et al demonstrated the effective promotion of DNA replication in eukaryotic organisms **5** by replacing the benzene group of 4 with pyridine [5]. Likewise, metabolic steadiness of sulfone-based nicotinamide phosphoribosyltransferase inhibitor **6** is enriched 160-fold when its benzene ring is replaced with pyridine

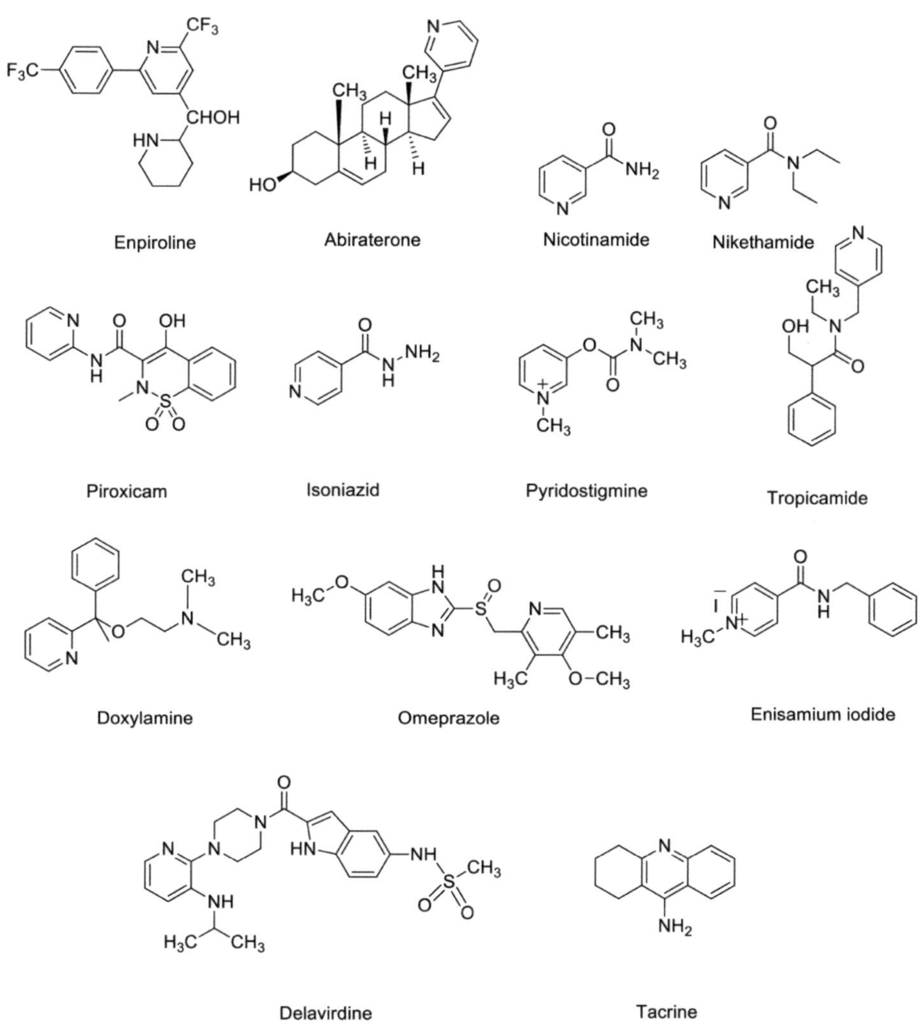

Figure 3.
Some commercially available drugs which contain pyridine rings.

in 7 [6]. A pyridine ring in a compound is also adept of increasing its cell permeability. Hong et al observed that a pyridine-containing positive allosteric modulator **9** with 190-fold the cellular penetrability of **8** (**Figure 2**). It is thus valid to say that incorporation of nitrogen-containing heterocyclic moiety greatly disturbs the physicochemical parameters of the bioactive molecule [7].

Some drugs available in the market which contain pyridine rings (**Figure 3**), such as enpiroline for malaria [8], abiraterone for prostate cancer [9], nicotinamide for vitamin B deficiency [10], nikethamide for a respiratory stimulant [11], piroxicam for inflammatory [12], isoniazid to treat active TB infections [13], pyridostigmine to improve muscle strength in patients with a certain muscle disease [14], tropicamide to dilate the pupil and help with examination of the eye [15], doxylamine for the short-term treatment of insomnia [16], omeprazole to treat gastric and duodenal ulcers [17], delavirdine for an antiviral against HIV/ AIDS [18], enisamium iodide for influenza [19], and tacrine for an oral acetylcholinesterase inhibitor previously used for the prevention of Alzheimer's disease [20].

3. Some pyridine scaffolds isolated from natural sources and their pharmacological importance

Trigonelline **10** was first isolated from the fenugreek seeds, which is used as a spice in South Asian regions. Trigonelline, a plant harmone that is extensively spread in plants and also exists in many animal species, such as bryozoans, arthropods, coelenterates, cnidarians, mollusks, crustaceans, echinoderms, marine poriferans, marine fishes, and mammals. The constituents of trigonelline presents in the pods of various fabaceae species and coffee. It also presents in mammalian urine after administration of nicotinic acid. The pharmacological activities of trigonelline have been more thoroughly screened than fenugreek's other components, particularly for diabetes and central nervous system disease [21]. Trigonelline has neuroprotective, hypoglycemic, memory-improving, hypolipidemic, antimigraine, antibacterial, sedative, antitumor, and antiviral activities, and it has been shown to decrease diabetic auditory neuropathy and platelet formation. It acts by affecting β-cell regeneration, insulin secretion, activities related to glucose metabolism, free radical scavenging, axonal extension, and neuron impulsiveness.

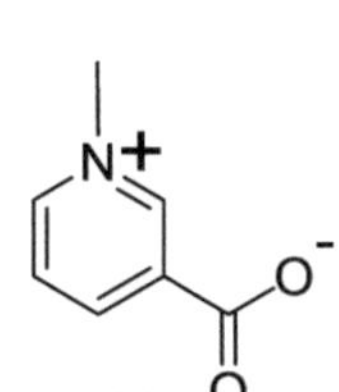

10

Trigonelline

The dried leaves of *Nicotiana tabacum* are named as tobacco. The tobacco was used by native American Indians about 8000 years, where the dried leaves were smoked in tube rituals for healing and ritualistic purposes [22, 23]. The compound Nicotine **1** was identified from dried leaves of *N. tabacum* leaves by Posselt and

Reimann [24]. Pictet and Cr´epieux established the structure through total synthesis in 1895 [25]. Nicotine is also present (albeit in lower amounts) in other species of the Solanaceae plant family, such as tomatoes, green peppers, and potatoes. At present, tobacco is cultivated in over many countries worldwide, where it is used to make cigars and as the source of nicotine for replacement therapy (NRT). The physiological studies of nicotine in a variety of cell systems and in animals have been evaluated by many researchers.

Nicotine stimulates the ion exchange channels to activate the discharge of neurotransmitters including serotonin (5-HT), dopamine, acetylcholine (ACh), norepinephrine, β-endorphins, γ-aminobutyric acid (GABA), and glutamate into the mesolimbic area, the corpus striatum, and the frontal cortex.

Picciotto and Zoli have explained that knocking out the α4β2 subunit gene in rats abolished the effects of nicotine and the discharge of dopamine. In associated studies, the α3β4-nAChR is occupied in the cardiovascular effects of nicotine and the α7-nAChR is tangled in memory, learning, and sensory gating [26]. Some other studies revealed that consumption of nicotine decreases the risk of Parkinson's disease (e.g. neurodegenerative disease) and anxiety and depression. In recent times, preliminary evaluations have described lower rates of SARS-CoV-2 contamination among smokers [27–30]. Various structurally related natural products to nicotine have also been identified from a variety of sources; many reviews on their biological activities are available.

Nicotinic acid **2** offers alkaloids with the pyridine moiety in the laboratory preparation. This nucleus presents in such alkaloids as nicotine, nornicotine, anabasine, ricine, anatabine, and arecoline. Furthermore, many alkaloids contain the pyridine ring as part of their total skeleton [31]. For example, anabasine is isolated from nicotinic acid and lysine [32]. Alkaloids with the pyridine ring occur in plants such as tobacco (*N. tabacum*), castor (*Ricinus communis*), and betel nuts (*Areca catechu*). The sesquiterpene-derived nucleus isolates partly from nicotinic acid and partly from the acetate biochemical pathway. There are more than 200 alkaloids identified in this group as potential compounds.

Demole & Demole isolated two terpenoid-based alkaloids from Burley tobacco (*Nicotiana tabacum*), 1,3,6,6-tetramethyl-5,6,7,8-tetrahydroisoquinolin-8-one **11** and 3,6,6-trimethyl-5,6-dihydro-7H-pyrindan-7-one **12** (**Figure 4**). Remarkably, **11** may be obtained from the glands of the *Castor fiber*, or by a synthetic method. Compound **11** has also been used to improve the flavor of tobacco [33].

Ricinine **13** is a familiar 2-pyridone derivative that occurs in the castor bean *Ricinus communis*. Nowadays, interest of the researchers has been focused on the relationship between the ricinine biogenesis and the pyridine nucleotide cycle [34]. The isomeric mixtures of pyridones ricinidine (**14**) and nudifluorine (**15**) have been isolated from the leaves of *Trewia nudiflora* (**Figure 5**) [35, 36].

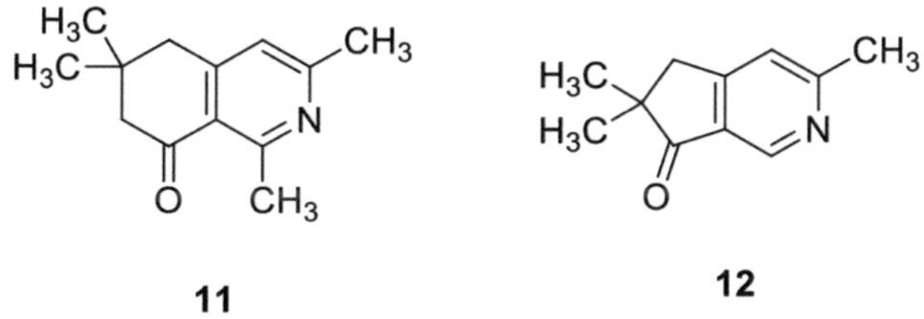

Figure 4.
Terpenoid-based alkaloids.

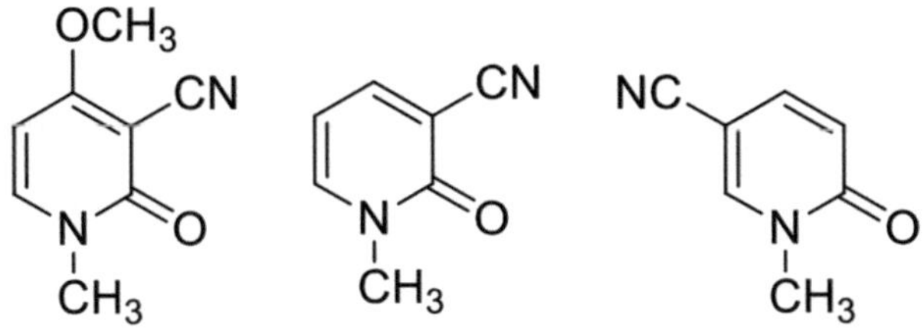

13 Ricinine; 14 Ricindine; 15 Nudifluorine

Figure 5.
2-Pyridone derivatives isolated from the different plant species.

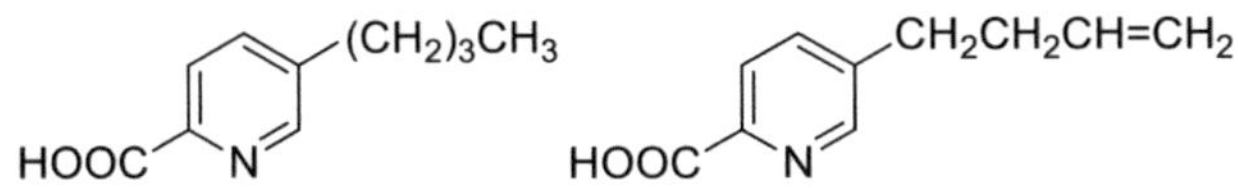

16 Fusaric acid; 17 Dehydrofusaric acid

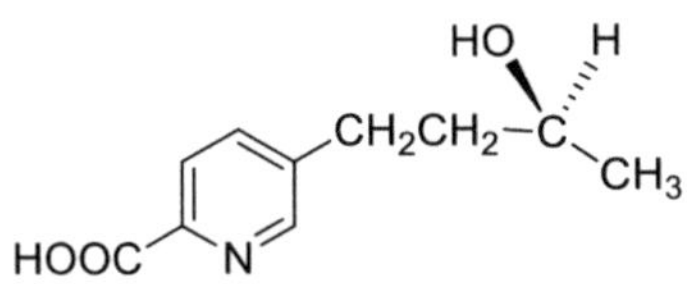

18 (+)-S-fusarinolic acid

Figure 6.
Fusaric acids from the mycelium species.

Fusaric acid (**16**) a systemic wilt toxin present especially in cotton plants [37, 38], was formed by various species of Fursaria and other fungi [39]. Dehydrofusaric acid (**17**) and (+)-S-fusarinolic acid (**18**) (**Figure 6**), metabolites of fusaric acid, have been attained from the mycelium of different *Fusaria, S. cerevisiae,* and *Gibberella fujikurvi* [39–41].

Ceropegia Juncea is described to be an important orgin of traditional ayurvedic practices [42]. The ethanolic extract of the plant was found to show significant biological activities in animal study, such as analgesic, antipyretic, antiulcer, hepato-protective, local anesthetic, mast-cell stabilizing, hypotensive, and tranquilizing activities. In 1991, Thirugnanasambantham et. al. reported Cerpegin **19** , a pyridine alkaloid, from the stem of the plant *Ceropegia Juncea* [43, 44]

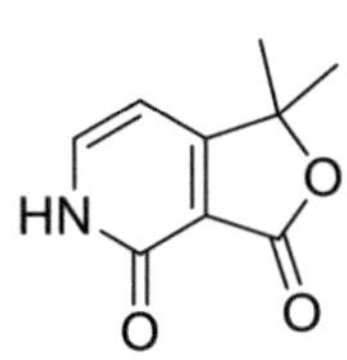

19

Cerpegin

(-)-Cytisine **20** and its derivatives are of great attention as pharmacological outfits and as vital drugs for the ailments of an extensive variety of conditions, from eating disorders, nicotine and alcohol dependence, stress, schizophrenia and Parkinson's diseases. (-)-Cytisine itself is used as a support to give up tobacco smoking, even though it is not very effective and proper physical alteration might well make it more so. The only linked compound in current medicinal use from cytosine, though not firmly a cytisine derivatives, is the anti-smoking drug varenicline [45]. Several researchers recommend that some cytisinoids display assured as hunger reducers, stress relief medicine, or drugs to treat neurodegenerative diseases [46].

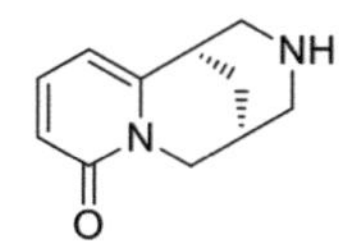

20

Cytisine (1R, 5S)

Actinomytes from soil and marine are a potent source for diverse compounds in the drug discovery. Wataru Aida et al isolated pyridine-containing natural compounds, such as fuzanins A (**21**), B (**22**), C (**23**), and D (**24**). The compounds were isolated from the Kitasatospora sp. IFM10917. The structure of each compound was proven by the source of spectroscopic and chemical analysis. Out of these, Fuzanin D (**24**) demonstrated cytotoxicity against human colon carcinoma DLD-1 cells (IC50, 41.2 mM) (**Figure 7**) and adequate inhibition of Wnt signal transcription besides with low cytotoxicity at 25 mM when it was screened for its Wnt signal inhibitory activity using a luciferase reporter gene assay in SuperTOP-Flash transfected cells [47].

Germana Esposito and the co-workers [48] isolated 13 novel nitrogen compounds from the Indonesian sponge *Acanthostrongylophora ingens*, and their chemical structures were established using NMR spectroscopy and HR-ESI-mass spectroscopy. All isolated compounds were evaluated in standard bioactivity assays, including antibacterial, antikinases, and amyloid β-42 assays. The most fascinating bioactivity outcome was acquired with the compound acanthocyclamine A (**25**), which shown

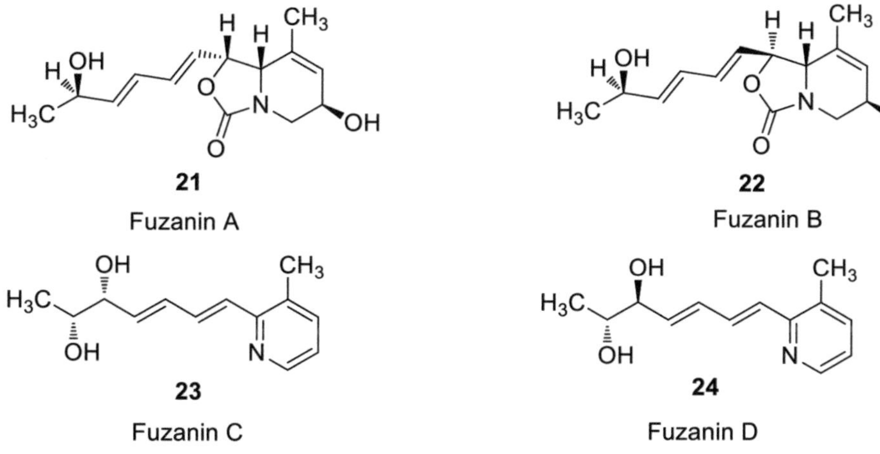

Figure 7.
Isolated from the culture extract of Kitasatospora sp. IFM10917.

for the exact *Escherichia coli* antibacterial action and as a result on amyloid β-42 assembly stimulated by aftin-5 and zero toxicity at the dose of 26 μM. These outcomes focus the potentiality of a bipiperidine skeleton as a favorable scaffold for inhibiting or decreasing the creation of amyloid β-42, a significant competitor in the beginning of Alzheimer's disease.

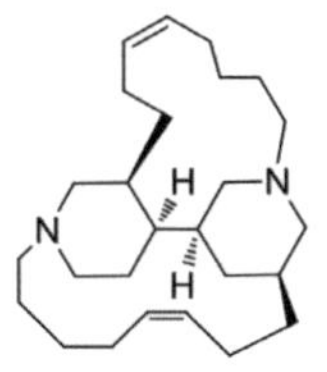

25

Acanthocyclamine A

Xin Wei et al reported three pyridine-type alkaloids, (-)-vincapyridines A–C (**26-28**), besides with two known alkaloids namely nauclefine **29** and vincamajoreine **30** (**Figure 8**) have been isolated from the stem of *Vinca major* grown in Pakistan. All the isolated compounds were assessed for their cytotoxicity against glioma initiating cell lines (GITC-3# and GITC-18#), glioblastoma cell lines (U-87MG and T98G), and lung cancer cell line A-549, but anyone entities was active at 20 μg/mL concentration [49].

Recently, Dumaa Mishig et al have isolated seven pyridine alkaloids (**31–37**), from the plants of *Caryopteris mongolica* Bunge. According to SciFinder and Reaxys

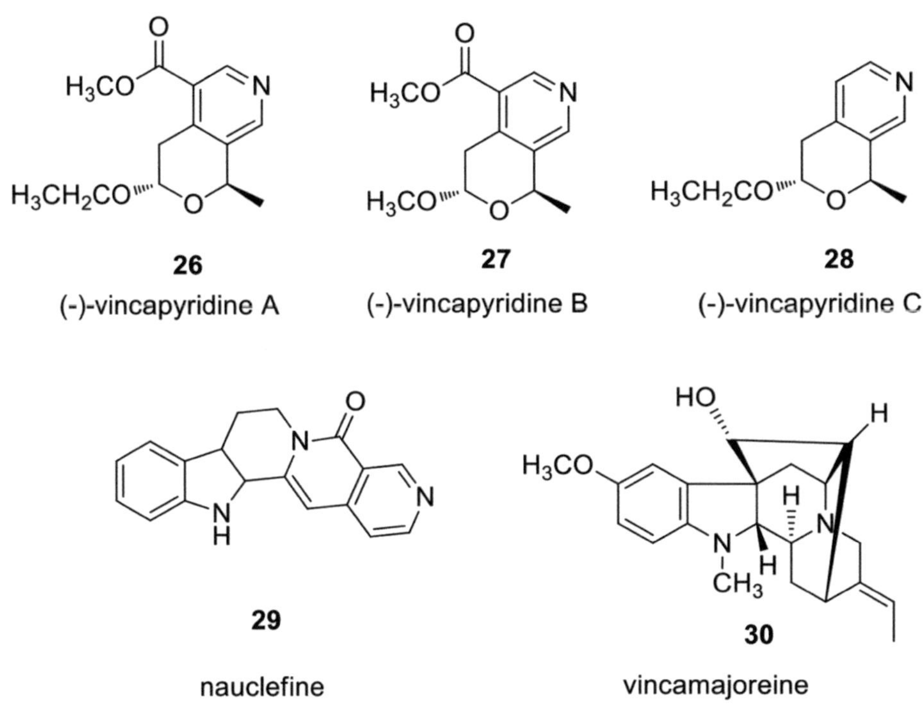

Figure 8.
Pyridine-type alkaloids isolated from Vinca major.

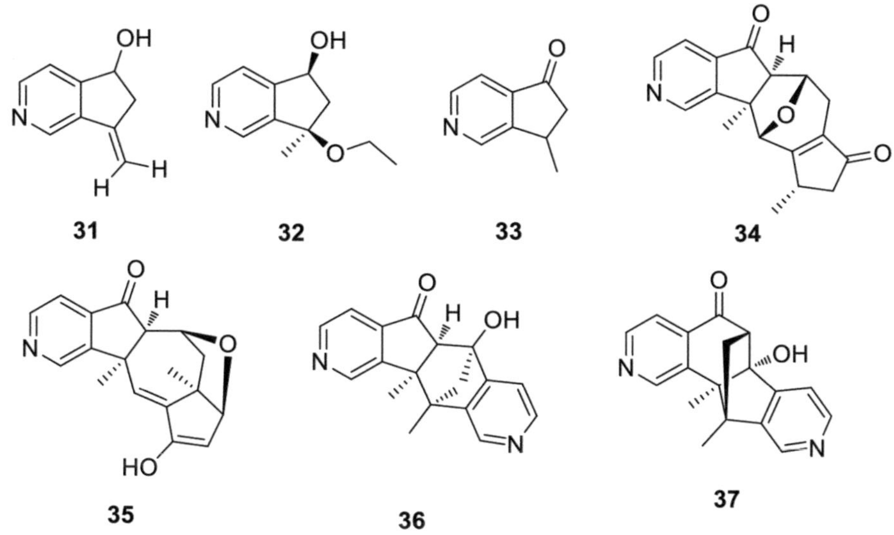

Figure 9.
New compounds obtained from the aerial parts of C. mongolica.

database search, the compounds **32, 34, 35, 36**, and **37** (**Figure 9**) represent new chemical structures. The chemical structures of these compounds were elucidated by 1H NMR, 13C NMR, and 2D NMR (COSY, HSQC, HMBC, and NOESY) and mass spectroscopic methods [50].

Noranabasamine (**38**) is an alkaloid that has been isolated from the Dendrobatidae amphibian—*Phyllobates terribilis* [51]. Noranabasamine is basically related to the analogous plant alkaloid anabasamine, which is known to inhibit acetylcholine esterase and exhibits anti-inflammatory activity. (S)-Anabasamine (**39**) was found in the poisonous semi-shrub *Anabasis aphylla* of Central Asia [52]. After administration of anabasamine to rats, hepatic alcohol dehydrogenase was improved and levels of ethanol were decreased in the blood stream [53]. In addition, the adrenal-regulated production of tryptophan pyrrolase was induced in the liver of those rats that were administered anabasamine.

All the earlier investigation with (S)-noranabasamine (**38**) and (S)-anabasamine (**39**) generally focused on the isolation of this alkaloid from other related alkaloids

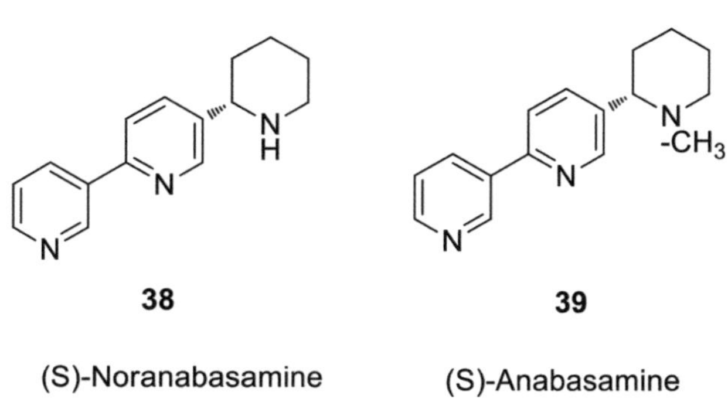

Figure 10.
Poisonous compounds isolated from the skin of amphibians.

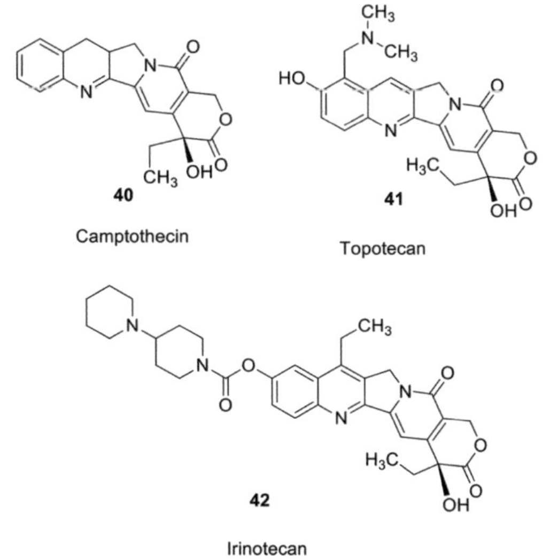

Figure 11.
Isolated and semisynthetic compounds of camptothecin.

found in amphibian skin and plants specimen (**Figure 10**). The mild concentrations in plants and amphibians, the difficulty in extraction, and the less existence in nature make these compounds smart goals for synthesis.

Camptothecin **40**, identified from the Chinese horticulture tree *Camptotheca acuminate* Decne , that belongs to Nyssaceae family was subjected to further clinical trials by National Cancer Institute in the 1970s but was stopped because of severe bladder toxicity [54]. Topotecan **41** and irinotecan **42** are semi-synthetic compounds of camptothecin for the healing of ovarian cancers and colorectal cancers, respectively (**Figure 11**).

Ageladine-A (**43**) is the first example of this family which contains 2-amino-imidazolopyridine. Ageladine-A was isolated from the combined extract of the sponge and purified by ODS flash chromatography, gel filtration, and ODS HPLC. Ageladine-A showed antiangiogenic activity [55].

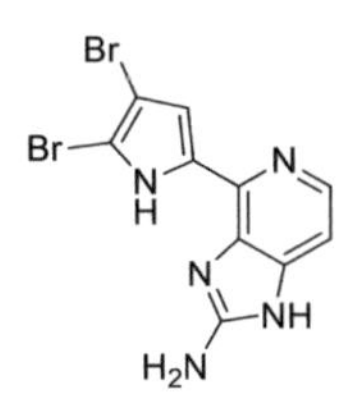

43

Ageladine-A

Aaptamine (**44**) from Aaptos aaptos [56] possesses α-adrenoceptor blocking activity in the isolated rabbit aorta. Amphimedine (**45**), a fused pentacyclic yellow aromatic alkaloid from a Pacific sponge Amphimedon spp. [57], is a cytotoxic agent.

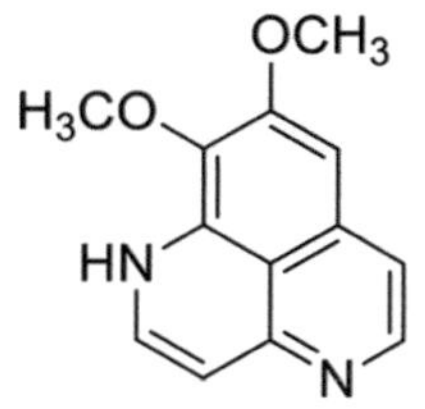

44

Aaptamine

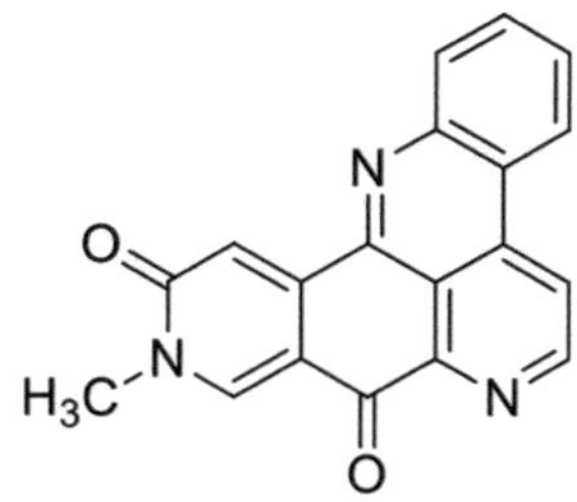

45

Amphimedine

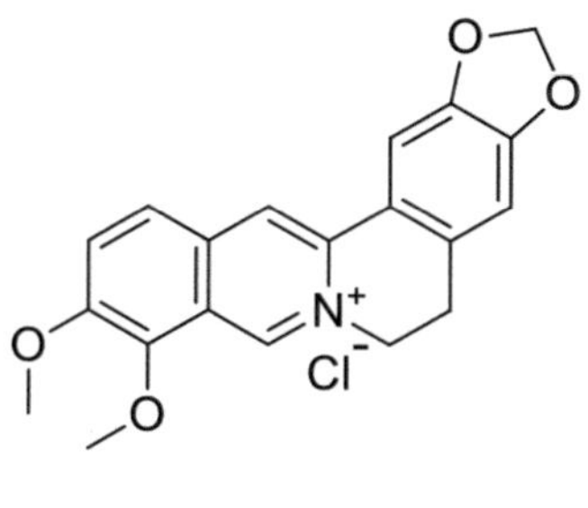

46

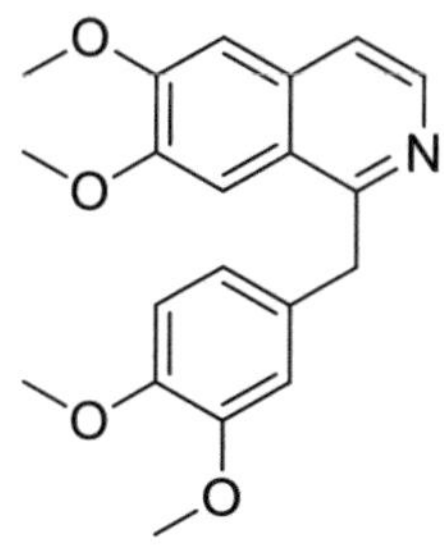

47

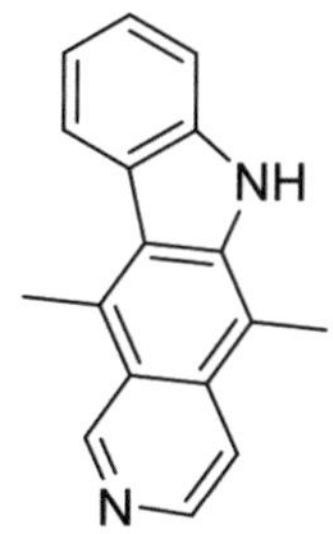

48

Figure 12.
Berberine, papaverine, and ellipticine.

Berberine **46** is a comparatively nontoxic alkaloid found in several plants, including goldenseal (*Hydrastis canadensis*), barberry (*Berberis vulgaris*), Oregon grape (*Berberis aquifolium*), and goldthread (*Coptis trifolia*). It has a long past and is most commonly used as an antibacterial agent [58, 59]. Papaverine **47** is used as a vasodilator under the trade name Para-Time® SR and is used as oral medicine to treat erectile dysfunction (**Figure 12**) [60]. Ellipticine **48** is used in cancer treatment, as it is alleged to act through DNA intercalation and inhibition of topoisomerase II [61].

4. Conclusion

The nitrogen containing heterocyclic compounds, especially pyridine scaffolds tangled into the various natural product compounds. The isolated as well as synthesized pyridine compounds exhibited various pharmacological properties due to their diverse physiochemical properties like water solubility, weak basicity, chemical stability, hydrogen bond-forming ability, protein-binding capacity, cell permeability, and size of the molecules attracted the attention of medicinal chemists for the past few years. In this chapter, we addressed some important pyridine-based compounds and their pharmacological applications. Natural product research is a mandatory tool for exploring bioactive compounds with unique properties and mode of action to face the future challenges.

Acknowledgements

We dedicate this chapter to our respectful Prof. (Late). P. Ramesh, Department of Natural Products Chemistry, Madurai Kamaraj University, Madurai. India.

Conflict of interest

The authors declare no conflict of interest.

Author details

Edayadulla Naushad[1*] and Shankar Thangaraj[2]

1 Department of Chemistry, Vel Tech Rangarajan Dr. Sagunthala R&D Institute of Science and Technology, Chennai, India

2 Department of Environmental Science, Sri Paramakalyani Centre for Environmental Science, Manonmaniam Sundaranar University, Tamilnadu, India

*Address all correspondence to: edayam2004@gmail.com

References

[1] Jampilek J. Heterocycles in medicinal chemistry. Molecules. 2019;**24**:3839. DOI: 10.3390/molecules24213839

[2] Ashok Kumar SK, Shah SK, Kazi S, et al. Pyridine: the scaffolds with significant clinical diversity. RSC Advances. 2022;**12**:15385-15406. DOI: 10.1039/D2RA01571D

[3] Lin SX, Curtis MA, Sperry J. Pyridine alkaloids with activity in the central nervous system. Bioorganic & Medicinal Chemistry. 2020;**28**:115820. DOI: 10.1016/j.bmc.2020.115820

[4] Pennington LD, Moustakas DT. The necessary nitrogen atom: A versatile high-impact design element for multiparameter optimization. Journal of Medicinal Chemistry. 2017;**60**:3552-3579. DOI: 10.1021/acs.jmedchem.6b01807

[5] Vanotti E, Amici R, Bargiotti A, et al. Cdc7 kinase inhibitors: Pyrrolopyridinones as potential antitumor agents. 1. Synthesis and Structure–Activity Relationships. Journal of Medicinal Chemistry. 2008;**51**: 487-501. DOI: 10.1021/jm700956r

[6] Zheng X, Bauer P, Baumeister T, et al. Structure-based identification of ureas as novel nicotinamide phosphoribosyltransferase (Nampt) inhibitors. Journal of Medicinal Chemistry. 2013;**56**:4921-4937. DOI: 10.1021/jm400186h

[7] Hong SP, Liu KG, Ma G, et al. Tricyclic thiazolopyrazole derivatives as metabotropic glutamate receptor 4 positive allosteric modulators. Journal of Medicinal Chemistry. 2011;**54**:5070-5081. DOI: 10.1021/jm200290z

[8] Basco L, Gillotin C, Gimenez F, Farinotti R, Bras J. In vitro activity of the enantiomers of mefloquine, halofantrine and enpiroline against Plasmodium falciparum. British Journal of Clinical Pharmacology. 1992;**33**:517-520. DOI: 10.1111/j.1365-2125.1992.tb04081.x

[9] Castellan P, Marchioni M, Castellucci R, et al. Abiraterone acetate for early stage metastatic prostate cancer: Patient selection and special considerations. Therapeutics and Clinical Risk Management. 2018;**14**:2341-2347. DOI: 10.2147/TCRM.S159824

[10] Raghuramulu N, Srikantia SG, Narasinga Rao BS, Gopalan C. Nicotinamide nucleotides in the erythrocytes of patients suffering from pellagra. The Biochemical Journal. 1965;**96**:837-839. DOI: 10.1042/bj0960837

[11] Westlake EK, Campbell EJM. Effects of aminophylline, nikethamide, and sodium salicylate in respiratory failure. British Medical Journal. 1959;**1**:274-276

[12] Lister BJ, Poland M, DeLapp RE. Efficacy of nabumetone versus diclofenac, naproxen, ibuprofen, and piroxicam in osteoarthritis and rheumatoid arthritis. The American Journal of Medicine. 1993;**95**:S2-S9

[13] Hsu KHK. Thirty years after isoniazid: Its impact on tuberculosis in children and adolescents. Journal of the American Medical Association. 1984;**251**:1283-1285. DOI: 10.1001/jama.1984.03340340023018

[14] Andersen JB, Engeland A, Owe JF, Gilhus NE. Myasthenia gravis requiring pyridostigmine treatment in a national population cohort. European Journal of Neurology. 2010;**17**:1445-1450. DOI: 10.1111/j.1468-1331.2010.03089.x

[15] Bostock C, McDonald C. Antimuscarinics in older people: Dry mouth and beyond. Dental Update. 2016;**43**:186-191

[16] Friedman H, Greenblatt DJ, Scavone JM, et al. Clearance of the antihistamine doxylamine.Reduced in elderly men but not in elderly women. Clinical Pharmacokinetics. 1989;**16**:312-316

[17] Walan A, Bader JP, Classen M, et al. Effect of omeprazole and ranitidine on ulcer healing and relapse rates in patients with benign gastric ulcer. The New England Journal of Medicine. 1989;**320**:69-75. DOI: 10.1056/NEJM198901123200201

[18] Wang Z, Vince R. Design and synthesis of dual inhibitors of HIV reverse transcriptase and integrase: Introducing a diketoacid functionality into delavirdine. Bioorganic & Medicinal Chemistry. 2008;**16**:3587-3595. DOI: 10.1016/j.bmc.2008.02.007

[19] Te Velthuis AJW, Zubkova TG, Shaw M, et al. Enisamium reduces influenza virus shedding and improves patient recovery by inhibiting viral RNA polymerase activity. Antimicrobial Agents and Chemotherapy. 2021;**65**:e02605-e02620. DOI: 10.1128/AAC.02605-20

[20] Ahmed M, Rocha JBT, Corrêa M, et al. Inhibition of two different cholinesterases by tacrine. Chemico-Biological Interactions. 2006;**162**:165-171. DOI: 10.1016/j.cbi.2006.06.002

[21] Zhou J, Chan L, Zhou S. Trigonelline: A plant alkaloid with therapeutic potential for diabetes and central nervous system disease. Current Medicinal Chemistry. 2012;**19**:3523-3531. DOI: 10.2174/092986712801323171

[22] Yildiz D. Nicotine, its metabolism and an overview of its biological effects. Toxicon. 2004;**43**:619-632. DOI: 10.1016/j.toxicon.2004.01.017

[23] Stratton K, Shetty P, Wallace R, et al. Clearing the Smoke: Assessing the Science Base for Tobacco Harm Reduction. Washington (DC): Institute of Medicine (US) Committee to Assess the Science Base for Tobacco Harm Reduction; 2001

[24] Posselt W, Reimann L. Chemical investigation of tobacco and the preparation of the characteristic active principle of this plant. Magazin f Pharmazie. 1828;**24**:138-161

[25] Pictet A, Crepieux P. Ueber phenyl-und pyridylpyrrole und die constitution des nicotins. Berichte derDeutschen Chemischen Gesellschaft. 1895;**28**:1904-1912

[26] Picciotto MR, Zoli M, Rimondini R, et al. Acetylcholine receptors containing the beta2 subunit are involved in the reinforcing properties of nicotine. Nature. 1998;**391**:173-177. DOI: 10.1038/34413

[27] Tindle HA, Newhouse PA, Freiberg MS. Beyond smoking cessation: Investigating medicinal nicotine to prevent and treat COVID-19. Nicotine & Tobacco Research. 2020;**22**:1669-1670

[28] Farsalinos K, Barbouni A, Niaura R. Systematic review of the prevalence of current smoking among hospitalized COVID-19 patients in China: Could nicotine Be a therapeutic option? Internal and Emergency Medicine. 2020;**15**:845-852

[29] Farsalinos K, Niaura R, Le Houezec J, et al. Nicotine and

SARS-CoV-2: COVID-19 may be a disease of the nicotinic cholinergic system. Toxicology Reports. 2020;7:658-663

[30] Simon X, Curtis MA, Sperry J. Pyridine alkaloids with activity in the central nervous system. Bioorganic & Medicinal Chemistry. 2020;**28**:115820

[31] Aniszewski T. Alkaloids-Secrets of life. Alkaloid Chemistry, Biological significance, Applications and Ecological Role. First ed. Amsterdam, Netherlands: Elsevier Science; 2007. pp. 85-92

[32] Funayama S, Cordell GA. Chapter 4 - Alkaloids Derived from Lysine. In: Alkaloids A Treasury of Poisons and Medicines. London, England: Academic Press; 2015. pp. 129-140

[33] Demole E, Demole C. A Chemical Study of Burley Tobacco Flavour (Nicotiana tabacum L.) V. Identification and Synthesis of the Novel Terpenoid Alkaloids 1, 3, 6, 6-Tetramethyl-5, 6, 7, 8-tetrahydro-isoquinolin-8-one and 3, 6, 6-Trimethyl-5,6-dihydro-7H -2-pyrindin-7-one. Helvetica Chimica Acta. 1975;**58**:523-531. DOI: 10.1002/hlca.19750580223

[34] Saravana Kumar P, Li Y, He M, Yuvaraj P, Balakrishna K, Ignacimuthu S. Rapid Isolation of Ricinine, a Pyridone Alkaloid from Ricinus communis (L.) with Antifungal Properties. Journal of Biologically Active Products from Nature. 2022;**12**:33-41. DOI: 10.1080/22311866.2021.2021985

[35] Ganguly SN. Isolation of ricinidine from plant source. Phytochemistry. 1970;**9**:1667-1668

[36] Mukherjee M, Chatterjee A. Structure and synthesis of nudiflorine : A new pyridone alkaloid. Tetrahedron. 1966;**22**:1461-1466. DOI: 10.1016/S0040-4020(01)99443-8

[37] Venkataraman K, Ramarao AV. Insect pigments derived from hydroxyanthraquinones. In: Rangaswami S, Subbarao NV, editors. Some Recent Developments In the Chemistry of Natural Products. India, New Delhi: Prentice-Hall; 1972. p. 153

[38] Kalyanasundaram R, Venkata Ram CS. Production and systemic translocation of fusaric acid in Fusarium infected cotton plants. The Journal of Indian Botanical Society. 1956;**35**:7-10

[39] Pitel DW, Vining LC. Accumulation of dehydrofusaric acid and its conversion to fusaric and 10-hydroxyfusaric acids in cultures of *Gibberella fujikuroi*. Canadian Journal of Biochemistry. 1970;**48**:623-630

[40] Steiner K, Graf U, Hardegger E. Welkstoffe und Antibiotika. 39. Mitteilung [1]. Fusarinolsäure. Helvetica Chimica Acta. 1972;**54**:845-851. DOI: 10.1002/hlca.19710540309

[41] Frederiks JC. Über L (+) - Fusarinolsäure, ein neus wechselprodukt von Fusarein Verhandlungen der Naturforschenden Gesellschaft in Basel. 1971:84-87

[42] Basak S. Cerpegin alkaloid and its analogues: Chemical synthesis and pharmacological profiles. Journal of the Indian Chemical Society. 2018;**95**:1175-1190

[43] AdibattiN,ThirugnanasambanthamP, Kulothungan C, et al. A pyridine alkaloid from *Ceropegia juncea*. Phytochemistry. 1991;**30**:2449-2450

[44] Sukumar E, Gopal RH, Rao RB, ViswanathanS,ThirugnanasambanthamP. Pharmacological actions of cerpegin, a novel pyridine alkaloid from *Ceropegia juncea*. Fitoterapia (Milano). 1995;**66**:403-406

[45] Zatonski W, Cedzynska M, Tutka P, West R. An uncontrolled trial of cytisine (Tabex) for smoking cessation. Tobacco Control. 2006;**15**:481-484. DOI: 10.1136/tc.2006.016097

[46] Perez EG, Mendez-Galvez C, Cassels BK. Cytisine: A natural product lead for the development of drugs acting at nicotinic acetylcholine receptors. Natural Product Reports. 2012;**29**:555-567. DOI: 10.1039/C2NP00100D

[47] Aida W, Ohtsuki T, Li X, Ishibashi M. Isolation of new carbamate- or pyridine-containing natural products, fuzanins A, B, C, and D from *Kitasatospora* sp. IFM10917. Tetrahedron. 2009;**65**:369-373. DOI: 10.1016/j.tet.2008.10.040

[48] Esposito G, Mai LH, Longeon A, Mangoni A, Durieu E, Meijer L, et al. A collection of bioactive nitrogen-containing molecules from the marine sponge *Acanthostrongylophora ingens*. Marine Drugs. 2019;**17**:472. DOI: 10.3390/md17080472

[49] Tokuyama T, Daly JW. Steroidal alkaloids (batrachotoxins and 4β-hydroxybatrachotoxins), "indole alkaloids" (calycanthine and chimonanthine) and a piperidinyldipyridin. Tetrahedron. 1983;**39**:41-47. DOI: 10.1016/S0040-4020(01)97627-6

[50] Mishig D, Gruner M, Lübken T, et al. Isolation and structure elucidation of pyridine alkaloids from the aerial parts of the Mongolian medicinal plant Caryopteris mongolica Bunge. Scientific Reports. 2021;**11**:13740. DOI: 10.1038/s41598-021-93010-4

[51] Sadykov AS. Doklady Akademii Nauk Uzbekskoi SSR. 1967;**24**:34-35

[52] Muzaev S. Doklady Akademii Nauk Uzbekskoi SSR. 1982;**9**:47-48

[53] Muzaev S. Doklady Akademii Nauk Uzbekskoi SSR. 1977;**7**:60-61

[54] Rahier NJ, Thomas CJ, Hecht SM. Camptothecin and its analogs. In: Cragg GM, Kingston DGI, Newman DJ, editors. Anticancer Agents from Natural Products, ed 2. Boca Raton: CRC/Taylor & Francis; 2012. pp. 5-25

[55] Fujita M, Nakao Y, Matsunaga S, Seiki M, et al. Ageladine A: An Antiangiogenic Matrixmetalloproteinase Inhibitor from the Marine Sponge Agelas nakamurai. Journal of the American Chemical Society. 2003;**125**:15700-15701. DOI: 10.1021/ja038025w

[56] Nakamura H, Kobayashi J, Ohizumi Y, Hirata Y. Isolation and structure of aaptamine a novel heteroaromatic substance possessing α-blocking activity from the sea sponge Aaptos aaptos. Tetrahedron Letters. 1982;**23**:5555-5558. DOI: 10.1016/S0040-4039(00)85893-1

[57] Schmitz FJ, Agarwal SK, Gunasekera SP, Schmidt PG, Shoolery JN. Journal of the American Chemical Society. 1983;**105**:4835-4836. DOI: 10.1021/ja00352a052

[58] Birdsall T, Kelly G. Berberine: Therapeutic potential of an alkaloid in several medicinal plants. Alternative Medicine Review. 1997;**2**:94-103

[59] Taylor CE, Greenough WB. Control of diarrheal diseases. Annual Review of Public Health. 1989;**10**:221-244. DOI: 10.1146/annurev.pu.10.050189.001253

[60] Kalsi JS, Cellek S, Muneer A, Kell PD, Ralph MS. Current oral treatments for erectile dysfunction. Expert Opinion on Pharmacotherapy. 2002;**3**:1613-1629. DOI: 10.1517/14656566.3.11.1613

[61] Stiborova MJ, Sejbal L, Borek-Dohalska D, Aimova J, Poljakova K, et al. The anticancer drug ellipticine forms covalent DNA adducts, mediated by human cytochromes P450, through metabolism to 13-hydroxyellipticine and ellipticine N2-oxide. Cancer Research. 2004;**64**:8374-8380

Chapter 5

Pyridine Heterocycles in the Therapy of Oncological Diseases

Lozan T. Todorov and Irena P. Kostova

Abstract

Oncological diseases pose a major challenge for modern medicine. Heterocyclic compounds play a vital role in modern medical and pharmaceutical science as most medicinal substances incorporate them. Nitrogen-containing heterocycles serve as the basis of numerous drugs and, therefore, are deeply involved in the design and synthesis of promising new therapeutic agents. Pyridine or pyrimidine scaffolds, with a number of substituents attached, comprise a large portion of FDA-approved drugs. They are chemically stable in the human body, manifest an affinity for DNA via hydrogen bonding, and present an opportunity for the development of novel anticancer agents. A large number of pyridine-based molecules are synthesized and tested for anticancer activity each year. The present chapter aims to introduce the most current synthetic approaches, published in scientific literature, and would also elaborate on structure-activity relationships described therein.

Keywords: pyridine, anticancer, biological activity, synthetic approaches, structure-activity relationship

1. Introduction

Oncological diseases pose a major problem worldwide in terms of societal, healthcare, financial, and economic impact with the number of cancer cases continually rising. The research for novel anticancer drugs comprises a significant portion of contemporary research and development in the field of medicine and pharmacy. Nitrogen heterocycles are a component of 59% of FDA-registered drugs [1] as of 2014. Among them, pyridine is the second most commonly incorporated nitrogen heterocycle. Pyridine-containing drugs are quite heterogeneous in terms of chemical structure, pharmacokinetics and pharmacodynamics – antihistamines (chlorpheniramine, brompheniramine), antiarrhythmic (disopyramide), antihypercholesterolaemic (cerivastatin), antitubercular (isoniazid, ethionamide), antibiotic (telithromycin), antiretroviral for AIDS treatment (indinavir), and anticancer (crizotinib, abiraterone) to name a few.

A multitude of natural substances contains pyridine. They tend to be involved in a number of physiological processes, among which is cancer pathogenesis. The pyridine ring is a chemically stable heterocyclic structure. Its nitrogen atom is able to participate in hydrogen bonding, which allows pyridines to bind to DNA and exhibit anticancer effects [2]. Pyridine can play the role of pharmacophore and can also serve as a stable basis for the synthesis of novel anticancer drugs. The present chapter aims to inform the reader in a brief and concise manner on the latest developments in the search for pyridine-based anticancer drugs, their mechanisms of action, and the most utilized synthetic approaches. Herein are included the most common types of novel, pyridine-based compounds, found in the scientific literature that do not involve fused ring structures. They are represented by molecular hybrids that the authors have classified into the following groups in terms of structure:

- Coumarin-pyridine hybrids
- Chalcone-pyridine hybrids
- Combretastatin-pyridine hybrids
- Terpyridines and terpyridines isosteres

Additionally, the authors are also presenting data on biological activity, types of cancer cell lines being suppressed, and pharmacodynamic action of the molecules discussed, should such information be available.

2. Pyridine derivatives recently approved for anticancer treatment

A number of pyridines have recently been registered for anticancer treatment [3]. The list predominantly includes kinase inhibitors (apalutamide, pexidartinib, lorlatinib, acalabrutinib, abemaciclib, neratinib, and alpelisib) – drugs that inhibit cellular kinases. Kinases are a family of enzymes that participate in cellular metabolism, signaling, replication, and survival. Inhibiting them suppresses vital cellular functions, therefore, targeting cancer-specific kinases suppresses tumor growth. Ivosidenib and enasidenib serve as isocitrate dehydrogenase (IDH) inhibitors. IDH is involved in energy production and includes two subtypes (IDH1 and IDH2). Mutations in IDH1 and IDH2 can cause changes in DNA gene expression including expression of oncogenes [4]. Inhibition of these enzymes could impair cancer growth. Benetoclax is a Bcl-2 inhibitor. Bcl-2 is a protein that suppresses cell death (apoptosis) [4]. Overexpression of Bcl-2 can prevent or significantly delay cell death – a typical characteristic of cancer. These drugs have been approved by FDA within the period 2017–2019. Considering the extremely stringent approval process of novel medicinal molecules, such a large number of newly-approved anticancer agents underscores both the extreme intensity of scientific exploration for novel anticancer treatments as well as the important role of the pyridine structure plays in drug research.

3. Coumarin-pyridine hybrids

The coumarin (benzopyran-2-on) structure (**Figure 1**) is considered an important bioactive scaffold, included in numerous drugs currently in use [5].

Coumarins are derived both naturally and synthetically. The specific structure of the coumarin scaffold allows coumarin derivatives to interact with a large variety of receptors and enzymes. They are currently being clinically utilized as anticoagulants and antithrombotic agents with relatively low toxicity. Naturally occurring

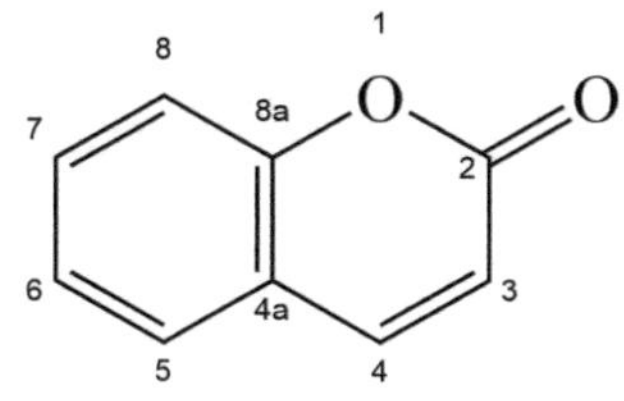

Figure 1.
Chemical structure of coumarin.

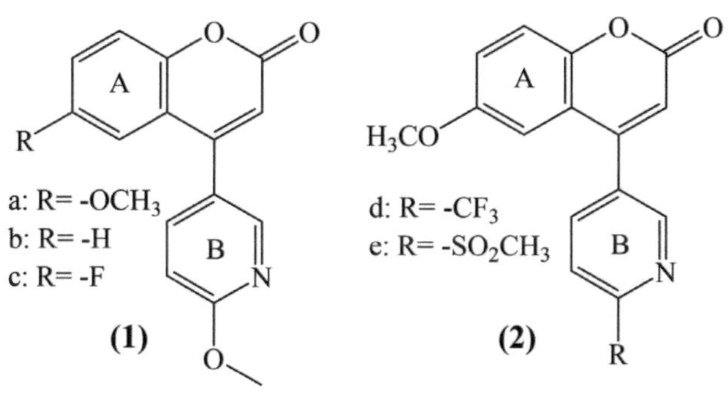

Figure 2.
Structure of the 4-arylcoumarin isosteres.

and synthetic derivatives have shown promise as antimicrobial, anti-inflammatory, anticancer, antioxidant, and MAO-B inhibitory agents [6]. They can exhibit cytostatic and cytotoxic activities against a significant number of cancer cell lines [7]. Adding a variety of functional groups and creating molecular hybrids is a promising direction for the development of novel medicinal molecules, aimed at alleviating a wide variety of maladies. Hybridization of coumarin derivatives with pyridines is a field of intense study in anticancer drug research [8–10].

4-Arylcoumarins are known for their cytotoxic and antiproliferative properties [11]. They can be viewed as structural analogs of the promising antiproliferative molecule combretastatin A-4 (CA-4), yielding very similar effects. For more information on CA-4 and its characteristics, please see Section 5. Pyridine isosteres of that class of compounds have been synthesized and tested for antiproliferative activity (**Figure 2**).

Pyridine derivatives manifest moderate activity against the A549 lung adenocarcinoma cell line [12]. Variants a and b significantly disrupt microtubule formation. Adding an electron-donating group in 6th place of ring A increases antiproliferative activity (**Figure 2**). Substituting with an electron-withdrawing group, such as a fluorine atom, in that same place decreases biological activity. Substituting the para-situated methoxy group in ring B only decreases the effect (**Figure 2**). The basics of the synthetic approach to yield 4-arylcoumarins are schematically presented in **Figure 3**.

Research and development of novel anticancer drugs are most often targeted toward a specific mechanism of action. A number of potential PI3K lipid kinase inhibitors have been synthesized by hybridization of coumarins and pyridines.

PI3K are enzymes, involved in the regulation of cellular growth, replication, and survival, as well as the mediation of protein kinase B (universally known as Akt). Upregulation of PI3K and Akt signaling is associated with tumor growth and tumor cell migration. The aforementioned substances have been tested for PI3K and Akt

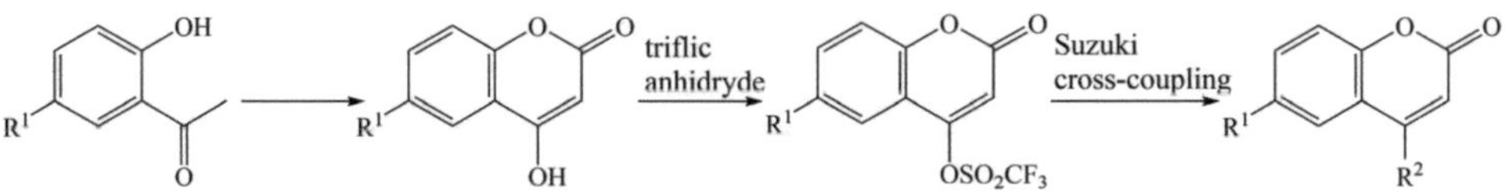

Figure 3.
Brief representation of the synthesis of 4-arylcoumarines.

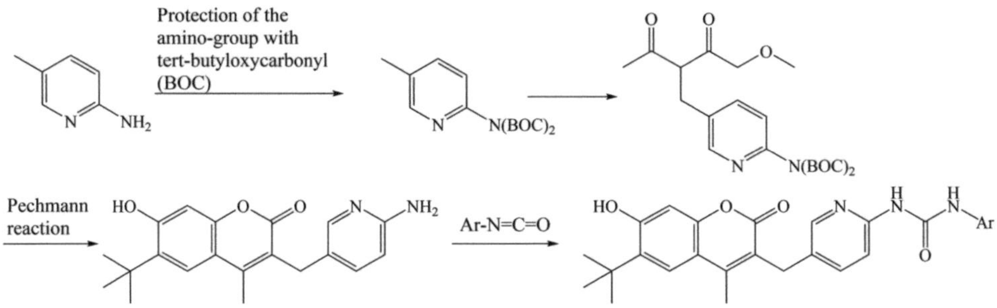

Figure 4.
Synthetic approach for generating some PI3K kinase inhibitors.

inhibition as well as antiproliferative activity against K562 (myelogenous leukemia), HeLa (cervical carcinoma), A549, and MCF-7 (adenocarcinoma) cancer strains [13]. A brief schematic of the synthesis is presented in **Figure 4**.

The member with difluoro-substituted phenyl ring (**Figure 5**) has the strongest effect on all observed cell lines.

All 3,4-disubstituted members exhibit a similar degree of antiproliferative effect. Another member, with monochloro substituted phenyl ring (**Figure 5**) has been found to significantly inhibit both PI3K and Akt and to initiate apoptosis in the K562 cell line.

A number of hybrid molecules have been synthesized using a novel approach [14]. The final step of the synthesis is conducted in two different media – in refluxing ethanol or under microwave heating. Microwave heating proves to be more energy-efficient, quicker, and produces significantly higher yields. **Figure 6** represents the basic synthesis of the most potent substance which exhibits promising activity against HCT-116 (colorectal carcinoma) and MCF-7 cell lines.

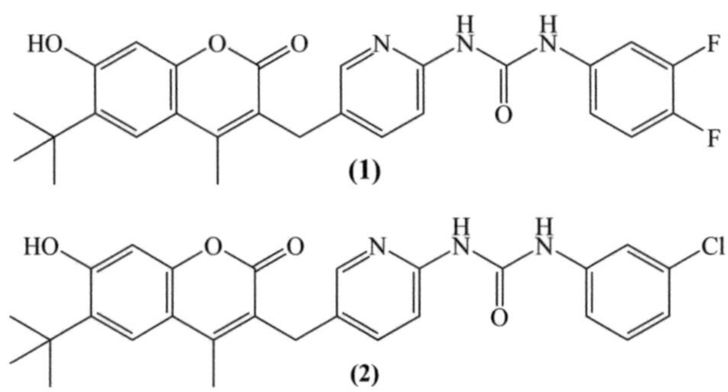

Figure 5.
The most active PI3K inhibitors against various cancer cell lines.

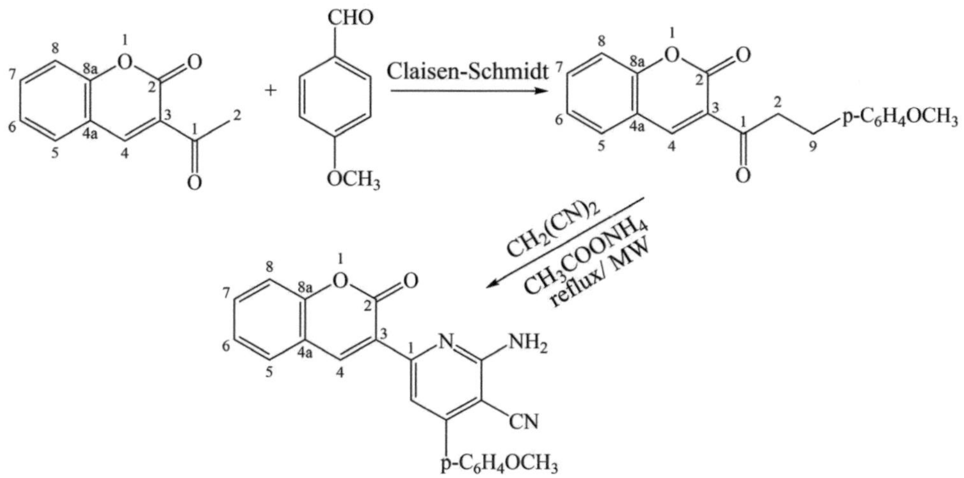

Figure 6.
Novel synthesis of coumarin-pyridine hybrid compounds.

4. Chalcone-pyridine hybrids

Chalcones are natural products from the flavonoid family, found in abundance in plants. Chalcone (**Figure 7**) is a molecular scaffold, characterized by uncomplicated chemistry, easy synthesis, and a large number of hydrogen atoms that, when substituted, can yield a huge selection of derivatives, exhibiting multiple physiological effects – antioxidant [15], antidiabetic [16], antihypertensive [17], anticancer [18], and many others.

They are known to inhibit cell proliferation, acting as antitumor agents both in vitro and in vivo. The antiproliferative properties of chalcones have been known for more than two decades [19]. Chalcones tend to bind to the so-called colchicine binding site in tubulin – a building block of microtubules. Microtubules are essential structures in all eukaryotic cells, responsible for keeping the structural integrity of cells, cell division, and many others [20]. Disrupting their synthesis is the mechanism of action of a number of antineoplastic drugs [21]. Attaching a pyridine moiety to the chalcone skeleton would be a way to complement the observed anticancer activity.

A promising design approach for the synthesis of chalcone-pyridine derivatives would be replacing one of the benzene rings with pyridine. A number of such molecules have been generated and then tested for antiproliferative activities and tubulin polymerization suppression [22]. α-(4-pyridyl) ketones and the necessary aldehydes undergo an aldol reaction to yield a number of chalcone-pyridine hybrids. The aforementioned step in the synthesis of the most potent member is presented in **Figure 8**.

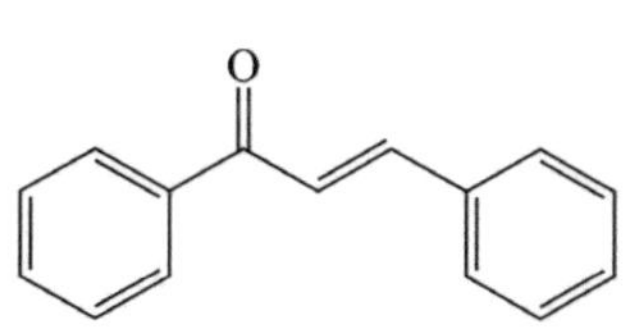

Figure 7.
The chalcone molecular scaffold.

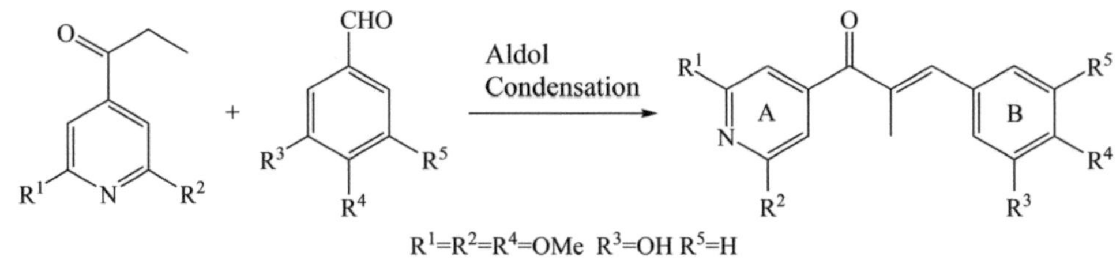

Figure 8.
Chalcone synthesis via aldol condensation.

All generated substances prove to be effective against K652 cells. The most potent one (**Figure 8**) is almost as effective as combretastatin A-4. It acts as a microtubule-destabilizing agent with an IC_{50} lower than that of CA-4. It connects with the colchicine binding site with 88% potency at 5 μM concentration, arresting the cell cycle of K562 at the G2/M phase and inducing apoptosis in a concentration-dependent manner.

The α-positioned methyl moiety to the carbonyl group tends to improve activity. The exposed hydroxyl at the meta-position of ring B (R^3) is important for the biological activity – changing it to methoxy decreases the observed effect. Adding electron-donating groups to ring A increases the effect, while adding electron-withdrawing groups (such as chlorine atoms) decreases the activity.

Aldol condensation has also been applied to generate a number of pyridinium bromide salts that have manifested promising antiproliferative activity against MCF-7, HeLa, U-87MG (malignant glioblastoma), and HEK293 (kidney) cell lines [23]. A brief summary of the synthesis of the two most active members is presented in **Figure 9**.

In terms of the structure-activity relationship, adding a strongly electron-donating functional group at the para-position of the phenyl radical R increases biological activity. Interestingly, adding the strongly electron-withdrawing nitro group also improves the antiproliferative properties. Replacing the radical R with a coumarin substituent (potentially anticancer-bearing) nullifies the anticancer effect.

Another class of substances that have been synthesized incorporates pyridine nucleus not as a substitute of one of the chalcone phenyl rings, but as a substituent [24]. They have been tested for their antiproliferative effect and colchicine-binding ability. The synthesis of the most active compounds is shown in **Figure 10**.

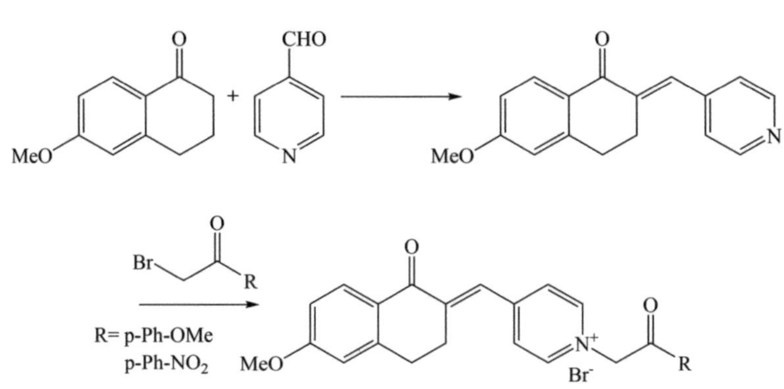

Figure 9.
Pyridinium bromide salts' synthesis.

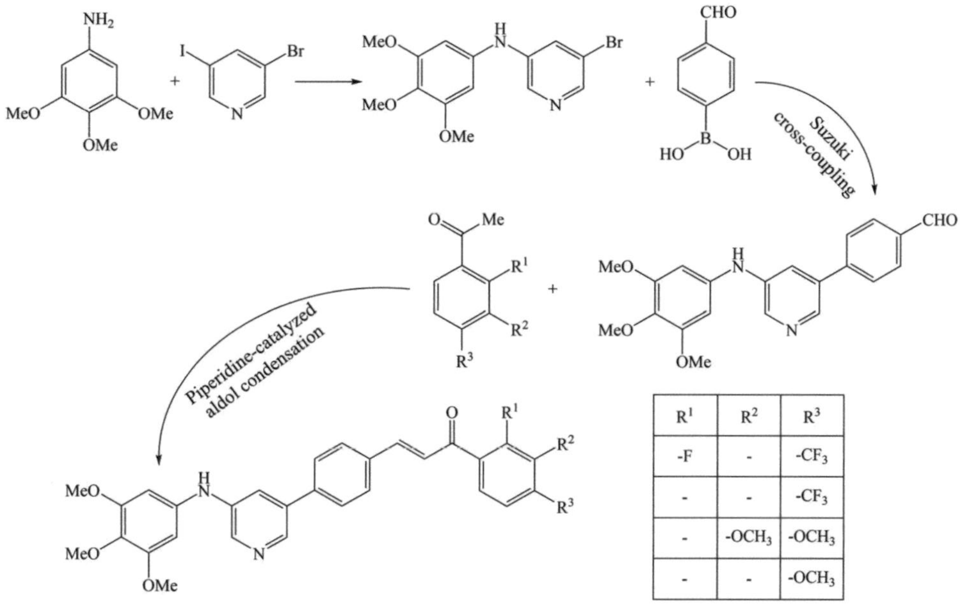

R^1	R^2	R^3
-F	-	$-CF_3$
-	-	$-CF_3$
-	$-OCH_3$	$-OCH_3$
-	-	$-OCH_3$

Figure 10.
Synthesis of pyridine substituted chalcones.

As in the previous case, adding electron-withdrawing groups, particularly in para-position, to the chalcone phenyl ring increases biological activity. Adding electron-donating groups (methoxy) to the same position has the same effect on ACHN (renal adenocarcinoma), MCF-7, and A549 cancer cell lines. The novel compounds have been docked in silico to the tubulin receptor, yielding promising results in terms of microtubule disruption.

5. Combretastatin: Pyridine hybrids

Combretastatins are a family of stilbenes, derived from the bark of the African Willow tree [25]. Combretastatin A-4 (**Figure 11**) in particular is an effective, selective inhibitor of tubulin polymerization by binding to the colchicine binding site. Thus it inhibits microtubule growth and acts as an antivascular and antimitotic agent, preventing cellular multiplication, changing endothelial cell structure, and resulting in tumor necrosis [26].

The cis-orientation of rings A and B is crucial for combretastatin A-4's cytotoxicity [27]. CA-4's application has been limited by its low solubility in aqueous media.

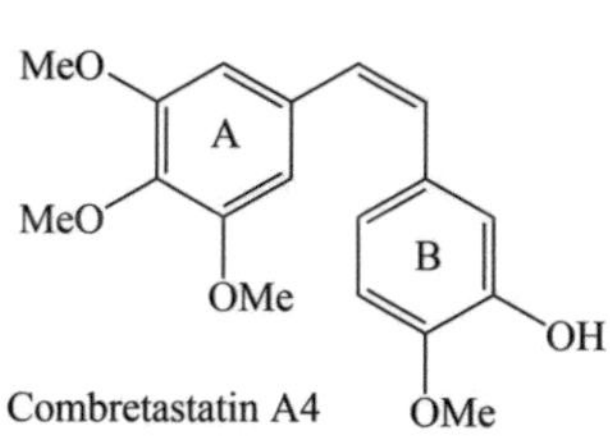

Combretastatin A4

Figure 11.
Structure of combretastatin A4.

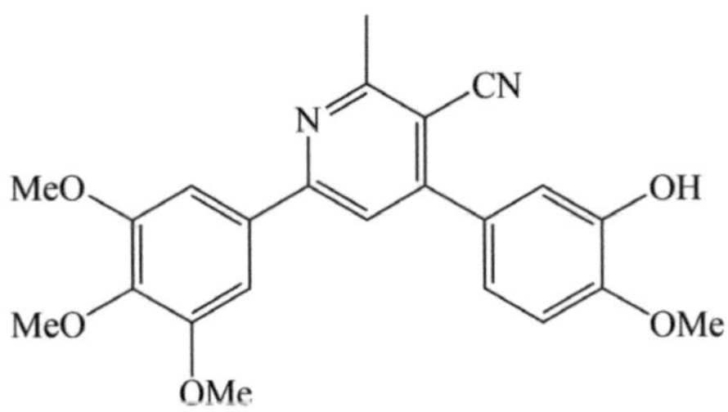

compound 1

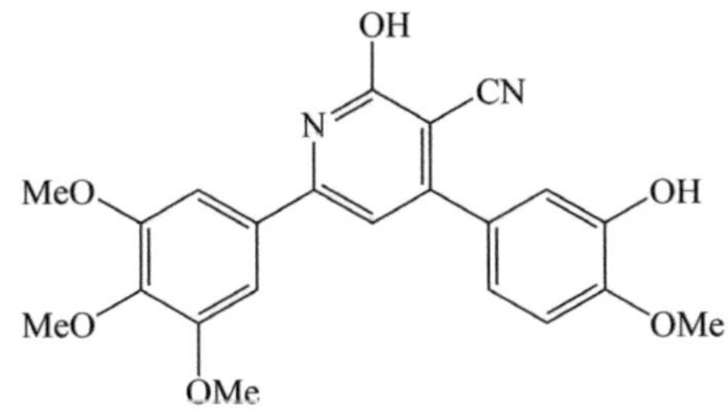

compound 2

Figure 12.
CA-4 analogs – 2,4-diphenyl-substituted pyridines.

Modification of its molecular structure (changing the aromatic rings and replacing the stilbene bridge) to increase its bioavailability, while maintaining its physiological effect has been a source of numerous investigations [28–30].

A number of combretastatin A-4 analogs with pyridine aromatic rings as a linker have been synthesized [31]. Two examples are presented in **Figure 12**.

Compound 1 manifests moderate cytotoxicity against MCF-7 cancer cells. Replacing the methyl group in its pyridine cycle with a hydroxyl group causes negation of the observed effect (compound 2). The antiproliferative effect associated with these 2,4-diphenyl-substituted pyridine structures is not very clearly manifested.

Interesting observations have been made with similar compounds, utilizing a pyridine linker between the two phenyl rings [32]. Among dozens of substances, three exhibit notable anticancer activity (**Figure 13**).

In terms of the structure-activity relationship, when the phenyl rings are at a para position from each other in the pyridine linker, cytotoxicity is low. Meta-position improves biological activity. The best results are observed with a 2,6-diphenyl substituted pyridine linker. 3,4,5-trimetoxy substituted ring A does not contribute significantly to biological activity. Compound 3 is the only one from a multitude of members, bearing such substituent, that yields promising results. It is an almost full analog of CA-4 – the stilbene linker is replaced with a 2,6-disubstituted pyridine. On the other hand, a 2,4-dimethoxy substituted ring A causes significant suppression against several cell lines – MDA-MB-231 (breast cancer), A549, and HeLa. Any other

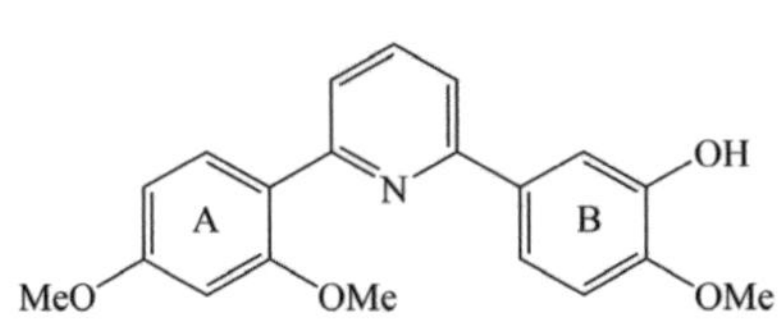

compound 2

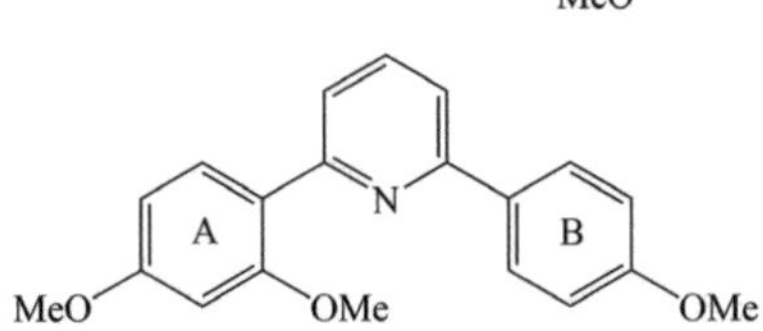

compound 1

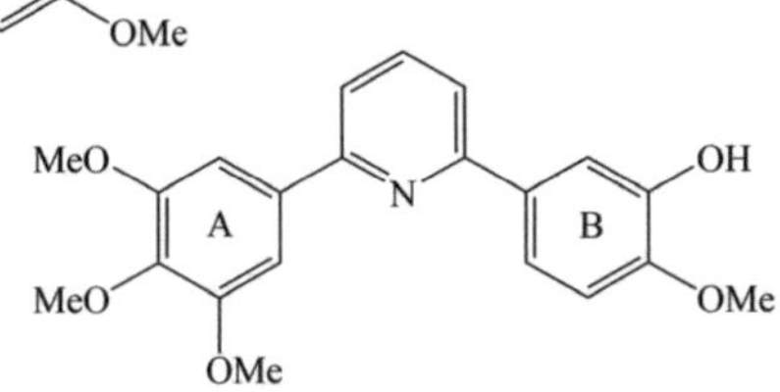

compound 3

Figure 13.
CA-4 analogs - 2,6-diphenylsubstitited pyridines.

type of dimethoxy substitution (e.g., 3,4-; 2,5-, etc.) decreases the antiproliferative effect. 3,4,5-trimethoxy substitution in ring B also weakens the biological effect. With 2,4-dimethoxysusbtituted ring A, 3-monomethoxy and 4-monomethoxy substituted ring B offer high antiproliferative effect, while 2-monomethoxy offers lesser activity. Thus, compounds 1, 2, and 3 potently inhibit cell survival and growth, arrest the cell division cycle and bind to the colchicine site to a degree, similar to combretastatin A4.

6. Terpyridine derivatives

Terpyridine is a known ligand in a variety of complexes [33]. Its structural analogs tend to bind to and intercalate in nucleic acids [34, 35]. α-Terpyridine (**Figure 14**) and its isosteres have manifested significant topoisomerase I and II inhibitory activity as well as notable cytotoxicity against a variety of cancer cell lines [36, 37]. Topoisomerases are a family of enzymes that catalyze changes in the topological state of the DNA double helix. They are involved in DNA replication and transcription, hence impairment of their function inhibits cellular replication – a way to suppress rapid tumor growth.

Terpyridines can be derived by way of the Kröhnke pyridine synthesis [38], represented in **Figure 15**.

Two families of terpyridine isosteres have been synthesized and tested for topoisomerase inhibitory activity and cytotoxicity – molecules with four aryl groups (furyl, thienyl, and pyridyl) and molecules with three aryl groups (**Figure 16**).

Three-ringed terpyridine members manifest low topoisomerase inhibitory activity and cytotoxicity. Some 2,4,6-trisubstituted members exhibit significant biological activity (listed in **Table 1**).

Notably, topoisomerase I inhibiting substances do not suppress topoisomerase II and topoisomerase II inhibiting substances do not suppress topoisomerase I. Interestingly, topoisomerase inhibitors manifest low toxicity toward a variety of cancer cell lines – MCF-7, HeLa, DU145 (prostate cancer), and HCT15 (colorectal

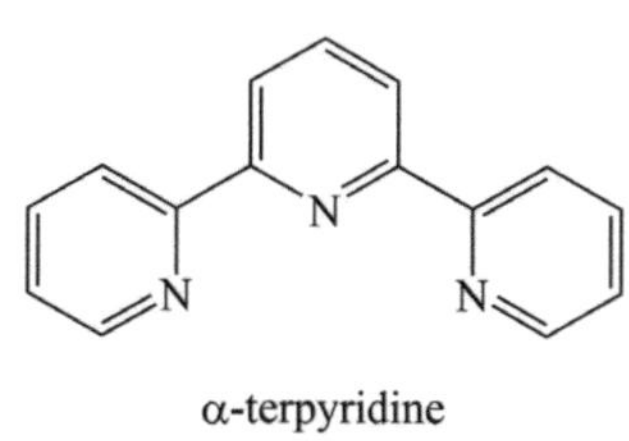

Figure 14.
Chemical structure of α-terpyridine.

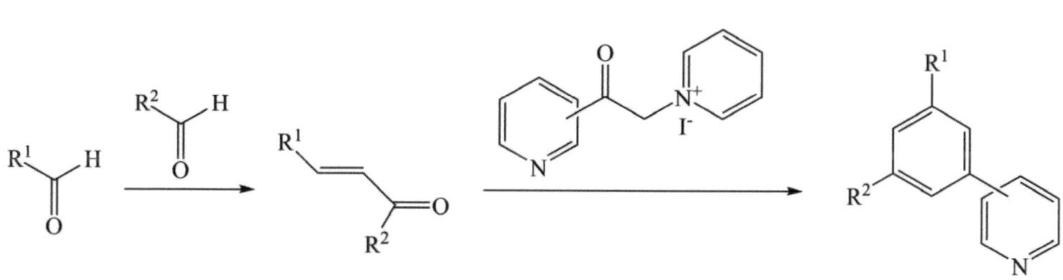

Figure 15.
Schematic representation of the synthesis of terpyridines and their isosteres.

R²=2/3/4-pyridyl
R¹/R³=2/3-furanyl/thienyl
(2,4,6-triaryl substituted)

R¹/R²= 2/3/4-pyridyl and/or
2/3-furanyl/thienyl
(2,6-diaryl substituted)

Figure 16.
Structures of the investigated terpyridines.

Moiety:	R^1	R^2	R^3	Biological activity
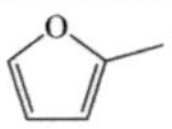 = a	a	g	c	Topoisomerase I inhibitor
 = b	a	g	d	Topoisomerase I inhibitor
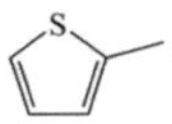 = c	c	e	d	Topoisomerase I inhibitor
 = d	c	g	d	Topoisomerase I inhibitor
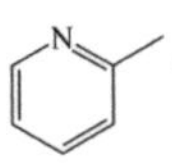 = e	a	g	d	Topoisomerase II inhibitor
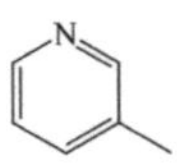 = f	c	g	f	Topoisomerase II inhibitor
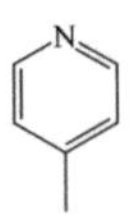 = g	a	g	b	Topoisomerase II inhibitor
	c	g	c	High cytotoxicity
	c	g	a	High cytotoxicity
	c	f	d	High cytotoxicity
	c	f	a	High cytotoxicity
	d	g	c	High cytotoxicity
	d	g	d	High cytotoxicity

Table 1.
Biological effect of various terpyridine isosteres with four aryl groups.

cancer). At the same time, some trisubstituted terpyridines did not behave as enzyme inhibitors but despite that are highly cytotoxic. In terms of molecular structure 2-furyl and 2-thienyl moieties in 2nd place, 4-pyridyl in 6th place, and 2/3-thienyl in 4th place seem to have the greatest impact on biological activity.

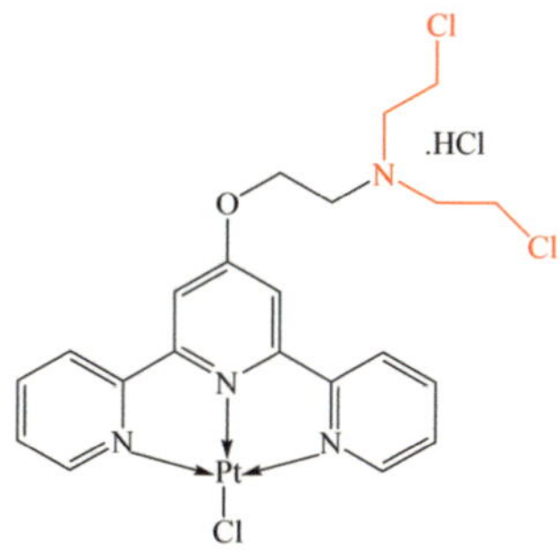

Figure 17.
An example of terpyridine-platinum complex. The ligand incorporates a "nitrogen mustard" moiety (in red), linked to the central pyridine ring. That molecular "tail" increases antiproliferative activity and DNA-binding of both the ligand itself and its platinum complex.

Terpyridines are being intensely studied in the field of oncology not so much for their intrinsic antiproliferative properties but for their ability to chelate metal ions. Recent data show that chelating copper ions with terpyridine ligands produce coordination compounds with high cytotoxicity and G0/G1 cell cycle phase inhibitory activity [39]. Experiments have demonstrated that complexes of terpyridines manifest antiproliferative activity in the nanomolar range against a large variety of cancer cell lines – MCF-7, A549, HCT-116, U-251 (glioblastoma), and PANC-1 (pancreatic carcinoma). At the same time, the observed IC_{50} doses against normal human fibroblasts (NHDF) are about 10–15 times higher, demonstrating good selectivity and potentially lower toxicity toward healthy human tissues. Numerous terpyridine complexes with platinum (**Figure 17**), palladium, and lanthanides have also recently been synthesized [40–43], bearing promising protein-binding, DNA-binding, and antiproliferative activities.

7. Conclusions

The pyridine heterocycle is an important chemical structure, ubiquitously utilized within the field of modern pharmaceutical science, research, and development. Its characteristic physicochemical properties (chemical stability, participation in hydrogen bonding, and numerous hydrogen atoms that can be substituted) make it an attractive molecular basis for synthesis of medicinal substances. Its nitrogen atom makes it a useful pharmacophore, imbuing potential drug molecules with novel pharmacological effects. Attaching it to extant compounds can modify their pharmacokinetics, pharmacodynamics, and physiological effect. The authors' aim is that the present chapter reveals to the reader the important role pyridine chemistry plays in the field of oncology. Pyridine-based compounds are being intensely researched in the hope of inventing novel oncological drugs that combine significant anticancer cytotoxicity with an improved safety profile and a targeted mechanism of action. Within the past several years a large number of novel pyridine anticancer molecules have been synthesized, yielding some very promising results. Substances of both natural and synthetic origin have been generated and/or modified, synthetic approaches have been refined and interesting and potentially important structure-activity relationships have been revealed. Hopefully, the authors have been able to present the subject of pyridines in oncology to the reader's satisfaction, both informing them as well as sparking an interest in this rapidly evolving area of research.

Conflict of interest

The authors declare no conflict of interest.

Author details

Lozan T. Todorov* and Irena P. Kostova
Department of Chemistry, Medical University – Sofia, Faculty of Pharmacy, Sofia, Bulgaria

*Address all correspondence to: ltodorov@pharmfac.mu-sofia.bg

References

[1] Vitaku E, Smith DT, Njardarson JT. Analysis of the structural diversity, substitution patterns, and frequency of nitrogen heterocycles among US FDA approved pharmaceuticals: Miniperspective. Journal of Medicinal Chemistry. 2014;**57**(24):10257-10274

[2] Özkay Y, Işıkdağ İ, İncesu Z, Akalın G. Synthesis of 2-substituted-N-[4-(1-methyl-4, 5-diphenyl-1H-imidazole-2-yl) phenyl] acetamide derivatives and evaluation of their anticancer activity. European Journal of Medicinal Chemistry. 2010;**45**(8):3320-3328

[3] Sahu R, Mishra R, Kumar R, Mazumder A, Kumar A. Pyridine moiety: Recent advances in cancer treatment. Indian Journal of Pharmaceutical Sciences. 2021;**83**(2):162-185

[4] Chao DT, Korsmeyer SJ. BCL-2 family: Regulators of cell death. Annual Review of Immunology. 1998;**16**

[5] Riveiro ME, De Kimpe N, Moglioni A, Vazquez R, Monczor F, Shayo C, et al. Coumarins: Old compounds with novel promising therapeutic perspectives. Current Medicinal Chemistry. 2010;**17**(13):1325-1338

[6] Sandhu S, Bansal Y, Silakari O, Bansal G. Coumarin hybrids as novel therapeutic agents. Bioorganic & Medicinal Chemistry. 2014;**22**(15):3806-3814

[7] Marshall M, Kervin K, Benefield C, Umerani A, Albainy-Jenei S, Zhao Q, et al. Growth-inhibitory effects of coumarin (1, 2-benzopyrone) and 7-hydroxycoumarin on human malignant cell lines in vitro. Journal of Cancer Research and Clinical Oncology. 1994;**120**(1):S3-S10

[8] Fayed EA, Sabour R, Harras MF, Mehany A. Design, synthesis, biological evaluation and molecular modeling of new coumarin derivatives as potent anticancer agents. Medicinal Chemistry Research. 2019;**28**(8):1284-1297

[9] Zhang L, Xu Z. Coumarin-containing hybrids and their anticancer activities. European Journal of Medicinal Chemistry. 2019;**181**:111587

[10] Rawat A, Reddy AVB. Recent advances on anticancer activity of coumarin derivatives. European Journal of Medicinal Chemistry Reports. 2022;**100038**

[11] Bailly C, Bal C, Barbier P, Combes S, Finet J-P, Hildebrand M-P, et al. Synthesis and biological evaluation of 4-arylcoumarin analogues of combretastatins. Journal of Medicinal Chemistry. 2003;**46**(25):5437-5444

[12] Mutai P, Breuzard G, Pagano A, Allegro D, Peyrot V, Chibale K. Synthesis and biological evaluation of 4 arylcoumarin analogues as tubulin-targeting antitumor agents. Bioorganic & Medicinal Chemistry. 2017;**25**(5):1652-1665

[13] Ma C-C, Liu Z-P. Design and synthesis of coumarin derivatives as novel PI3K inhibitors. Anti-Cancer Agents in Medicinal Chemistry (Formerly Current Medicinal Chemistry-Anti-Cancer Agents). 2017;**17**(3):395-403

[14] El-Naggar AM, Hemdan MM, Atta-Allah SR. An efficient one-pot synthesis of new Coumarin derivatives as potent anticancer agents under microwave irradiation. Journal of Heterocyclic Chemistry. 2017;**54**(6):3519-3526

[15] Iqbal H, Prabhakar V, Sangith A, Chandrika B, Balasubramanian R. Synthesis, anti-inflammatory and antioxidant activity of ring-A-monosubstituted chalcone derivatives. Medicinal Chemistry Research. 2014;**23**(10):4383-4394

[16] Kaushal R, Kaur M. Bio-medical potential of chalcone derivatives and their metal complexes as antidiabetic agents: A review. Journal of Coordination Chemistry. 2021;**74**(4-6):725-742

[17] Mahapatra DK, Bharti SK. Therapeutic potential of chalcones as cardiovascular agents. Life Sciences. 2016;**148**:154-172

[18] Modzelewska A, Pettit C, Achanta G, Davidson NE, Huang P, Khan SR. Anticancer activities of novel chalcone and bis-chalcone derivatives. Bioorganic & Medicinal Chemistry. 2006;**14**(10):3491-3495

[19] Ducki S, Forrest R, Hadfield JA, Kendall A, Lawrence NJ, McGown AT, et al. Potent antimitotic and cell growth inhibitory properties of substituted chalcones. Bioorganic & Medicinal Chemistry Letters. 1998;**8**(9):1051-1056

[20] Howard J, Hyman AA. Dynamics and mechanics of the microtubule plus end. Nature. 2003;**422**(6933):753-758

[21] Dumontet C, Jordan MA. Microtubule-binding agents: A dynamic field of cancer therapeutics. Nature Reviews Drug Discovery. 2010;**9**(10):790-803

[22] Xu F, Li W, Shuai W, Yang L, Bi Y, Ma C, et al. Design, synthesis and biological evaluation of pyridine-chalcone derivatives as novel microtubule-destabilizing agents. European Journal of Medicinal Chemistry. 2019;**173**:1-14

[23] Gondru R, Saini R, Vaarla K, Singh S, Sirassu N, Bavantula R, et al. Synthesis and characterization of chalcone-pyridinium hybrids as potential anti-cancer and anti-microbial agents. ChemistrySelect. 2018;**3**(5):1424-1431

[24] Madhavi S, Sreenivasulu R, Yousuf Ansari M, Jawed Ahsan M, Ramesh RR. Synthesis, biological evaluation and molecular docking studies of pyridine incorporated chalcone derivatives as anticancer agents. Letters in Organic Chemistry. 2016;**13**(9):682-692

[25] Pettit GR, Singh SB, Niven ML, Hamel E, Schmidt JM. Isolation, structure, and synthesis of combretastatins A-1 and B-1, potent new inhibitors of microtubule assembly, derived from Combretum caffrum. Journal of Natural Products. 1987;**50**(1):119-131

[26] Chaplin D, Pettit G, Hill S. Anti-vascular approaches to solid tumour therapy: Evaluation of combretastatin A4 phosphate. Anticancer Research. 1999;**19**(1A):189-195

[27] Woods J, Hadfield JA, Pettit G, Fox BW, McGown AT. The interaction with tubulin of a series of stilbenes based on combretastatin A-4. British Journal of Cancer. 1995;**71**(4):705-711

[28] Ohsumi K, Nakagawa R, Fukuda Y, Hatanaka T, Morinaga Y, Nihei Y, et al. Novel combretastatin analogues effective against murine solid tumors: Design and structure– activity relationships. Journal of Medicinal Chemistry. 1998;**41**(16):3022-3032

[29] Liou J-P, Chang Y-L, Kuo F-M, Chang C-W, Tseng H-Y, Wang C-C, et al. Concise synthesis and structure–activity relationships of combretastatin A-4 analogues, 1-aroylindoles and 3-aroylindoles, as novel classes of potent antitubulin agents. Journal of Medicinal Chemistry. 2004;**47**(17):4247-4257

[30] Jian X-E, Yang F, Jiang C-S, You W-W, Zhao P-L. Synthesis and biological evaluation of novel pyrazolo [3, 4-b] pyridines as cis-restricted combretastatin A-4 analogues. Bioorganic & Medicinal Chemistry Letters. 2020;**30**(8):127025

[31] Shringare SN, Chavan HV, Bhale PS, Dongare SB, Mule YB, Patil SB, et al. Synthesis and pharmacological evaluation of combretastatin-A4 analogs of pyrazoline and pyridine derivatives as anticancer, anti-inflammatory and antioxidant agents. Medicinal Chemistry Research. 2018;**27**(4):1226-1237

[32] Zheng S, Zhong Q, Mottamal M, Zhang Q, Zhang C, LeMelle E, et al. Design, synthesis, and biological evaluation of novel pyridine-bridged analogues of combretastatin-A4 as anticancer agents. Journal of Medicinal Chemistry. 2014;**57**(8):3369-3381

[33] Mukkala VM, Helenius M, Hemmilä I, Kankare J, Takalo H. Development of luminescent europium (III) and terbium (III) chelates of 2, 2′: 6′, 2 ″-terpyridine derivatives for protein labelling. Helvetica Chimica Acta. 1993;**76**(3):1361-1378

[34] Carter PJ, Cheng C-C, Thorp HH. Oxidation of DNA and RNA by oxoruthenium (IV) metallointercalators: Visualizing the recognition properties of dipyridophenazine by high-resolution electrophoresis. Journal of the American Chemical Society. 1998;**120**(4):632-642

[35] Jennette K, Lippard S, Vassiliades G, Bauer W. Metallointercalation reagents. 2-hydroxyethanethiolato (2, 2′, 2 ″-terpyridine)-platinum (II) monocation binds strongly to DNA by intercalation. Proceedings of the National Academy of Sciences. 1974;**71**(10):3839-3843

[36] Son J-K, Zhao L-X, Basnet A, Thapa P, Karki R, Na Y, et al. Synthesis of 2, 6-diaryl-substituted pyridines and their antitumor activities. European Journal of Medicinal Chemistry. 2008;**43**(4):675-682

[37] Basnet A, Thapa P, Karki R, Na Y, Jahng Y, Jeong B-S, et al. 2, 4, 6-Trisubstituted pyridines: Synthesis, topoisomerase I and II inhibitory activity, cytotoxicity, and structure–activity relationship. Bioorganic & Medicinal Chemistry. 2007;**15**(13):4351-4359

[38] Adib M, Tahermansouri H, Koloogani SA, Mohammadi B, Bijanzadeh HR. Kröhnke pyridines: An efficient solvent-free synthesis of 2, 4, 6-triarylpyridines. Tetrahedron Letters. 2006;**47**(33):5957-5960

[39] Malarz K, Zych D, Kuczak M, Musioł R, Mrozek-Wilczkiewicz A. Anticancer activity of 4′-phenyl-2, 2′: 6′, 2 ″-terpyridines–behind the metal complexation. European Journal of Medicinal Chemistry. 2020;**189**:112039

[40] Adams M, Sullivan MP, Tong KK, Goldstone DC, Hanif M, Jamieson SM, et al. Mustards-derived Terpyridine–platinum complexes as anticancer agents: DNA alkylation vs coordination. Inorganic Chemistry. 2021;**60**(4):2414-2424

[41] Kacar O, Adiguzel Z, Yilmaz VT, Cetin Y, Cevatemre B, Arda N, et al. Evaluation of the molecular mechanisms of a palladium (II) saccharinate complex with terpyridine as an anticancer agent. Anti-Cancer Drugs. 2014;**25**(1):17-29

[42] Li C, Xu F, Zhao Y, Zheng W, Zeng W, Luo Q, et al. Platinum (II) terpyridine anticancer complexes possessing multiple mode of DNA interaction and EGFR inhibiting activity. Frontiers in Chemistry. 2020;**8**:210

[43] Hussain A, Gadadhar S, Goswami TK, Karande AA, Chakravarty AR. Photoactivated DNA cleavage and anticancer activity of pyrenyl-terpyridine lanthanide complexes. European Journal of Medicinal Chemistry. 2012;**50**:319

Chapter 6

The Expanding Role of Pyridine Derivatives as Privileged Scaffolds in Cardiac Ionic Channels

Yasodha Krishna Janapati, Sunithasree Cheweti, Bojjibabu Chidipi, Medidi Srinivas and Sunil Junapudi

Abstract

Pyridine-based ring systems are heterocycle-structured subunits that are being abundantly employed in drug design, primarily because of their tremendous effect on pharmacological activity, which has resulted in the discovery of various broad-spectrum medicinal compounds. Pyridine derivatives are employed to treat multiple medical illnesses, including prostate cancer, AIDS, tuberculosis, angina, ulcer, arthritis, urinary tract analgesic, Alzheimer's disease, and cardiovascular diseases. This chapter emphasized the currently available synthetic pyridine derivatives, including nimodipine, ciclopirox, efonidipine, nifedipine, milrinone, and amrinone, effects on cardiac ionic channels and their mechanisms of action for the cure. Pyridine derivatives regulate several voltage-gated ion channel behaviors, including sodium (Na_v), calcium (Ca_v), and potassium (K_v) channels, and are set as a therapeutic approach. Particularly, calcium-channel blockers are the most common action of medicines with a dihydropyridine ring and are often used to treat hypertension and heart-related problems. Finally, this chapter gives the prospects of highly potent bioactive molecules to emphasize the advantages of using pyridine and dihydropyridine in drug design. This chapter discusses pyridine derivatives acting on cardiac ionic channels to combat CVS diseases. The book chapter describes the importance of pyridine derivatives as a novel class of medications for treating cardiovascular disorders.

Keywords: pyridine derivatives, privileged, scaffolds, cardiac ions

1. Introduction

1.1 The physiological role of Pyridine derivatives

Heterocycles are vital in the pharmaceutical sectors, which are an integral part of the essential roof of life processes, that is, DNA and RNA [1–3]. Recently, 90% of newly produced and commercialized medicines integrate heterocyclic compounds [4]. Pyridine and dihydropyridine are 6-membered heterocyclic rings with a wide variety of therapeutic potential in cardiovascular diseases, ulcers, HIV, antibacterial activity, etc. [5–9]. Pyridines are typically found in plants with the alkaloids, such as nicotine, anabasine, and trigonelline [10]. In the biochemical process, nicotinamide adenine dinucleotide (NAD) redox reactions are reduced to NADH, and a dihydropyridine ring is present in NAHD. We can also notice dihydropyridine ring in NADPH structure which reduced from the $NADP^+$ [11]. The food and drug administration (FDA) has approved 14% of drugs containing pyridine and dihydropyridine scaffolds [10].

1.2 Natural and commercial drugs with pyridine and dihydropyridine scaffolds

Pyridine and dihydropyridine are versatile chemicals used to make libraries with various functional groups and therapeutic objectives. The existence of pyridine or dihydropyridine heterocycles significantly impacts pharmacological properties. For instance, the pyridine ring in a medication boosts physiological properties, potency, metabolic stability, permeability, and binding to the protein [12]. There is a myriad of commercially accessible medications that include pyridine rings on the market which we listed in the below table.

Name of drugs	Disease	References
Pyridine derivative		
Abiraterone	Prostate cancer	[13]
Delavirdine	Antiviral against HIV/ AIDS	[14]
Doxylamine	Allergies	[15]
Enpiroline	Malaria	[16]
Isoniazid	Tuberculosis	[17]
Nicotinamide	Pellagra	[18]
Nikethamide	Respiratory stimulant	[19]
Omeprazole	Ulcers	[20]
Piroxicam	Arthritis	[21]
Pyridostigmine	Myasthenia gravis	[22]
Tacrine	Alzheimer's	[23]
Tropicamide	Antimuscarinics	[24]
Nicorandil	Vasodilator	[25]
Metyrapone	NSAID	[26]
Bromazepam	Anxiety	[27]
Etoricoxib	NSAID	[28]
Tenoxicam	Rheumatoid arthritis and osteoarthritis	[29]
Droxicam		[30]
Ampiroxicam	Anti-inflammatory	[31]
Lornoxicam	rheumatoid arthritis	[32]
Clonixin	Arthritis, migraine, and tissue disorders	[33]
Phenazopyridine	Urinary tract infections and analgesic activity	[34]
Pitavastatin	Lowering cholesterol	[35]

Name of drugs	Disease	References
Ceftaroline fosamil, tedizolid, ceftazidime, delafloxacin	Antibiotic	[36–39]
Ethionamide	Tuberculosis	[40]
Nevirapine, tipranavir, indinavir	HIV/AID	[41, 42]
Axitinib, sorafenib, regorafenib, alpelisib	Cancer treatment	[43–47]
Lorlatinib, acalabrutinib	Lung cancer	[48, 49]
Abemaciclib, neratinib	Breast cancer	[50, 51]
Nedocromil	cure allergic conjunctivitis	[52]
Betahistine	Ménière's disease	[53]
Amifampridine	Lambert-Eaton myasthenic syndrome (LEMS)	[54]
Chlorpheniramine	antihistaminic	[55]
Pyridoxine	Deficiency of vitamin B_6 and peripheral neuritis	[56]
Amlexanox	Asthma and Rhinitis	[57]
Carbinoxamine	Rhinitis and vasomotor rhinitis	[58]
Doxylamine	Allergies	[59]
Brompheniramine	Cough, and Nasal congestion	[60]
Nedocromil	Allergic conjunctivitis	[61]
Nedocromil	Allergic conjunctivitis	[61]
Rupatadine	Allergic rhinitis	[62]
Acrivastine	Rhinitis	[63]
Indacaterol	Asthma	[64]
Triprolidine	Antihistamine	[65]
Bepotastine	Itching	[66]
Niacin	Pellagra and Hypertriglyceridemia	[67]
Pyrithion	Dandruff and Seborrheic Dermatitis	[68]
Nicotine	Symptoms of nicotine and Smoking cessation	[69]
Lemborexant, zolpidem	Insomnia	[70–72]
Quinine, chloroquine	Malaria	[73, 74]
Diiodohydroxyquinoline	Amebiasis	[75]
Telithromycin	Pneumonia	[76]
Trovafloxacin	Chlamydia, and Gonorrhea	[77, 78]
Imiquimod	Warts	[79]
Ubrogepant	Migraine	[52]
Chromium picolinate	Regulation of insulin function	[80]
Chromium nicotinate	Chromium deficiency	[81]
Dihydropyridine ring–containing drug		
Ciclopirox	Ringworm and athlete's foot	[82]
Doravirine	HIV/AIDS	[83]
NADH	nutraceutical	[84]

Name of drugs	Disease	References
Cabotegravir	HIV1	[85]
Huperzine a	Alzheimer's disease	[86]
Nifedipine pyridine and dihydropyridine ring systems	Raynaud's syndrome	[87]
Milrinone and amrinone	Vasodilators	[88, 89]

2. Pyridine and dihydropyridine scaffolds with cardiovascular action

Torsemide with pyridine is an approved medicine that stimulates diuresis and reduces the patient's blood pressure [90]. Most dihydropyridine rings act as calcium-channel blockers, most commonly used to treat high blood pressure and cardiovascular disorders [91, 92]. The dihydropyridine ring-containing drugs are nilvadipine, nifedipine, amlodipine, azelnidipine, clevidipine, and felodipine [10]. Nimodipine helps cure vasospasm and subarachnoid hemorrhage [93, 94]. Levamlodipine, isradipine, nicardipine, benidipine, felodipine, nisoldipine, nitrendipine, and clevidipine are used to treat hypertension [95–102]. Efonidipine is specially used to treat hypertension and angina [103]. Torasemide is also a cure for renal and liver diseases other than heart failure and hypertension [104]. Quinidine is used to treat atrial fibrillation and flutter [105]. Papaverine used as vasodilator [106].

The nifedipine drug is also used to treat diseases premature birth and Raynaud's syndrome [87]. Milrinone and amrinone are FDA-approved vasodilators containing pyridine and dihydropyridine ring systems [88, 89].

Examples of a few pyridines and dihydropyridine derivatives of cardiovascular action drugs are shown in **Figure 1**.

3. Pyridine derivatives regulation of cardiac ion channel behaviors is established as a therapeutic strategy

3.1 Cardiac ion channels

Ion channels are pore-forming membrane proteins that permit ions to pass through the channels. The selective permeability of ion channels on the cell membrane causes the heart to produce an action potential. The ion channels reduce the activation energy required for ion movement across the lipophilic cell membrane. Ion channels are established within the membrane of all excitable cells and various intracellular organelles. In search for new drugs, ion channels are a recurrent target [107].

All elements of cardiac function, including rhythmicity and contractility, rely on ion channels. Ion channels are unavoidably important therapeutic targets for heart pathology, such as atrial fibrillation or angina [108].

3.2 Cardiac action potential and ion channels

The cardiac action potential is characterized by a rapid shift in membrane potential (voltage) across the cell membrane of heart cells. The passage of ions between the interior and exterior of cells via proteins known as ion channels generates the cardiac

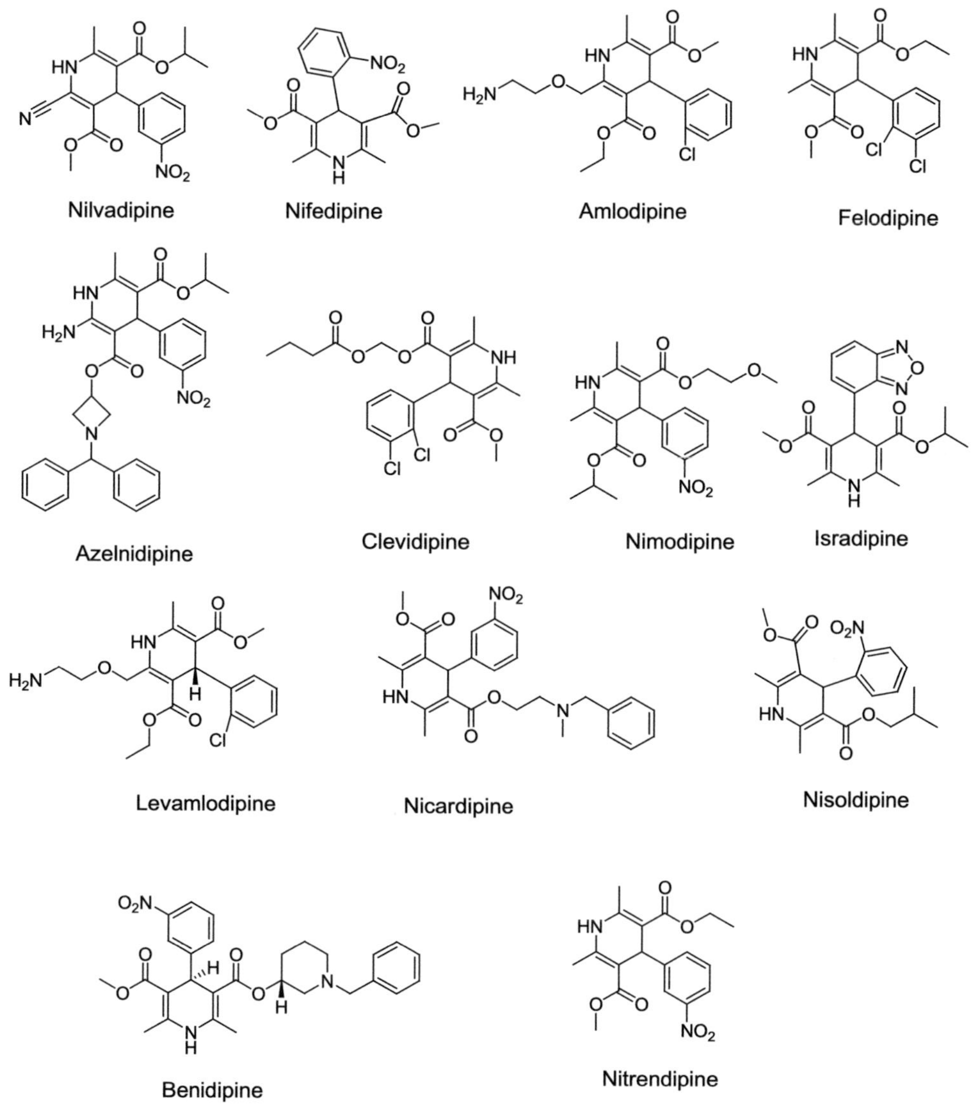

Figure 1.
Pyridine and dihydropyridine derivatives of cardiovascular action drugs.

action potential [109]. Ion channels have unique structures and are composed of numerous proteins situated in the cell membrane [107]. Identifying the ion channels that create the action potential is accomplished by examining the molecular basis of hereditary cardiac arrhythmias.

Normal atrioventricular and ventricle contraction requires the fast stimulation or activation of cardiac cell clusters. An activation mechanism must authorize rapid heart rate variations and respond to changes in autonomic tone. These responsibilities are executed by generating the cardiac action potential [107]. The five phases of the cardiac action potential are depicted in **Figure 2** [107]:

1. In healthy functioning cardiac cells, phase 4 (resting potential) is around −90 mV.

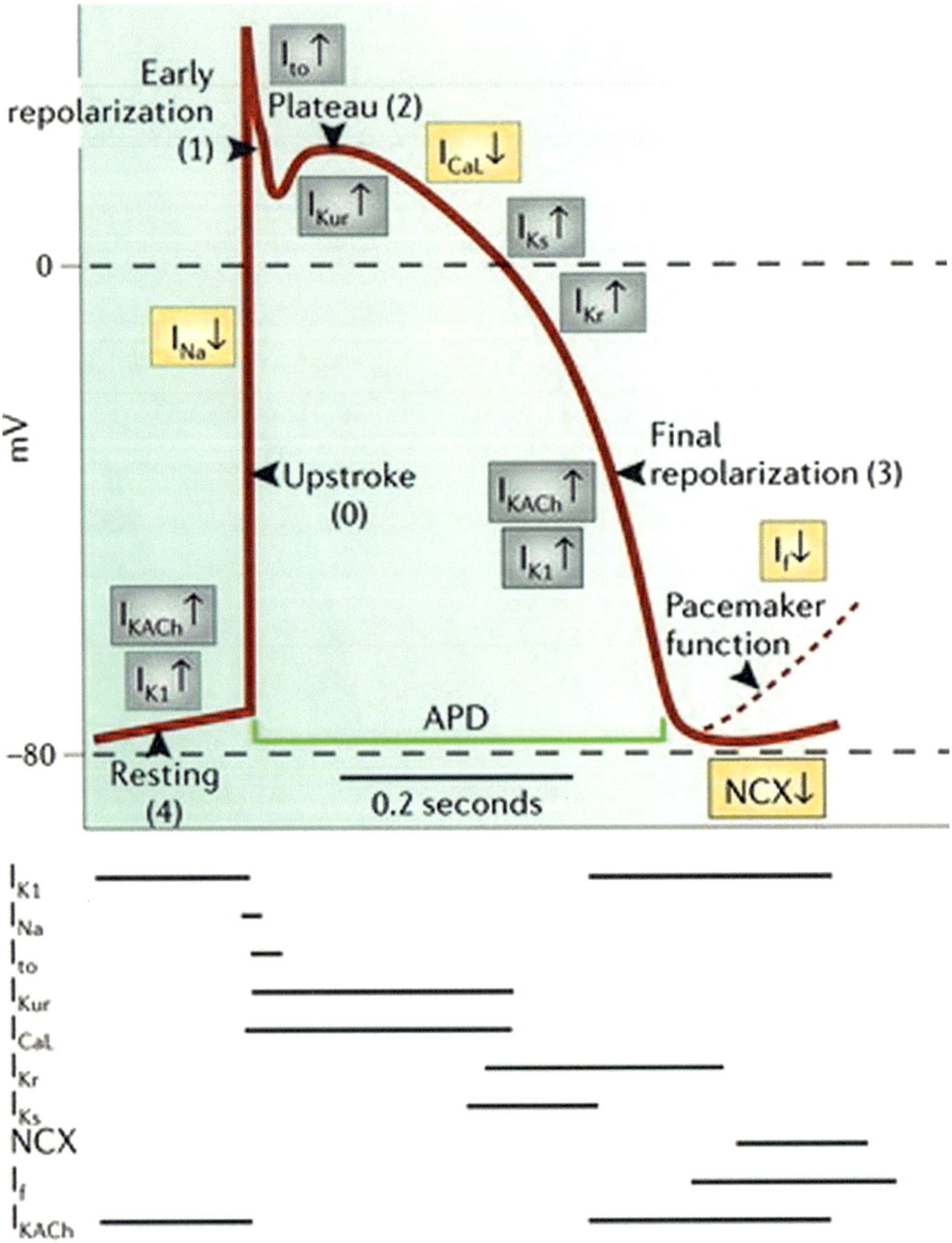

Figure 2.
Membrane currents that provide a standard action potential.

2. Phase 0 is known as the rapid depolarization phase. The membrane potential shifts toward the charge. This phase is central to the rapid cardiac impulse propagation (conduction velocity, θ = 1 m/s).

3. Phase 1 is characterized by fast repolarization. This phase of the action potential establishes the potential for phase 2.

4. The most prolonged phase is phase 2, a plateau phase. It distinguishes excitable cells and indicates the time of calcium entry into the cell.

5. Phase 3 is the rapid repolarization phase, during which the membrane potential is restored to its resting value [110].

The five phases of the action potential are resting (4), upstroke (0), early repolarization (1), plateau (2), and final repolarization. A broken line represents a fall in potential toward the end of phase 3 in pacemaker cells, such as the sinus node. The inward currents I_{Na}, I_{Ca}, and the sodium-calcium exchanger are illustrated in yellow boxes (NCX). It is electrogenic and can produce both inward and outward currents. Gray boxes represent I_{KAch}, I_{K1}, Ito, I_{Kur}, I_{Kr}, and I_{Ks}. The action potential duration (APD) is typically between 200 and 400 milliseconds [111].

The start of the action potential and the variances observed throughout the heart show that ion channels dispersed on the cell membrane have selective permeability. Ion channels minimize the activation energy required for ion transport across the lipophilic cell membrane [107].

Ion channels have two primary characteristics: ion permeation and gating [112]. The passage of ions via an open channel is described by ion permeation. The classification of ion channels is based on the selective permeability of ion channels to specific ions (e.g., Na^+, K^+, and Ca^{2+} channels). Size, valency, and hydration energy are essential factors of selectivity. Ion channels do not function as simple fluid-filled pores but provide multiple binding sites for ions as they traverse the membrane. Most ion channels are singly occupied during permeation; specific K^+ channels may be multiply occupied. The bulk of ion channels has a nonlinear current–voltage relationship. The size of the current depends on the direction of ion migration into or out of the cells for the same absolute degree of change in voltage. This is known as rectification, an essential trait of K^+ channels; they carry minimal outward current at positive (depolarized) potentials. The fundamental mechanism of rectification differs depending on the kind of ion channel. The mechanism of significant inward rectification displayed by many K^+ channels is blocked by the internal Mg^+ and polyvalent cations [113].

Ion channel gating, which explains how they open and close, is their second characteristic. Ion channels can also be categorized into categories based on their gating mechanisms, including voltage-dependent, ligand-dependent, and mechanosensitive gating. Voltage-gated ion channels modify their conductance in response to variations in membrane potential. The gating mechanism used by ion channels is typically voltage-dependent [109].

Changes in the electrical membrane potential close to the channel cause a set of transmembrane proteins called voltage-gated ion channels to open and close. The channel proteins' shape is altered by the membrane potential, which also regulates how they open and close. Ions must diffuse through the membrane through transmembrane protein channels because they are unable to generally flow through cell membranes. They are essential for enabling an immediate and coordinated depolarization in response to triggering voltage changes in excitable tissues, such as neurons and muscle cells [114]. The opening and closing of the channels are activated by changing ion concentration, and hence charge gradient between the sides of the cell membrane [115].

3.3 Voltage-gated sodium (Na_v)

Na_v channels are integral membrane proteins that change conformation in response to membrane potential depolarization, open a transmembrane pore, and convey sodium ions inward to initiate and propagate action potentials. Na_v is responsible for the rising phase of action potentials in excitable cells, such as neurons, myocytes, and certain types of glia. These channels cycle through three states: resting, active, and inactive. Even though the ions would not be able to move through the channels in their resting or inactive states, there is a variation in their structural conformation. When the membrane potential of a cell change, a modest but noteworthy number of Na^+ ions migrate into the cell down their electrochemical gradient, further depolarizing the cell. Therefore, the more the Na^+ channels get localized in a section of a cell's membrane, the more excitable and quickly propagating the action potential of that portion of the cell will be [116].

3.4 Voltage-gated calcium (Ca_v)

There are two voltage-gated Ca_v channels within the cardiac muscle: L-type calcium channels ("L" for Long-lasting) and T-type calcium channels ("T" for Transient, i.e., short). L-type channels are more numerous and densely populated within ventricular cell t-tubule membranes. On the other hand, T-type channels are located primarily within atrial and pacemaker cells but to a smaller extent than L-type channels. Higher positive membrane potentials activate L-type channels, take longer to open, and remain open for a longer time than T-type channels. This implies that T-type channels contribute more to depolarization (phase 0), whereas L-type channels contribute more to plateauing (phase 2) [117].

3.5 Voltage-gated potassium (K_v)

K_v is the most widely distributed ion channel type found in all living organisms. They are transmembrane channels specific for potassium and sensitive to voltage changes in the cell's membrane potential. During action potentials, they play a crucial role in returning the depolarized cell to a resting stage. Potassium channels are found in most cell types and control various cell functions [112].

The two main K^+ channels in cardiac cells are inward rectifiers and voltage-gated potassium channels.

Potassium channels that internally correct (K_{ir}) encourage the entry of K^+ into cells. However, the significance of this potassium influx increases when the membrane potential is lower than the equilibrium potential for K^+ (~ -90 mV). The amount of potassium entering the cell through the K_{ir} reduces as the membrane potential moves in a more positive direction, as it does when an adjacent cell stimulates the current flow. K_{ir} is therefore in charge of preserving the resting membrane potential and starting the depolarization phase. However, the channel starts to let K^+ leave the cell when the membrane potential continues to move in a more positive direction. The K_{ir} can also help with the last phases of the repolarization because of this outward influx of potassium ions at the more positive membrane potentials [118].

Depolarization activates voltage-gated K_v channels. These channels generate currents, such as the transient out potassium current I_{to1}. This current is made up of two parts. Both components activate quickly. However, $I_{to,\,fast}$ deactivates faster than $I_{to,\,slow}$. These currents contribute to the action potential's early repolarization phase (phase 1) [118].

The delayed rectifier potassium channels are yet another variety of voltage-gated potassium channels. These channels transport potassium currents that cause the action potential's plateau phase. They are named according to how quickly they activate: IK_s that activate slowly, IK_r that activate quickly, and IK_{ur} that activate extremely quickly [119].

4. Pyridine derivatives an ion channels modulator

The pyridine ring system can be found in a variety of natural products and pharmaceutically relevant molecules. Many of these compounds have fascinating and distinctive pharmacological characteristics that have often encouraged their production and reactivity. This chapter highlights recent advances in the regulation of several ion channel behaviors, such as voltage-gated sodium (Na_v), calcium (Ca_v), and potassium (K_v) channels by the Pyridine derivatives [120].

4.1 Regulation of voltage-gated calcium (Ca_v) ion channel by pyridine derivatives

Calcium channel blockers (CCBs) are unique drugs that prevent calcium from moving through calcium channels. They all have a similar mode of action, but are not interchangeable and can have diverse physiologic consequences. Calcium channel blockers are divided into dihydropyridines [DHPs] such as nifedipine and non-DHPs such as verapamil and diltiazem. These families bind to calcium channels at various binding locations, which could explain the clinical discrepancies. Non-dihydropyridines are more myocardial selective and tend to lower the heart rate, while dihydropyridines are more vascular selective [121]. Calcium channel blockers all relax atrial smooth muscle and cause peripheral vasodilation, decreasing blood pressure.

Furthermore, because calcium is directly implicated in cardiac contraction, lowering intracellular calcium concentrations via calcium channel blocking can reduce ventricular contractility. However, DHP CCBs do not exhibit this negative ionotropic effect, since they are more effective peripheral vasodilators than verapamil and diltiazem [122]. Because of their cardiac inotropic and vasomotor properties, DHPs are frequently employed as medicines. Many members of this class are commercially important cardio protectants, vasodilators, and calcium antagonists [123]. This possible peripheral vasodilation causes a baroreceptor-mediated increase in sympathetic tone, which mitigates the DHPs' negative inotropic action. In patients with heart failure and systolic dysfunction, it is recommended to avoid and use calcium channel blockers [non-dihydropyridines] with negative inotropic effects with caution [124]. Verapamil and diltiazem, unlike DHPs, lower the sinoatrial (SA) node conduction rate (negative chronotropes) and slow atrioventricular (AV) conduction (negative chromotropes) [125]. The rationale for employing non-DHPS (verapamil and diltiazem) for the treatment of supraventricular tachyarrhythmias (SVTS) and atrial fibrillation is to slow the rate of conduction via the AV node [126]. The DHP CCBs do not slow conduction across the AV node and are thus ineffective in treating SVT.

Furthermore, they do not disable the SA node's automaticity. Indeed, DHP CCBs may cause a rise in heart rate due to reflex tachycardia induced by powerful peripheral vasodilation. This effect is particularly noticeable with nifedipine quick release [127]. To emphasize immediate release, when used for acute blood pressure lowering, nifedipine has been linked to increased morbidity (myocardial ischemia and infarction), particularly in individuals with coronary artery disease (CAD) [125]. When taken for acute blood pressure reduction, immediate-release nifedipine has been linked to higher morbidity (myocardial ischemia and infarction), particularly in individuals with CAD [128]. Nifedipine was the chosen drug for hypertension crises because of its quick onset of action.

On the other hand, immediate-release nifedipine is no longer considered safe or efficacious for this indication. Sustained-release nifedipine formulations are less dangerous and do not cause strong reflex reactions to tachycardia. It is also worth noting that reflex tachycardia is not concerned with DHP CCBs with a delayed onset of action, such as amlodipine and felodipine.

To summarize, there are numerous distinctions between DHP and non-DHP CCBs. The non-DHPs are notable for being negative chronotropes, inotropes, and dromotropes. They should be taken with caution in individuals with heart failure and with drugs that have comparable hemodynamic effects. DHP CCBs are the most commonly used medications in individuals with hypertension and angina because they affect cardiac conduction [129].

4.2 Regulation of voltage-gated sodium (Na_v) ion channel behaviors by pyridine derivatives

Action potentials are initiated by voltage-gated sodium channels in neurons, cardiac muscle, and other electrically excitable cells. Sodium channel blockers are utilized in local anesthetic and in treating epilepsy, bipolar disorder, chronic pain, and cardiac arrhythmia. Pyridine, having the chemical formula C_5H_5N, is an essential heterocyclic organic molecule. The presence of a pyridine derivative, such as nicotinamide, as a nitrogen base distinguishes pyridine nucleotides (PNs). In addition to their role as soluble electron carriers, pyridine nucleotides [NAD(P)(H)] influence ion transport mechanisms. According to new research, pyridine nucleotides [NAD(P)(H)] influence ion transport processes in addition to their role as soluble electron carriers. PNs are vital in various physiological responses, including stress, energy metabolism, and cell survival/death in cardiovascular cells. The development of congestive heart failure may be influenced by oxidative stress in the myocardium (HF) [130]. Cells include an antioxidant system comprising GSH and thioredoxin (Trx) and reducing enzymes, such as superoxide dismutase and catalase, to protect against excessive ROS [131]. PNs function in regulating cellular redox status by acting as electron donors for both negative and positive oxidative stress regulators. Pyridine nucleotide regulation of ion channels may be essential for integrating cell ion transport to energetics and sensing oxygen levels or metabolite availability. Aside from these regulatory activities, current research has demonstrated that pyridine nucleotides also influence the activity of ion channels by acting as ligands or substrates of accessory subunits that modify channel gating. The modulation of K_{Na}/SLO2 channels by $NAD(P)^+$ shows that their activity may be linked to the cell's metabolic condition. This form of control may be especially relevant during ischemia–reperfusion, and other circumstances in which $NAD(P)^+$ buildup may promote K^+ efflux through these channels. High intracellular $NAD(P)^+$ levels would also increase the sensitivity of these channels to intracellular sodium [132].

Moreover, it has been proposed that in ischemic cardiac myocytes, increased $[Na^+]_i$ levels activate K_{Na}, and an increase in this current shortens ADP and promotes calcium overload [133, 134]. As a result, regulating these channels with pyridine nucleotides would allow them to adapt to both the metabolic and ionic circumstances present in the ischemic heart. Interestingly, despite the lack of direct proof, it has been claimed that SLO2 channels exist in the cardiac mitochondria [135]. Pyridine nucleotide control of these channels could present the preservation of the relationship between metabolism and ion transport in modern mitochondria and their prokaryotic progenitors. Although these findings are exciting, more research is needed to understand how intracellular changes in pyridine nucleotides influence SLO channels' activity and physiological relevance.

4.3 Regulation of voltage-gated potassium (K_v) ion channel activity by pyridine derivatives

Potassium channels are a diverse and widespread type of ion channel. They primarily regulate the cell's resting membrane potential and reduce the level of excitement. The current invention relates to novel pyridine and quinoline derivatives, pharmaceutical compositions incorporating them, and their use in treating ion channel disorders, such as potassium channel dysfunction. Potassium (K_v) channels also interact with pyridine nucleotide-binding proteins. These channels are essential in numerous physiological functions. They regulate the membrane potential of excitable cells and affect the shape and frequency of the action potential. These channels are also involved in the regulation of neurotransmitter release and cell volume [136, 137], proliferation, [138] and apoptosis [139]. They are also important in T-cell differentiation, activation, and cytokine generation [140]. These channels' activity affects baseline and agonist-stimulated vasomotor tone, and the membrane hyperpolarization generated by K_v channel activation governs the vasodilation [141]. Oxygen-sensitive variations in K_v channel activity drive hypoxic pulmonary vasoconstriction in small resistance arteries (HPV) [142, 143]. As a result, aberrant K_v channel activity has been linked to cardiac arrhythmias, pulmonary hypertension, epilepsy, and aberrant immunological responses [141, 144, 145]. The many functions of K_v channels are related to their various structures. The ion-conducting pore of K_v channels is produced by four membrane spanning subunits, assembled in a homotetrameric or heterotetrameric fashion. Twelve distinct Kv channel proteins have been reported so far [146, 146]. Several K_v families' pore-forming subunits interact in situ with accessory subunits that help channel construction and influence channel function, such as K_v family proteins, that interact with the cytosolic domains of Kv1 and Kv4 channel proteins [148]. Pyridine nucleotide function at the binding location N-type inactivation by NADPH, removal of inactivation by $NADP^+$, and membrane trafficking are the functions of voltage-gated potassium (K_v) ion channels' ancillary subunit-K_v [149]. Changes in the amount of cofactor binding, which passively replicates the physiological levels of these nucleotides, could modulate the gating of the K_v-K_v assembly. Thus, increased intracellular NAD(P)H levels would promote inactivation, but increased NAD(P) + levels would eliminate inactivation. Membrane voltage may influence catalysis via K_v contact with the cytosolic T1 domain or the C-terminus of K_v channels. The C-terminus of the shaker channel linked to Kv2 is in intimate contact with the K_v active site, according to the electron microscopic single particle analysis [150]. This analysis demonstrates that the K_v channel's inner helices, which are anticipated to move considerably during gate opening and closing, are directly connected to the channel's C-terminus. This suggests that the conformation and orientation of the Kv C-terminus relative to the subunits may change as a function of membrane voltage. $K_v1.5$'s C-terminal peptide interacts more avidly with NADPH than $NADP^+$ bound K_v2, and its deletion prevents differential regulation of $K_v1.5$ + Kv2 and $K_v1.5$ + K_v3 currents by reduced and oxidized nucleotides, despite the fact that the role of the K_v C-terminus in enhancing voltage sensitivity to K_v catalysis has not been studied [151]. Despite these observations, the general physiological function of the K_v C-terminus is unknown. The C-terminus of $K_v1.1$, unlike the C-terminus of $K_v1.5$, does not affect channel control by K_v1 coupled to pyridine nucleotides [152]. Although pyridine nucleotides have been shown to regulate K_v currents in heterologous systems, the physiological importance of this regulatory axis has yet to be determined. Even though K_v channels are involved in numerous physiological processes, their function is heavily influenced by posttranslational modification and subunit assembly. Pyridine nucleotide regulation may give additional control by linking the activity of these channels to changes in metabolic activity of the cell's redox state. For example, hypoxic depolarization of pulmonary artery smooth muscle cells (PASMCs), which underpins the HPV phenomenon, has been linked to the $K_v1.5$ inhibition [153]. The fact that $K_v1.5$ is oxygen sensitive when produced in PASMCs but not in other cell types suggest that factors other than the pore-forming subunits of the channels may be necessary for the channel's oxygen sensitivity [154]. The ability of pyridine nucleotide-binding K_v proteins to modulate K_v current might theoretically confer oxygen sensitivity to Kv1.5 channels. $K_i\beta$ is abundantly expressed in PASMCs, and its expression is substantially higher in the distal than the proximal bovine pulmonary artery, indicating a potential function in oxygen sensing and HPV infection [155]. Furthermore, the $K_v1.5$-$K_v1.3$ channels are the primary components of I_{Kv} in PASMC, and these channels are variably controlled by oxidized and reduced pyridine nucleotides in COS-7 cells [153, 155]. As a result, an increase in the NADPH:$NADP^+$ ratio during hypoxia may activate $K_v1.5$-$K_v1.3$ currents at more negative membrane potentials, whereas the current is blocked at higher positive membrane potentials, where inactivation becomes more pronounced. This activity has only been observed in hypoxic canine PASMC and not in other species [156]. This species difference could be attributed to variations in K_v expression. While inhibition may be related to K_v2, which does not impact K_v inactivation but shifts the voltage dependence of activation, hypoxia may increase K_v currents, whilst inhibition may be related to K_v2. However, it would be anticipated that a rise in the NADPH:$NADP^+$ ratio would result in a shift in the activation threshold, that is, more hyperpolarizing than depolarizing [147]. Therefore, more research is needed to implicate K_v in HPV and to determine the role of distinct K_v subunits in regulating the oxygen sensitivity of K_v channels.

5. Clinical approaches of Pyridine derivatives

A glance at the US FDA database reveals that pyridine and dihydropyridine drugs constitute nearly 14% and 4% of N-heterocyclic drugs are approved for the treatment of various diseases.

5.1 Pioglitazone

Pilot research was conducted to compare the effects of pioglitazone on cardiac function and oxidative stress in patients with type II diabetes and insulin resistance undergoing elective percutaneous coronary intervention to placebo [157]. In cardiac insulin resistance, pioglitazone corrects mitochondrial dysfunction [158], PPARgamma activation which is associated with improving cardiovascular risk were observed in many clinical investigations. The change in cardiovascular or metabolic markers and mRNA will be isolated from circulating mononuclear cells to investigate the degree of activation of the immune system, which is another measurement of the atherosclerosis risk [159]. It also have myocardial protection in atherosclerosis and coronary heart disease [160]. Pioglitazone reduces left ventricular mass in people with type II diabetes who have ischemic heart disease [161]. Pioglitazone treatment or physical training alone enhance the hearts in HIV patients with metabolic syndrome. The combination of physical training and pioglitazone treatment results on in reducing insulin resistance and subsequently improving cardiac metabolism, and enhancing heart function in the type II diabetes population with cardiovascular risk [162].

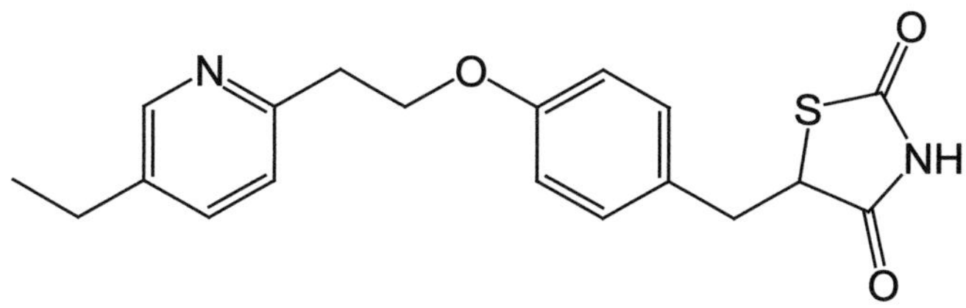

Pioglitazone

5.2 Niacin

Niacin plays a key role in regulating atherosclerotic plaque inflammation. It has a protective effect on endothelial progenitor cells and microparticles, and it is vigorously used in chronic statin therapy to treat atherosclerotic disease on chronic statin therapy. The effects of niacin on vascular health were assessed using fluorodeoxyglucose-PET/CT and circulating endothelial progenitor cells and microparticles [163]. Niacin reduces the elevation of triglycerides and HDL [164]. Extended-release niacin/laropiprant has a significant effect in patients with the atherosclerotic disease compared to placebo. Dilatation of arterial walls improved in statin therapy assessed by the brachial vasoreactivity [165].

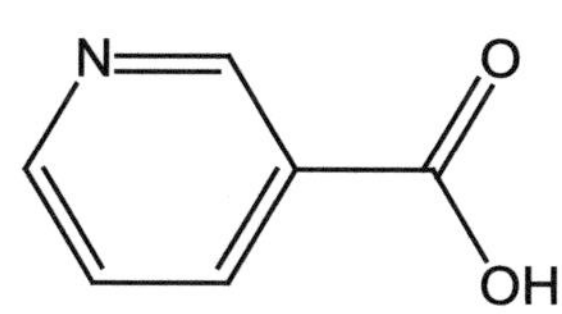

Niacin

5.3 Nicorandil

Nicorandil is recommended as a second-line treatment for the angina treatment [166]. Still, randomized-controlled trials going on to evaluate the benefit of nicorandil for patients with chronic total occlusion [167]. The treatment of oral nicorandil to reduced cardiac death after coronary revascularization in hemodialysis patients [168].

Nicorandil, a combination of nitrates, is an ATP-sensitive K+ channel activator that reduces infarct size in animal models. Moreover, a prospective and randomized, multicenter study was conducted by the Japan-working groups of acute myocardial infarction for the reduction of necrotic damage by activating K-ATP channel to determine potential use of nicorandil. The treatment of Nicorandil for acute myocardial infarction, reduces myocardial infarct size and improves regional wall motion [169]. The infarct size in ST-segment elevation myocardial infarction patients undergoing primary percutaneous coronary intervention treated by nicorandil before and after the reperfusion with those standard therapy treated by percutaneous coronary intervention [170].

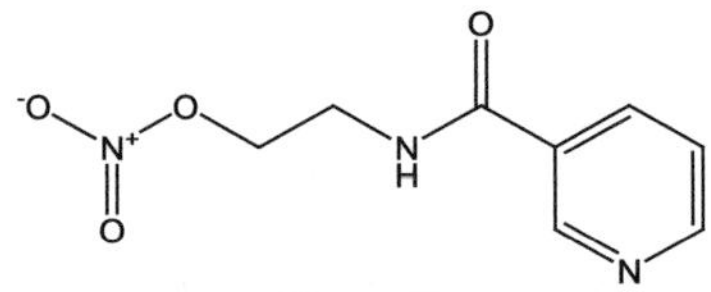

5.4 Riociguat

Riociguat has been shown pharmacodynamics affects in patients with pulmonary hypertension and heart failure with remodeled ejection fraction [171].

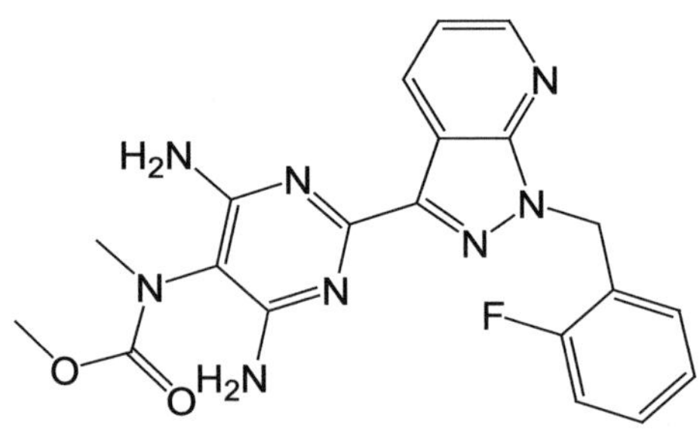

Riociguat

5.5 Vericiguat

Vericiguat (BAY1021189) is currently being developed to treat heart failure, which is a condition where the heart has unable to pump blood throughout the body. Patients with heart failure frequently also have renal impairment, which prevents the kidneys from properly filtering the blood [172]. Many investigators found the pharmacodynamic drug-drug interaction and the safety and tolerability of Isosorbide Mononitrate and Vericiguat in patients with stable coronary artery disease [173]. In Phase III clinical trials, the optimal dose of soluble guanylate cyclase stimulator BAY1021189 per day by orally preserved ejection fraction in the heart failure [174].

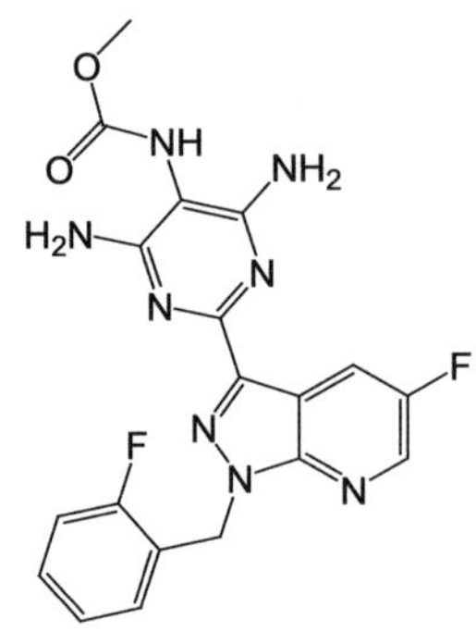

Vericiguat

5.6 Mangafodipir

Mangafodipir is also known as manganese dipyridoxyl diphosphate, and its lipophile metabolite manganese dipyridoxyl diethylene diamide has a catalytic antioxidants and iron chelators properties. In preclinical studies, these agents reduce injuries induced by oxidative stress in cancer chemotherapy and reperfusion/reoxygenation of ischemic/hypoxic myocardium. The treatment of Mangafodipir, decreased the size of the myocardial infarct by 55% in a in vivo myocardial infarct pig model. Most likely, mangafodipir promotes recovery of downregulated pathways and guards against fatal reperfusion damage [175].

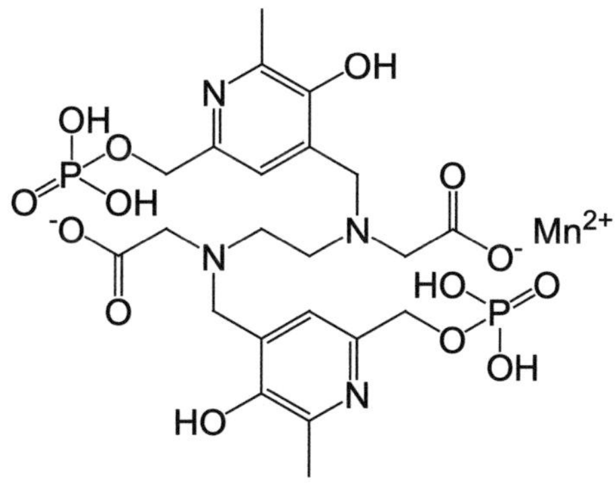

Mangafodipir

5.7 Tezosentan

Tezosentan has shown efficacy, and safety profile in patients with acute heart failure [176, 177].

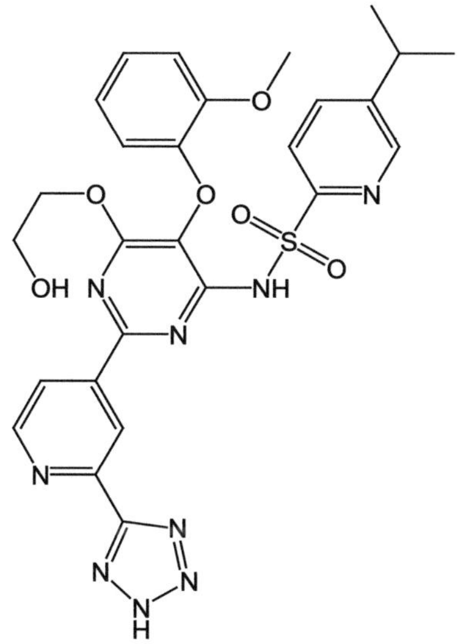

Tezosentan

Estrogens, dextrothyroxine, nicotinic acid, and clofibrate are used to treat coronary artery disease. These drugs cause more toxicity [178].

6. Conclusion

The book chapter describes the importance of pyridine derivatives as a novel class of medications for treating cardiovascular disorders. Pyridine derivatives are known to be ion channel modulators and change the action potential by changing voltage-gated potassium, sodium, and calcium ion channel activity. This chapter presents a critical study of many medications and research on designing and developing various pyridine and dihydropyridine-based derivatives. They have been classified based on their pharmacological activity. Every specific structural aspect relevant to exclusive activities has also been considered. The central pyridine core is more significantly tractable for producing anti-infectious and anticancer medicines. Dihydropyridine derivatives primarily regulate the dihydropyridine protein, also known as calcium channels. Dihydropyridine ring-containing drugs, including nimodipine, ciclopirox, efonidipine, nifedipine, milrinone, and amrinone, primarily function as calcium channel blockers, and are used to treat hypertension and heart issues.

The structure, application, and diversity of pyridine- and dihydropyridine-containing compounds will expand in the future decade, with tremendous potential for new cardiovascular, anti-inflammatory, anti-infectious, neurogenic, and anticancer therapies incorporating the two heterocycles. Because of the enormous structural diversity of pyridine- and dihydropyridine-containing compounds, the present literature just scratches the surface of potential therapeutic applications. In conclusion, paired with a broader chemical space, pyridine and dihydropyridine-containing compounds will aid medicinal chemists in designing bioactive molecules for specific targets.

Author details

Yasodha Krishna Janapati[1], Sunithasree Cheweti[2], Bojjibabu Chidipi[3], Medidi Srinivas[4] and Sunil Junapudi[4*]

1 School of Pharmacy and Health Sciences, United States International University – Africa, Nairobi, Kenya

2 Hindu College of Pharmacy, Nagarjuna University, Guntur, Andhra Pradesh, India

3 Molecular Pharmacology and Physiology, Morsani College of Medicine, University of South Florida, Tampa, FL, USA

4 Department of Pharmaceutical Chemistry, Geethanjali College of Pharmacy, Medchalmalkajgiri, Telanga, India

*Address all correspondence to: suniljunapudi@gmail.com

References

[1] Wang S, Yuan XH, Wang SQ, Zhao W, Chen XB, Yu B. FDA-approved pyrimidine-fused bicyclic heterocycles for cancer therapy: Synthesis and clinical application. European Journal of Medicinal Chemistry. 2021;**214**:113218. DOI: 10.1016/j.ejmech.2021.113218

[2] Jadhav M, Sankhe K, Bhandare RR, Edis Z, Bloukh SH, Khan TA. Synthetic strategies of pyrimidine-based scaffolds as aurora kinase and polo-like kinase inhibitors. 2021;**26**(17):5170. DOI: 10.3390/molecules26175170

[3] Abbas N, GSP M, Dhiwar PS, Patel S, Devasahayam G. Fused and substituted pyrimidine derivatives as profound anti-Cancer agents. Anti-Cancer Agents in Medicinal Chemistry. 2021;**21**(7):861-893. DOI: 10.2174/18715206206662007 21104431

[4] Bull JA, Mousseau JJ, Pelletier G, Charette AB. Synthesis of pyridine and dihydropyridine derivatives by regio- and stereoselective addition to N-activated pyridines. 2012;**112**(5):2642-2713. DOI: 10.1021/cr200251d

[5] Wang L, Bharti KR, Pavlov PF, Winblad B. Small molecule therapeutics for tauopathy in Alzheimer's disease: Walking on the path of most resistance. European Journal of Medicinal Chemistry. 2021;**209**:112915. DOI: 10.1016/j.ejmech.2020.112915

[6] Jubete G, Puig de la Bellacasa R, Estrada-Tejedor R, Teixido J, Borrell JI. Pyrido[2,3-d]pyrimidin-7(8 H)-ones: Synthesis and biomedical applications. Molecules. 2019;**24**(22):4161. DOI: 10.3390/ molecules24224161

[7] Mammoliti O, Palisse A, Joannesse C, El Bkassiny S, Allart B, Jaunet A, et al. Discovery of the S1P2 antagonist GLPG2938 (1-[2-Ethoxy-6-(trifluoromethyl)-4-pyridyl]-3-[[5-methyl-6-[1-methyl-3-(trifluoromethyl) pyrazol-4-yl]pyridazin-3-yl]methyl] urea), a preclinical candidate for the treatment of idiopathic pulmonary fibrosis. Journal of Medicinal Chemistry. 2021;**64**(9):6037-6058. DOI: 10.1021/acs.jmedchem.1c00138

[8] Yu L, Ran K, Zeng J, Wan G, He X, Feng Z, et al. Design, synthesis and biological evaluations of a series of pyrido[1,2-a]pyrimidinone derivatives as novel selective FGFR inhibitors. European Journal of Medicinal Chemistry. 2021;**220**:113499. DOI: 10.1016/j.ejmech.2021.113499

[9] Ling Y, Hao ZY Liang D, Zhang CL, Liu YF, Wang Y. The expanding role of pyridine and Dihydropyridine Scaffolds in drug design. Drug Design, Development and Therapy. 2021;**15**:4289

[10] Sperry J, Lin SX, Curtis MA. Pyridine alkaloids with activity in the central nervous system. Bioorganic & Medicinal Chemistry. 2020;**28**(24):115820. DOI: 10.1016/j.bmc.2020.115820

[11] Ziegler M, Pollak N, Dölle C. The power to reduce: Pyridine nucleotides--small molecules with a multitude of functions. The Biochemist. 2007;**402**(2):205-218. DOI: 10.1042/BJ20061638

[12] Moustakas D, Pennington LD. The necessary nitrogen atom: A versatile high-impact design element for multiparameter optimization. Journal of Medicinal Chemistry. 2017;**60**(9):3552-3579. DOI: 10.1021/acs.jmedchem.6b01807

[13] Castellan P, Marchioni M, Castellucci R, et al. Abiraterone acetate for early stage metastatic prostate cancer: Patient selection and special considerations. 2018;**14**:2341. DOI: 10.2147/TCRM.S159824

[14] Vince R, Wang Z. Design and synthesis of dual inhibitors of HIV reverse transcriptase and integrase: Introducing a diketoacid functionality into delavirdine. 2008;**16**(7):3587-3595. DOI: 10.1016/j.bmc.2008.02.007

[15] Friedman H, Greenblatt DJ, Scavone JM, et al. Clearance of the antihistamine doxylamine. Clinical Pharmacokinetics. 1989;**16**(5):312-316. DOI: 10.1111/j.1365-2125.1992

[16] Basco LK, Gillotin C, Gimenez F, Farinotti R. In vitro activity of the enantiomers of mefloquine, halofantrine and enpiroline against plasmodium falciparum. British Journal of Clinical Pharmacology. 1992;**33**(5):517-520. DOI: 10.1111/j.1365-2125.1992. tb04081.x

[17] Hsu KHK. isoniazid. JAMA. 1984;**251**(10):1283-1285. DOI: 10.1001/jama.1984.03340340023018

[18] Gopalan C, Raghuramulu N, Srikantia S, Rao B. Nicotinamide nucleotides in the erythrocytes of patients suffering from pellagra. The Biochemical Journal. Sep 1965;**96**(3):837. DOI: 10.1042/bj0960837

[19] Campbell E, Westlake EK. Effects of aminophylline, nikethamide, and sodium salicylate in respiratory failure. 1959;**1**(5117):274. DOI: 10.1136/bmj.1.5117.274

[20] Walan A, Bader JP, Classen M, et al. Effect of omeprazole and ranitidine on ulcer healing and relapse rates in patients with benign gastric ulcer. The New England Journal of Medicine. 1989;**320**(2):69-75. DOI: 10.1056/NEJM198901123200201

[21] De Lapp R, Lister BJ, Poland M. Efficacy of nabumetone versus diclofenac, naproxen, ibuprofen, and piroxicam in osteoarthritis and rheumatoid arthritis, The American Journal of Medicine. 1993;**95**(2):S2-S9. DOI: 10.1016/0002-9343(93)90390-B

[22] Andersen JB, Engeland A, Owe JF. Myasthenia gravis requiring pyridostigmine treatment in a national population cohort. European Journal of Neurology. 2010;**17**(12):1445-1450. DOI: 10.1111/ j.1468-1331.2010.03089.x

[23] Ahmed M, JBT R, Corrêa M, et al. Inhibition of two different cholinesterases by tacrine. Chemico-Biological Interactions. 2006;**162**(2):165-171. DOI: 10.1016/j.cbi.2006.06.002

[24] McDonald C, Bostock C. Antimuscarinics in older people: Dry mouth and beyond. Dental Update. 2016;**43**(2):186-191. DOI: 10.12968/denu.2016.43.2.186

[25] Pöch G, Kukovetz WR, Holzmann S. Molecular mechanism of action of nicoran. Journal of Cardiovascular Pharmacology. 1992;**20**:S1-S7. DOI: 10.1097/00005344-199206203-00002

[26] Metabolism W-D, Harvey JL, Paine AJ, Maurel P. Effect of the adrenal 11-beta-hydroxylase inhibitor metyrapone on human hepatic cytochrome P-450 expression: Induction of cytochrome P-450 3A4. Drug Metabolism and Disposition. 2000;**28**(1):96-101

[27] Siegel M, Sigel E. Structure, function, and modulation of GABAA receptors. The Journal of Biological Chemistry. 2012;**287**(48):40224-40231. DOI: 10.1074/jbc.R112.386664

[28] Patrignani P, Capone ML, Tacconelli S, Di Francesco L, Sacchetti A, Sciulli MG. Pharmacodynamic of cyclooxygenase inhibitors in humans. Prostaglandins & Other Lipid Mediators. 2007;**82**(1-4):85-94. DOI: 10.1016/j.prostaglandins.2006.05.019

[29] Mishra J, Kothekar V, Sahi S, Srinivasan M, Mohan A. Recognition of cyclooxygenase-2 (COX-2) active site by NSAIDs: A computer modelling study. Indian Journal of Biochemistry & Biophysics. 2001;**38**(1-2):56-63

[30] Roser R, Esteve J, Farré AJ. Pharmacological profile of droxicam. 1988;**19**(1):49-54

[31] Weissman A, Carty TJ, Marfat A, Moore PF, Falkner FC, Twomey TM. Ampiroxicam, an anti-inflammatory agent which is a prodrug of piroxicam. Agents and Actions. 1993:**39**(3):157-165. DOI: 10.1007/BF01998969

[32] Stimmeder D, Berg J, Fellier H, Christoph T, Grarup J. The analgesic NSAID lornoxicam inhibits cyclooxygenase (COX)- 1/-2, inducible nitric oxide synthase (iNOS), and the formation of interleukin (IL)-6 in vitro. 1999;**48**(7):369-379. DOI: 10.1007/s000110050474

[33] Dekomfeld T, Finch JS. Clonixin: A clinical evaluation of a new oral analgesic. The Journal of Clinical Pharmacology and New Drugs. 1971;**11**(5):369-379

[34] Krause KH, Suter DM, Preynat-Seauve O, Tirefort D, Feki A. Phenazopyridine induces and synchronizes neuronal differentiation of embryonic stem cells. Journal of Cellular and Molecular Medicine. 2009;**13**(9b):3517-3527. DOI: 10.1111/j.1582-4934.2009.00660.x

[35] Muster H, Morgan RE, Campbell SE, Yu CY, Sponseller CA. Comparison of the safety, tolerability, and pharmacokinetic profile of a single oral dose of pitavastatin 4 mg in adult subjects with severe renal impairment not on hemodialysis versus healthy adult subjects. Journal of Cardiovascular Pharmacology. 2012;**60**(1):42-48. DOI: 10.1097/FJC.0b013e318256cdf0

[36] Rybak M, Steed ME. Ceftaroline: A new cephalosporin with activity against resistant gram-positive pathogens. 2010;**30**(4):375-389. DOI: 10.1592/phco.30.4

[37] Muller L, Roger C, Roberts JA. Clinical pharmacokinetics and pharmacodynamics of oxazolidinones. 2018;**57**(5):559-575. DOI: 10.1007/s40262-017-0601-x

[38] Bryson R, Richards DM. Ceftazidime. A review of its antibacterial activity, pharmacokinetic properties and therapeutic use. 1985;**29**(2):105-161. DOI: 10.2165/00003495-198529020- 00002

[39] Markham A. Delafloxacin: First global approval. 2017;**77**(13):1481-1486. DOI: 10.1007/s40265-017-0790-5

[40] Cooksey R, Morlock GP, Metchock B, Sikes D, Crawford JT. ethA, inhA. katG loci of ethionamide-resistant clinical *Mycobacterium tuberculosis* isolates. Antimicrobial Agents and Chemotherapy. 2003;**47**(12):3799-3805. DOI: 10.1128/AAC.47.12.37 99-3805.2003

[41] Ambrose Z, Herman BD, Sheen C-W, et al. The human immunodeficiency virus type 1 nonnucleoside reverse transcriptase inhibitor resistance mutation I132M confers hypersensitivity to nucleoside analogs. Journal of Virology. 15 Apr 2009;**83**(8):3826-3833. DOI: 10.1128/ JVI.01968-08

[42] Feig M, Wittayanarakul K, Hannongbua S. Accurate prediction of protonation state as a prerequisite for reliable MM-PB(GB)SA binding free energy calculations of HIV-1 protease inhibitors. Journal of Computational Chemistry. 2008;**29**(5):673-685. DOI: 10.1002/jcc.20821

[43] Cohen EE, Rosen LS, Vokes EE, et al. Axitinib is an active treatment for all histologic subtypes of advanced thyroid cancer: Results from a phase II study. Journal of Clinical Oncology. 2008;**26**(29):4708. DOI: 10.1200/JCO.2007.15.9566

[44] Druker B, MWN D. Specific targeted therapy of chronic myelogenous leukemia with imatinib. 2003;**55**(3):401-423. DOI: 10.1124/pr.55.3.4

[45] Flaherty KT. Chemotherapy and targeted therapy combinations in advanced melanoma. 2006;**12**(7):2366s-2370s. DOI: 10.1158/1078-0432.CCR-05-2505

[46] Hotta K, Ueyama J, Tatsumi Y et al. Lack of contribution of multidrug resistance-associated protein and organic anion-transporting polypeptide to pharmacokinetics of regorafenib. 2015;**35**(9):4681-4689

[47] Konstantinopoulos PA, Barry WT, Birrer M, et al. Olaparib and α- specific PI3K inhibitor alpelisib for patients with epithelial ovarian cancer: A dose-escalation and dose-expansion phase 1b trial. 2019;**20**(4):570-580. DOI: 10.1016/S1470-2045(18) 30905-7

[48] Shaw AT, Felip E, Bauer TM, et al. Lorlatinib in non-small-cell lung cancer with ALK or ROS1 rearrangement. 2017;**18**(12):1590-1599. DOI: 10.1016/S1470-2045 (17)30680-0

[49] Liu D, Wu J, Zhang M. Acalabrutinib (ACP-196): A selective second-generation BTK inhibitor. Journal of Hematology & Oncology. 2016;**9**(1):1-4. DOI: 10.1186/s13045-016-0250-9

[50] Gelbert LM, Cai S, Lin X, et al. Preclinical characterization of the CDK4/6 inhibitor LY2835219: In-vivo cell cycle-dependent/independent anti-tumor activities alone/in combination with gemcitabine. 2014;**32**(5):825-837. DOI: 10.1007/s106 37-014-0120-7

[51] Burstein HJ, Sun Y, Dirix LY, et al. Neratinib, an irreversible ErbB receptor tyrosine kinase inhibitor, in patients with advanced ErbB2-positive breast cancer. Journal of Clinical Oncology. 2010;**28**(8):1301-1307. DOI: 10.1200/JCO.2009.25.8707

[52] Martelletti P, Negro A. Gepants for the treatment of migraine. Expert Opinion on Investigational Drugs. 2019;**28**(6):555-567. DOI: 10.1080/13543784.2019.1618830

[53] Schilder A, Murdin L, Hussain K. Betahistine for symptoms of vertigo. Cochrane Database of Systematic Reviews. 2016;(6):CD010696. DOI: 10.1002/14651858.CD010696.pub2

[54] Stangel M, Lindquist S. Treatment options for Lambert Eaton myasthenic syndrome: Focus on use of amifampridine. Neuropsychiatric Disease and Treatment. 2011;7:341-349. DOI: 10.2147/NDT. S10464

[55] Tagawa M, Kano M, Okamura N, et al. Neuroimaging of histamine H1-receptor occupancy in human brain by positron emission tomography (PET). British Journal of Clinical Pharmacology. 2001;**52**(5):501-509. DOI: 10. 1046/j.1365-2125.2001.01471.x

[56] Woolliscroft G, Webbon PM. Use of flunixin advocated. The Veterinary

Record. 1984;**115**(2):45. DOI: 10.1136/vr.115.2.45-b

[57] Bell J. Amlexanox used for allergic asthma and rhinitis. 2005;**25**(9):555-566. DOI:10.2165/00044011-200525090-00001

[58] Mandil H, Ramadan AA. Spectrophotometric determination of carbinoxamine maleate in pharmaceutical formulations by ternary complex formation with Cu(II) and eosin. Analytical Biochemistry. 2006;**353**(1):133-137. DOI: 10.1016/j.ab.2006.02.020

[59] Rial MJ, Fernández-Nieto M, Rodrigo-Muñoz JM, Sastre B, Sastre J, Del Pozo V. Doxylamine allergy in a pregnant woman: Suitability of the Basophil activation test. Journal of Investigational Allergology and Clinical Immunology. 2018;**28**(6):433-434

[60] Onaran H, Bökesoy TA. Atypical Schild plots with histamine H1 receptor agonists and antagonists in the rabbit aorta. European Journal of Pharmacology. 1991;**197**(1-2):49-56. DOI: 10.1016/0014-2999(91)90363-u

[61] Monteseirin J, Chacon P, Vega A, et al. L-selectin expression on neutrophils from allergic patients. 2005;**35**(9):1204-1213. DOI: 10.1111/j.1365-2222.2005.02320.x

[62] Merlos M, Giral M, Balsa D, et al. Rupatadine, a new potent, orally active dual antagonist of histamine and platelet-activating factor (PAF). The Journal of Pharmacology and Experimental Therapeutics. 1997;**280**(1):114-121

[63] Tietze K, Mann KV, Crowe JP. Nonsedating histamine H1- receptor antagonists, 1989;**8**(5):331-44

[64] Molimard M, Naline E, Trifilieff A, Fairhurst RA, Advenier C. Effect of indacaterol, a novel long-acting 2-agonist, on isolated human bronchi. The European Respiratory Journal. 2007;**29**(3):575-581. DOI: 10.1183/09031936.00032806

[65] Picatoste F, Claro E, Arbonés L, García A. Phosphoinositide hydrolysis mediated by histamine H1-receptors in rat brain cortex. European Journal of Pharmacology. 1986;**123**(2):187-196

[66] Simons K, Simons FE. Histamine and H1-antihistamines: Celebrating a century of progress. The Journal of Allergy and Clinical Immunology. 2011;**128**(6):1139-1150. DOI: 10.1016/j.jaci.2011.09.005

[67] Kamanna M, Kamanna VS. Mechanism of action of niacin. The American Journal of Cardiology. 2008;**101**(8):S20-S26. DOI: 10.1016/j.amjcard.2008.02.029

[68] Reeder NL, Kaplan J, Xu J, et al. Zinc pyrithione inhibits yeast growth through copper influx and inactivation of iron-sulfur proteins. Antimicrobial Agents and Chemotherapy. 2011;**55**(12):5753-5760. DOI: 10.1128/AAC.00724-11

[69] Jackson KJ, Marks MJ, Vann RE, et al. Role of α5 nicotinic acetylcholine receptors in pharmacological and behavioral effects of nicotine in mice. The Journal of Pharmacology and Experimental Therapeutics. 2010;**334**(1):137-146. DOI: 10.1124/jpet.110.165738

[70] Walley T, Dündar Y, Dodd S, Strobl J, Boland A, Dickson R. Comparative efficacy of newer hypnotic drugs for the short-term management of insomnia. 2004;**19**(5):305-322. DOI: 10.1002/ hup.594

[71] Scammell TE, Mahoney CE, Mochizuki T. Dual orexin receptor antagonists increase sleep and cataplexy in wild type mice. 2020;**43**(6):zsz302. DOI: 10.1093/sleep/zsz302

[72] Lüscher C, Tan KR, Rudolph U. Hooked on benzodiazepines: GABAA

receptor subtypes and addiction. 2011;**34**(4):188-197. DOI: 10.1111/j.1365-2222.2005.02320.x

[73] Ferriprotoporphyrin I, Fitch CD. Phospholipids, and the antimalarial actions of quinoline drugs. 2004;**74**(16):1957-1972. DOI: 10.1016/j.lfs.2003.10.003

[74] Ceramai A, AFG S. Inhibition by chloroquine of a novel haem polymerase enzyme activity in malaria trophozoites. 1992;**355**(6356):167-169. DOI: 10.1038/355167a0

[75] Asrani CH, Damle SS, Ghotge VV, et al. Efficacy and safety of metronidazole versus a combination of metronidazole and diiodohydroxyquinoline for the treatment of patients with intestinal amebiasis: A primary care physician research group study. 1995;**56**(7):678-683. DOI: 10.1016/0011-393X(95)85137-2

[76] Green M, Farrell DJ, Shackcloth J, Barbadora KA. Streptococcus pyogenes isolates with high-level macrolide resistance and reduced susceptibility to telithromycin associated with 23S rRNA mutations. Antimicrobial Agents and Chemotherapy. 2006;**50**(2):817-818. DOI: 10.1128/AAC.50.2.817-818.2006

[77] Bébéar C, Bébéar CM, Grau O, Charron A, Renaudin H, Gruson D. Cloning and nucleotide sequence of the DNA gyrase (gyrA) gene from *Mycoplasma hominis* and characterization of quinolone-resistant mutants selected in vitro with trovafloxacin. 2000;**44**(10):2719-2727. DOI: 10. 1128/AAC.44.10.2719-2727.2000

[78] Maurice A, Gootz TD, Zaniewski RP, Haskell SL, Kaczmarek FS. Activities of trovafloxacin compared with those of other fluoroquinolones against purified topoisomerases and gyrA and grlA mutants of Staphylococcus aureus. Antimicrobial Agents and Chemotherapy. 1999;**43**(8):1845-1855. DOI: 10.1128/AAC.43.8.1845

[79] Burg G, Dummer R, Urosevic M, Kempf W, Hoek K, Hafner J. Imiquimod in basal cell carcinoma: How does it work. The British Journal of Dermatology. 2003;**149**:57-58. DOI: 10.1046/j.0366-077X.2003.05630.x

[80] Broadhurst CL, Domenico P. Clinical studies on chromium picolinate supplementation in Diabetes Mellitus—a review. 2006;**8**(6):677-687. DOI: 10.1089/dia.2006.8.677

[81] Sreejayan N, Hua Y, Clark S, Ren J. Molecular mechanisms of chromium in alleviating insulin resistance. The Journal of Nutritional Biochemistry. 2012:23. DOI: 10.1016/j.jnutbio.2011.11.001

[82] Hube B, Niewerth M, Kunze D, Seibold M, Schaller M, Korting HC. Ciclopirox olamine treatment affects the expression pattern of *Candida albicans* genes encoding virulence factors, iron metabolism proteins, and drug resistance factors. 2003;**47**(6):1805-1817. DOI: 10.1128/AAC.47.6.1805-1817.2003

[83] Menéndez-Arias L. Molecular basis of human immunodeficiency virus type 1 drug resistance: overview and recent developments. Antiviral Research. 2013;**98**(1):93-120. DOI: 10.1016/j.antiviral.2013.01.007

[84] Brenner C, Belenky P, Bogan KL. NAD+ metabolism in health and disease. Trends in Biochemical Sciences. 2007;**32**(1):12-19. DOI: 10.1016/j.tibs.2006.11.006

[85] Heneine W, Andrews CD. Cabotegravir long-acting for HIV-1 prevention. Current Opinion in HIV and AIDS. 2015;**10**(4):258-263. DOI: 10.1097/COH.0000000000000161

[86] Aisen P, Little JT, Walsh S. An update on huperzine a as a treatment for

Alzheimer's disease. 2008;**17**(2):209-215. DOI: 10.1517/13543784.17.2.209

[87] Vaast P, Dubreucq-Fossaert S, Houfflin-Debarge V, et al. Acute pulmonary oedema during nicardipine therapy for premature labour. European Journal of Obstetrics & Gynecology and Reproductive Biology. 2004;**113**(1):98-99. DOI: 10.1016/j.ejogrb.2003.05.004

[88] Hess M, Shipley JB, Tolman D, Hastillo A. Milrinone: Basic and clinical pharmacology and acute and chronic management. American Journal of Medical Science. 1996;**311**(6):286-291. DOI: 10.1097/00000441-199606 000-00011 100

[89] Makuuchi M, Kobayashi T, Sugawara Y, Ohkubo T, Imamura H. Effects of amrinone on hepatic ischemia-reperfusion injury in rats. Journal of Hepatology. 2002;**37**(1):31-38. DOI: 10.1016/s0168-8278(02)00084-3

[90] Díez J, Lopez B, González A, Hermida N, Laviades C. Myocardial fibrosis in chronic kidney disease: Potential benefits of torasemide. Kidney International. Supplement. 2008;**74**:S19-S23. DOI: 10.1038/ki.2008.512

[91] Elliott H, Toal CB, Meredith PA. Long-acting dihydropyridine calcium-channel blockers and sympathetic nervous system activity in hypertension: A literature review comparing amlodipine and nifedipine GITS. Blood Pressure. 2012;**21**(1):3-10. DOI: 10.31 09/08037051.2012.690615

[92] Wang JG, Kario K, Lau T, et al. Use of dihydropyridine calcium channel blockers in the management of hypertension in Eastern Asians: A scientific statement from the Asian Pacific Heart Association. 2011;**34**(4):423-430. DOI: 10.1038/hr.2010.259

[93] Hare W, Dong CJ, Guo Y, Agey P, Wheeler L. Nimodipine enhancement of α2 adrenergic modulation of NMDA receptor via a mechanism independent of Ca2+ channel blocking. Investigative Ophthalmology & Visual Science. 2010;**51**(8):4174-4180. DOI: 10.1167/iovs.09-4613

[94] Allen GS, Ahn HS, Preziosi TJ, et al. Cerebral arterial spasm–a controlled trial of nimodipine in patients with subarachnoid hemorrhage. The New England Journal of Medicine. 1983;**308**(11):619-624. DOI: 10.10 56/NEJM198303173081103

[95] February F, Johnson R, Dludla P, Mabhida S, Benjeddou M, Louw J. Pharmacogenomics of amlodipine and hydrochlorothiazide therapy and the quest for improved control of hypertension. 2019;**24**(3):343-357. DOI: 10.1007/s10741-018-09765-y

[96] Forrester T, Fletcher H, Roberts G, Mullings A. An open trial comparing isradipine with hydralazine and methyl dopa in the treatment of patients with severe pre-eclampsia. Journal of Obstetrics and Gynaecology Research. 1999;**19**(3):235-238. DOI: 10.1080/01443619964977

[97] Fairhurst A, Thayer SA, Welcome M, Chhabra A. Effects of dihydropyridine calcium channel blocking drugs on rat brain muscarinic and alpha-adrenergic receptors. Biochemical Pharmacology. 1985:34. DOI: 10.1016/0006-2952(85)90121-2

[98] Bruncko M. Dihydropyridine-based calcium channel blockers for the treatment of angina pectoris and hypertension. Bioactive Heterocyclic Compound Classes. 2012:135-151

[99] Szamburska O, Mielcarek J, Grobelny P. The effect of betacarotene

on the photostability of nisoldipine. 2005;**27**(3):167-172. DOI: 10.1358/mf.2005.27.3.890873

[100] Hofmann F, Regulla S, Schneider T, Nastainczyk W, Meyer HE. Identification of the site of interaction of the dihydropyridine channel blockers nitrendipine and azidopine with the calcium channel alpha 1 subunit. The EMBO Journal. 1991;**10**(1):45-49. DOI: 10. 1002/j.1460-2075.1991.tb07919.x

[101] Zhang JG, Dehal SS, Ho T et al. Human cytochrome p450 induction and inhibition potential of clevidipine and its primary metabolite h152/81. Drug metabolism and disposition. 2006;**34**(5):734-737

[102] Wan J, Liu Z, Zheng X, Yang X, Wang E. Affinity and specificity of levamlodipine-human serum albumin interactions: Insights into its carrier function. Biophysical. 2009;**96**(10):3917-3925. DOI: 10.1016/j.bpj.2008.12.3965

[103] Shigenobu K, Tanaka H. Efonidipine hydrochloride: A dual blocker of L- and T-type ca(2+) channels. 2002;**20**(1):81-92. DOI: 10.1111/j.1527-3466.2002.tb00084.x

[104] Vormfelde SV, Sehrt D, Toliat MR, et al. Genetic variation in the renal sodium transporters NKCC2, NCC, and ENaC in relation to the effects of loop diuretic drugs. 2007;**82**(3):300-309. DOI: 10.1038/sj.clpt.6100131

[105] Hanck DA, Sheets MF, Fozzard HA, Lipkind GM. Sodium channel molecular conformations and antiarrhythmic drug affinity. 2010;**20**(1):16-21. DOI: 10.1016/j.tcm.2010.03.002

[106] Swerdlow N, Weber M, Breier M, Ko D, Thangaraj N, Marzan DE. Evaluating the antipsychotic profile of the preferential PDE10A inhibitor, papaverine. 2009;**203**(4):723-735. DOI: 10.1007/s00213-008-1419-x

[107] Wit AL, Janse MJ. Electrophysiological mechanisms of ventricular arrhythmias resulting from myocardial ischemia and infarction. Physiological Reviews. 1989;**69**(4):1049-1169

[108] Almendral J, Zaballos M, Fernández I, Rodríguez L, García S, Varela O, Quintela O, Anadón MJ. Effects of intravenous lipid emulsions on the reversal of pacing-induced ventricular arrhythmias and electrophysiological alterations in an animal model of ropivacaine toxicity. 2022:1-0. DOI: 10.1080/15563650.2022.2080075

[109] Rudy Y, Luo CH. A model of the ventricular cardiac action potential. Depolarization, repolarization, and their interaction. Circulation research. 1991;**68**(6):1501-1526

[110] Goldberger AL. Fractal mechanisms in the electrophysiology of the heart. IEEE Engineering in Medicine and Biology Magazine. 1992;**11**(2):47-52

[111] Carlsson L, Nattel S. Innovative approaches to anti-arrhythmic drug therapy. Nature Reviews Drug Discovery. 2006;**5**(12):1034-1049

[112] Hille B. Ionic channels in excitable membranes. Current problems and biophysical approaches. Biophysical Journal. 1978;**22**(2):283-294. DOI: 10.1016/S0006-3495(78)85489-7

[113] Nichols G. Colin. Lopatin AN, Makhina EN, Potassium channel block by cytoplasmic polyamines as the mechanism of intrinsic rectification. 1994;**372**(6504):366-369

[114] Williams S. Purves D. Augustine GJ, Fitzpatrick D, Katz LC, La Mantia A, McNamara JO, Voltage-Gated Ion Channels. 2001.

[115] Catterall WA. structure and function of voltage-gated sodium

channels. 2000. DOI: 10.1016/S0896-6273(00)81133-2

[116] Catterall W, Yu FH. Voltage-gated sodium channel family. 2003;**4**(3):1-7. DOI: 10.1186/gb-2003-4-3-207

[117] Nargeot J. A tale of two (calcium) channels. 2000;**86**(6):613-615. DOI: 10.1161/01.res.86.6.613

[118] Hibino H, Inanobe A, Furutani K, Murakami S, Findlay IA, Kurachi Y. Inwardly rectifying potassium channels: Their structure, function, and physiological roles. 2010;**90**(1):291-366. DOI: 10.1152/physrev.00021.2009

[119] Snyders D. Structure and Function of Cardiac Potassium Channels. 1999;**42**(2):377-390. DOI: 10.1016/s0008-6363(99)00071-1

[120] Balasubramanian M. Formation of Completely or Partially Reduced Pyridines and Quinolines. Pyridines: From Lab to Production. 2013:413

[121] Opie LH. Pharmacological differences between calcium antagonists. European Heart Journal. 1997;**18**(A):71-79. DOI: 10.1093/eurheartj/18.suppl_A.71

[122] Bechem M, Schramm M. Calcium-Agonists. Journal of molecular and cellular cardiology. 1987;**19**:63-75. DOI: 10.1016/s0022-2828(87)80005-6

[123] Ramin M, Miri R, Javidnia K et al. Synthesis and in vitro dual calcium channel antagonist-agonist activity of some 1, 4-dihydro-2,6-dimethyl-3-nitro and cyano-4-(1-methyl-5- nitro-1H-imidazol-2-yl)-5-pyridinecarboxylates. DARU Journal of Pharmaceutical Sciences. 2015;**16**(4):263-270

[124] DeMaria A, Low RI, Takeda P, Mason DT. The effects of calcium channel blocking agents on cardiovascular function. The American Journal of Cardiology. 1982;**49**(3):547-553

[125] Okazaki H. Kawai CH, Konishi TO, Matsuyama EI. Comparative effects of three calcium antagonists, diltiazem, verapamil and nifedipine, on the sinoatrial and atrioventricular nodes. 1981;**63**(5):1035-1042

[126] Bauman JL, Phillips BG, Gandhi AJ, Sanoski CA. Just VL. Comparison of intravenous diltiazem and verapamil for the acute treatment of atrial fibrillation and atrial flutter. The Journal of Human Pharmacology and Drug Therapy. 1997;**17**(6):1238-1245

[127] Carlos FH, Alviar L, Devarapally S, Nadkarni GN, Romero J, Benjo AM, et al. Efficacy and safety of dual calcium channel blockade for the treatment of hypertension: A meta-analysis. American Journal of Hypertension. 2013;**26**(2):287-297

[128] Poole-Wilson PA, Lubsen J, Wagener G, Kirwan BA, de Brouwer S. Effect of long acting nifedipine on mortality and cardiovascular morbidity in patients with symptomatic stable angina and hypertension. Journal of Hypertension. 2005;**23**(3):641-648

[129] DeDea L. How do dihydropyridine and nondihydropyridine CCBs differ?. JAAPA. 2012;**25**(3):15

[130] Takeshita A, Ide T, Tsutsui H, Kinugawa S, Utsumi H, Kang D, et al. Mitochondrial electron transport complex I is a potential source of oxygen free radicals in the failing myocardium. Circulation research. 1999;**85**(4):357-363

[131] Segal A, Abo A, Pick E, Hall A, Totty N, Teahan CG. Ctivation of the NADPH oxidase involves the small GTP-binding protein p21rac1. Nature. 1991;**353**(6345):668-670

[132] Bhatnagar A, Kilfoil PJ, Tipparaju SM, Barski OA. Regulation of ion channels by pyridine nucleotides. Circulation research. 2013;**112**(4):721-741

[133] Shattock M, Mitani AT. Role of Na-activated K channel, Na-K-Cl cotransport, and Na-K pump in [K] e changes during ischemia in rat heart. The American Journal of Physiology-Heart and Circulatory Physiology. 1992;**263**(2):H333-H340

[134] Rodrigo G, Lawrence C. The Na+-activated K+ channel contributes to K+ efflux in Na+-loaded guinea-pig but not rat ventricular myocytes. 2001

[135] Nehrke K, Wojtovich AP, Sherman TA, Nadtochiy SM, Urciuoli WR, Brookes PS. SLO-2 is cytoprotective and contributes to mitochondrial potassium transport. PloS one. 2011;**6**(12):e28287

[136] Yellen G. The voltage-gated potassium channels and their relatives. Nature. 2002;**419**(6902):35-42

[137] Nerbonne JM. Molecular basis of functional voltage-gated K+ channel diversity in the mammalian myocardium. The Journal of Physiology. 2000;**525**(2):285-298

[138] Dubois B. Rouzaire-Dubois JM, Role of potassium channels in mitogenesis. Progress in biophysics and molecular biology. 1993;**59**(1):1-21

[139] Yuan J, Krick S, Platoshyn O, Sweeney M, Kim H. Activation of K+ channels induces apoptosis in vascular smooth muscle cells. The American Journal of Physiology. 2001;**280**(4):C970-C979

[140] Cahalan M, Chandy KG, Wulff H, Beeton C, Pennington M, Gutman GA. K+ channels as targets for specific immunomodulation. Trends in pharmacological sciences. 2004;**25**(5):280-289

[141] Liu Y, Gutterman DD, Miura H. Redox modulation of vascular tone: Focus of potassium channel mechanisms of dilation. Arteriosclerosis, thrombosis, and vascular biology. 2005;**25**(4):671-678

[142] Hampl V, Archer SL, Souil E, Dinh-Xuan AT, Schremmer B, Mercier JC, et al. Molecular identification of the role of voltage-gated K+ channels, Kv1. 5 and Kv2. 1, in hypoxic pulmonary vasoconstriction and control of resting membrane potential in rat pulmonary artery myocytes. The Journal of Clinical Investigation. 1998;**101**(11):2319-2330

[143] Michelakis E, Archer SL, London B, Hampl V, Wu X, Nsair A, et al. Impairment of hypoxic pulmonary vasoconstriction in mice lacking the voltage-gated potassium channel Kv1. 5. The FASEB Journal. 2001;**15**(10):1801-1803

[144] Sanguinetti M, Tristani-Firouzi M, Chen J, Mitcheson JS. Molecular biology of K+ channels and their role in cardiac arrhythmias. The American Journal of Medicine. 2001;**110**(1):50-59

[145] Yuan J, Mandegar M. Role of K+ channels in pulmonary hypertension. Vascular pharmacology. 2002;**38**(1):25-33

[146] Catterall WA. From ionic currents to molecular mechanisms: the structure and function of voltage-gated sodium channels. Neuron. 2000;**26**(1):13-25

[147] Choe S. Potassium channel structures. Nature Reviews Neuroscience. 2002;**3**(2):115-121

[148] Schwarz J, Pongs O. Ancillary subunits associated with voltage-dependent K+ channels. Physiological reviews. 2010;**90**(2):755-796

[149] Pongs O, Rettig J, Heinemann SH, Wunder F, Lorra C, Parcej DN, Oliver Dolly J. Inactivation properties of

voltage-gated K+ channels altered by presence of beta-subunit. Nature. 1994;**369**(6478):289-294

[150] Grigorieff N, Sokolova O, Accardi A, Gutierrez D, Lau A, Rigney M. Conformational changes in the C terminus of Shaker K+ channel bound to the rat Kvbeta2-subunit. Proceedings of the National Academy of Sciences. 2003;**100**(22):12607-12612

[151] Barski O, Tipparaju SM, Li XP, Kilfoil PJ, Xue B, Uversky VN, et al. Interactions between the C-terminus of Kv1.5 and Kvβ regulate pyridine nucleotide-dependent changes in channel gating. Pflügers Archiv: European Journal of Physiology. 2012;**463**(6):799-818

[152] Zhou M, Pan Y, Weng J, Cao Y, Bhosle RC. Functional coupling between the Kv1. 1 channel and Aldoketoreductase Kvβ1. The Journal of Biological Chemistry. 2008;**283**(13):8634-8642

[153] Aaronson P, Sylvester JT, Shimoda LA. Hypoxic pulmonary vasoconstriction. Physiological reviews. 2012;**92**(1):367-520

[154] Yuan J, Platoshyn O, Brevnova EE, Burg ED, Yu Y, Remillard CV. Acute hypoxia selectively inhibits KCNA5 channels in pulmonary artery smooth muscle cells. The American Journal of Physiology. 2006;**290**(3):C907-C916

[155] Bhatnagar A, Tipparaju SM, Saxena N, Liu SQ, Kumar R. Differential regulation of voltage-gated K+ channels by oxidized and reduced pyridine nucleotide coenzymes. The American Journal of Physiology. 2005;**288**(2):C366-C376

[156] Weir E, Post JM, Hume JR, Archer SL. Direct role for potassium channel inhibition in hypoxic pulmonary vasoconstriction. The American Journal of Physiology. 1992;**262**(4):C882-C890

[157] https://www.clinicaltrials.gov/ct2/show/study/NCT00771004.

[158] https://clinicaltrials.gov/ct2/show/NCT01588470.

[159] https://clinicaltrials.gov/ct2/show/NCT00479986.

[160] https://clinicaltrials.gov/ct2/show/NCT03011775.

[161] https://clinicaltrials.gov/ct2/show/NCT01947790.

[162] https://clinicaltrials.gov/ct2/show/NCT00656851.

[163] https://clinicaltrials.gov/ct2/show/NCT02003638.

[164] https://clinicaltrials.gov/ct2/show/NCT00590629.

[165] https://clinicaltrials.gov/ct2/show/NCT01126073

[166] https://clinicaltrials.gov/ct2/show/NCT05087797

[167] https://clinicaltrials.gov/ct2/show/NCT01396395

[168] https://clinicaltrials.gov/ct2/show/NCT00848562

[169] https://clinicaltrials.gov/ct2/show/NCT00212030

[170] https://clinicaltrials.gov/ct2/show/NCT03445728

[171] https://clinicaltrials.gov/ct2/show/NCT02744339

[172] https://clinicaltrials.gov/ct2/show/NCT04722484

[173] https://www.clinicaltrials.gov/ct2/show/NCT03255512

[174] https://clinicaltrials.gov/ct2/show/NCT01951638

[175] https://clinicaltrials.gov/ct2/show/NCT00966563

[176] https://www.clinicaltrials.gov/ct2/show/NCT00524433

[177] https://clinicaltrials.gov/ct2/show/NCT00525707.

[178] https://clinicaltrials.gov/ct2/show/NCT00000483.

Chapter 7

Fused Pyridine Derivatives: Synthesis and Biological Activities

Huseyin Istanbullu, Gulsah Bayraktar and Merve Saylam

Abstract

Five-membered heteroaromatic ring fused pyridine derivatives are of increasing interest in drug design and medicinal chemistry. The structural similarity of many drugs (especially antiviral and anticancer ones) with DNA bases such as adenine and guanine is a key factor to explain their effectiveness. Apart from these, it is also found in the structures of substances with antituberculosis, antibacterial, antifungal, anti-inflammatory, and antimalarial activities. Another advantage of this group of compounds is their positive contribution to solubility, polarity, lipophilicity, and hydrogen bonding capacity properties of the compounds they are incorporated into. In this chapter, various bioactivities of fused pyridine derivatives will be categorized and summarized.

Keywords: fused pyridine, medicinal chemistry, furopyridines, thiazolopyridine, triazolopyridine, oxadiazolopyridine

1. Introduction

Fused pyridine heterocyclic ring derivatives are frequently used structures in drug research. Due to the vastness of the chemical space of fused pyridine derivatives, the most common fused pyridine derivatives, namely furopyridines, thienopyridines, pyrrolopyridines, oxazolopyridines, isoxazolopyridines, oxadiazolopyridines, imidazopyridines, pyrazolopyridines, thiazolopyridines, isothiazolopyridines, triazolopyridines, thiadiazolopyridines, tetrazolopyridines, selenazolopyridines, and dithiolopyridines, with their bioactivities were selected to cover in this chapter.

2. Fused pyridine derivatives

2.1 Furopyridines

Furopyridine synthesis was firstly reported almost a century ago. Since furopyridines are isosteres of benzofuran and indole cores, they are frequently encountered in the chemical structure of compounds possessing various bioactivities such as antihypertensive and antimicrobial.

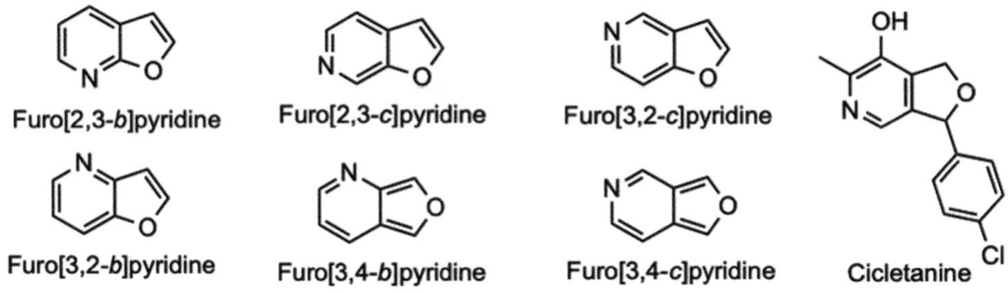

Figure 1.
Furopyridine isomeric structures and example drug molecule bearing furopyridine ring.

One of the first studies on furopyridine derivatives focused on anti-inflammatory, anti-aggregation, and anticoagulant activities [1, 2]. Sato et al. reported tetrahydrofuro[3,4-b]pyridine derivatives with coronary vasodilating activity [3]. Garay et al. examined the effect of furopyridines on the stimulation of K^+ movement across human red cells membrane [4].

On the other hand, cicletanine, a diuretic drug bearing furopyridine scaffold, used in the treatment of hypertension, also is a competitive histamine antagonist (**Figure 1**) [5, 6]. Clinical trial on its usage in hypertension with diabetes is ongoing (NCT02709031).

In addition to the activities mentioned before, there are several studies on furopyridine containing compounds with antimicrobial, anti-infective, and antiproliferative activities [7–14]. Also, furopyridine scaffold is present in a HIV protease inhibitor, L-754394 [15, 16]. Interestingly, it is also found in the structure of the antibiotic isolated from the fungus, *Cladobotryum varium* [17].

Compounds bearing furopyridine scaffold were reported in many studies as both core structure and substituent with kinase inhibitor properties, namely selective inhibitors of cdc-like kinases (CLKs), cyclin-dependent kinase (CDK2) inhibitors, and dk1, cdk2, Fyn, JNK3 kinase inhibitors [18–21].

On the other hand, furopyridine derivatives were reported possessing melanin-concentrating hormone (MCH1) receptor modulator activity and melatoninergic MT1 and MT2 receptor activity [22, 23].

In addition to these, inhibitor effect against angiogenetic targets on VEGFR2, Tie-2, and EphB4, mGluR5 noncompetitive antagonist activity, cannabinoid-1 receptor inverse agonist activity, σ receptor affinity, 5-HT1A agonists/5-HT3 antagonist activity, and 5-HT1F receptor agonist activity of various compounds bearing furopyridine fused ring were also reported [24–29].

2.2 Thienopyridines

The first report on bioactivity of thieno[3,2-b]pyridines focused on chemotherapy of parasites (*Entamoeba histolytica*) [30].

Thienopyridine ring system is an important structural element of anti-aggregation drugs (**Figure 2**). Ticlopidine, tetrahydrothieno[3,2-c]pyridine derivative, is the first reported drug with in vitro anti-inflammatory (carrageenan-induced edema) and inhibition of ADP-induced platelet aggregation activity in 1974 [2]. Then clopidogrel, having the same ring was reported in 1987 and is still on the market for antiplatelet

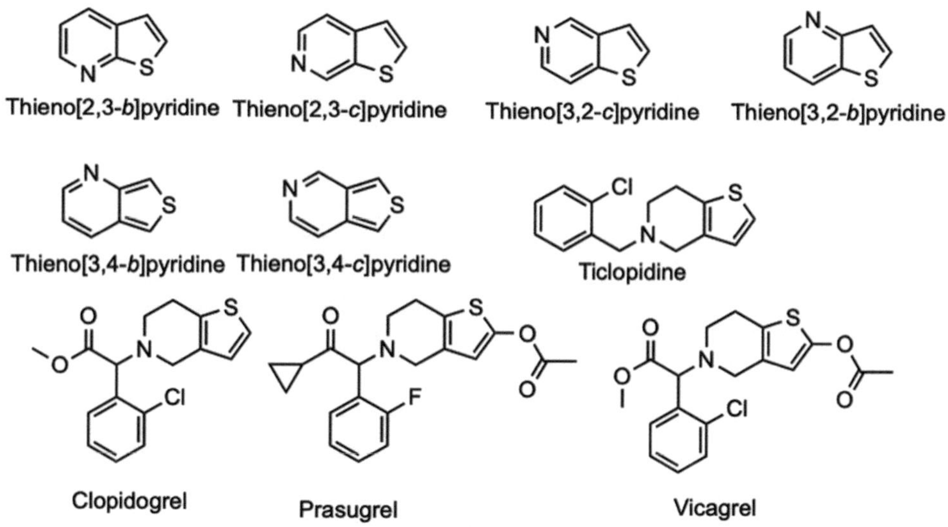

Figure 2.
Thienopyridine isomeric structures and example drug molecules bearing thienopyridine ring.

therapy [31]. Third drug of this class, prasugrel, was reported to the literature in 2000 [32]. Lastly, vicagrel was reported in 2011 to literature and is still undergoing clinical trials (NCT05162053) (**Figure 2**) [33].

On the other hand, compounds containing thienopyridine ring were reported having antimicrobial, anti-infective, antiviral, and antiproliferative effects [34–45].

Also, thienopyrimidine ring occurs either as core scaffold or a substituent in a group of kinase inhibitors such as VEGFR, EGFR, Src, Aurora, KDR, B-Raf, Pim kinases, check point 1 kinase (CHK1) IκB kinase-β (IKKβ), COT, and JAK2 inhibitors [46–56].

In addition to these, thienopyridine bearing structures are also associated with HMG-CoA reductase inhibitors, agonists for the luteinizing hormone receptor, histone lysine demethylase KDM5A Inhibitors, ubiquitin C-terminal hydrolase-L1 (UCH-L1) inhibitors, alkaline phosphatase (ALPase) activity, 5-HT1A agonists/5-HT3 antagonists, allosteric modulators of metabotropic Glu5 (mGlu5) and mGlu2 receptors, urotensin-II receptor antagonists, positive allosteric modulator targeting the M4 muscarinic acetylcholine receptor (M4 mAChR), selective inhibitors of *Plasmodium falciparum* glycogen synthase-3 (PfGSK-3), urea transporter inhibitors, and uridine diphosphate-galactose glycosyltransferase 8 (UGT8) inhibitor in the literature [28, 57–69].

2.3 Pyrrolopyridines

There are six isomeric structures of pyrrolopyridine ring, and azaindole term is also commonly used in the literature.

First reported bioactivity of pyrrolopyridine-bearing compound had been synthesized by Hooper et al. and had pyrrolo[3,2-b]pyridine scaffold with moderate antibacterial effect [70].

The first pyrrolo[2,3-b]pyridine-derived drug in literature is vemurafenib, a B-Raf enzyme inhibitor for the treatment of melanoma [71, 72]. On the other hand,

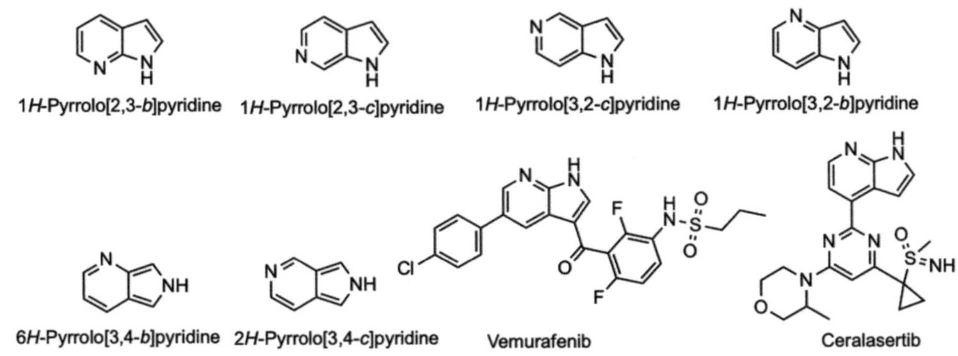

Figure 3.
Pyrrolopyridine isomeric structures and example drug molecules bearing pyrrolopyridine ring.

ceralasertib, a pyrrolo[2,3-b]pyridine-bearing compound, is under phase II trials as ATR kinase inhibitor for antineoplastic therapy (NCT04417062) (**Figure 3**) [73].

On the other hand, several studies were reported on pyrrolopyridine ring-derived compounds with antimicrobial, anti-infective, and antiviral activities [74–79].

Da Settimo et al. reported that pyrrolo[3,4-c]pyridine derivatives with local anesthetic activity and aldose reductase inhibitory properties [80].

Additionally, Kulagowski et al. found out that pyrrolo[2,3-b]pyridine derivatives showed selective D4 receptor antagonist activity [81].

As mentioned before, similar to thienopyridine ring, platelet aggregation inhibitor activity of pyrrolo[3,2-*c*]pyridine-derived scaffold was reported by Altomare et al. [82].

Moreover, antiproliferative activity of several pyrrolopyridine derivatives was investigated in many studies [83–91].

Apart from these, compounds bearing pyrrolopyridine moiety were found in various kinase inhibitors such as Met, insulin-like growth factor-1 receptor (IGF-1R), tyrosine, Aurora, Fes and Flt3 tyrosine kinases, Traf2 and Nck-interacting kinase (TNIK), Tau Tubulin Kinase 1 (TTBK1), JAK1 selective, BTK, DYRK1A, and RAF-1 dual inhibitor [92–103].

Lastly, many compounds containing fused pyrrolopyridine anologs were reported in the literature having several different bioactivities such as allosteric mGluR5 antagonist activity, diacylglycerol acyltransferase-2 inhibitors, antagonists of the G-protein-coupled chemoattractant receptor (CRTh2), in vivo TNF-a inhibitory activity, preventing protein phosphatase 2A (PP2A) inhibition, human neutrophil elastase (HNE) inhibitors, retinoic acid receptor-related orphan C2 (RORC2) inverse agonist, selective GluN2B negative allosteric modulators, 5-HT_{1F} receptor agonist, agonist of ORL-1(Opioid receptor-like) receptor, and cannabinoid 1 (CB1) and cannabinoid 2 (CB2) receptor agonist activity [104–116].

2.4 Oxazolopyridines, isoxazolopyridines, and oxadiazolopyridines

Oxazolopyridine derivatives, an aza analog of benzoxazole, have been studied extensively since the first report of their synthesis by Fraser and Tittensor in 1956 (**Figure 4**) [117]. Yet, the first bioactivity (anthelmintic and acaricidal activity) of compounds with oxazolopyridine moiety, namely oxazolo[4,5-b]pyridine, was reported nearly 20 years later by Rüfenacht et al., and then, oxazolo[5,4-b]

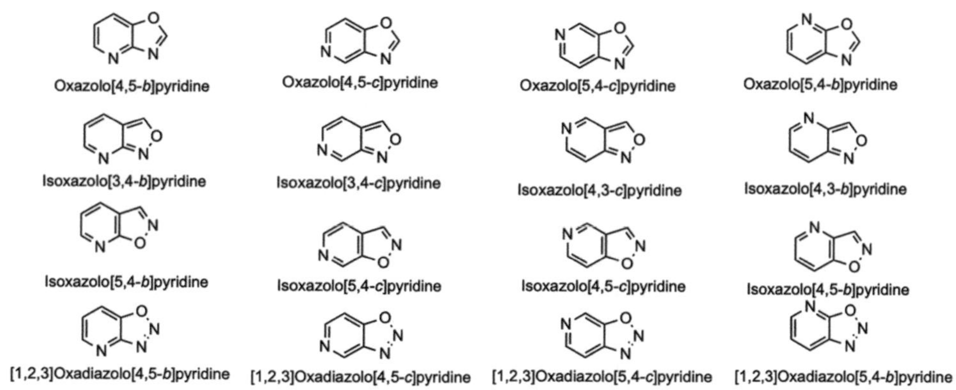

Figure 4.
Oxazolopyridine, isoxazolopyridine, and oxadiazolopyridine isomeric structures.

pyridine-bearing compounds were reported having carrageenan rat foot edema assay activity by Clark et al. [118, 119]. Later, antimicrobial, anti-infective, antiviral, and antiproliferative activities of several compounds having oxazololopyridine moiety were reported [120–124].

Additionally, various bioactivities such as fatty acid amide hydrolase (FAAH), topoisomerase II, monoamine oxidase B, GSK-3beta-, sphingomyelin synthase 2 inhibitory, SIRT1 activation, and histamine H3-receptor antagonistic activity of oxazololopyridine moiety-bearing compounds were reported in the literature [125–133].

Although the synthesis of isoxazolo[5,4-b]pyridines was reported in 1968 by Markillie, there has been a few bioactivity studies on isoxazolopyridine derivatives including GABAergic activity, HMG-CoA reductase inhibitory activity, anticancer activity, polo-like kinase inhibitor activity, and gamma-secretase modulator activity (**Figure 4**) [57, 134–138].

The synthesis of oxadiazolopyridine core was firstly reported by Bailey et al. in 1971 (**Figure 4**) [139]. Only antitumor activity and fluorescent properties of oxadiazolopyridine containing compounds were reported [140, 141].

2.5 Imidazopyridines

Imidazo[4,5-b]pyridine, the first synthesized imidazopyridine isomer, was synthesized by Takahashi and Yajima in 1946, and then analeptic activity of imidazopyridine was reported in 1965 [142, 143].

Imidazopyridines are one of the most studied fused pyridine ring systems; therefore, it is found in many drugs' structures (**Figure 5**). The various bioactivity profiles of these groups of compounds might be associated with the fact that imidazopyridines, also known as 3-deazapurines, are isosteres of purine ring.

Miroprofen, an imidazo[1,2-a]pyridine derived NSAID, has analgesic, antipyretic, and anti-inflammatory activity. Another imidazo[1,2-a]pyridine derivative, Zolpidem, is a hypnotic drug and positive GABA-A receptor modulator. Similarly, Alpidem, Necopidem, and Saripidem are other imidazo[1,2-a]pyridine containing anxiolytic drugs. Olprinone acts as a cardiotonic agent and is used in Japan. Zolimidine is a marketed anti-ulcerative drug. Minodronic acid, a bone resorption inhibitor and Sch 28080, gastric antisecretic compound, and H^+K^+-ATPase inhibitior are other imidazo[1,2-a]pyridine-bearing compounds [144–152].

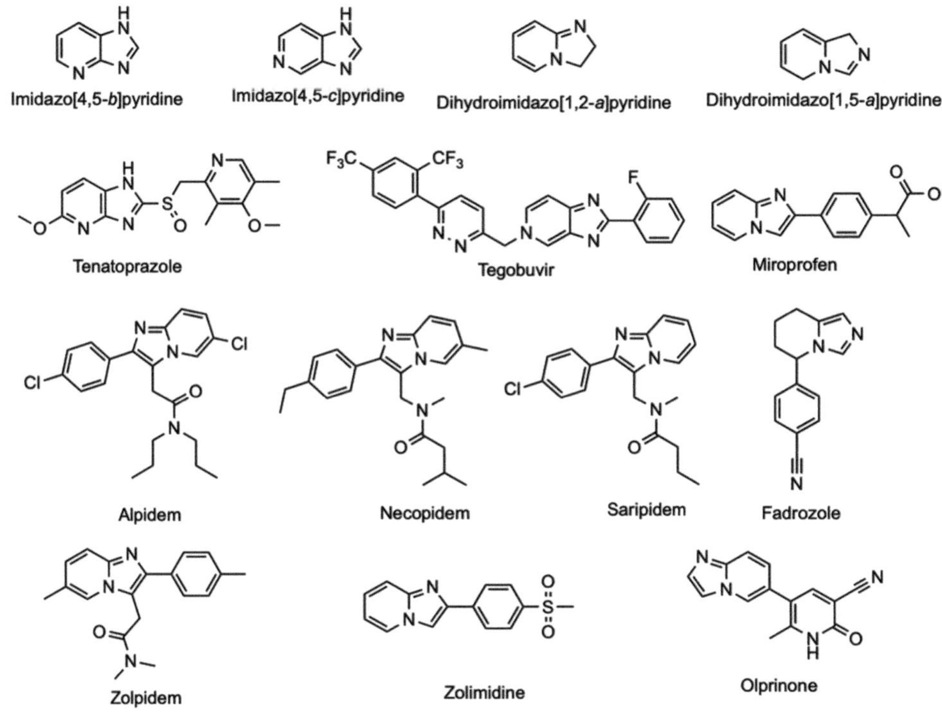

Figure 5.
Imidazopyridine isomeric structures and example drug molecules bearing imidazopyridine ring.

Imidazo[4,5-b]pyridine ring is occurred in various drugs including Vardax (sulmazole), a cardiotonic drug with positive inotropic activity, phosphodiesterase inhibition and adenosine receptor antagonist activity, and Rimegepant and Telcagepant, antimigraine drugs possessing CGRP receptor antagonists activity. Additionally, imidazo[4,5-b]pyridine-derived Tenatoprazole is reported with proton pump inhibitory activity and gastric acid secretion inhibitory properties in rats [153–158].

On the other hand, Tegobuvir, imidazo[4,5-c]pyridine-bearing compound, is used in prophylaxis and treatment of HCV infection, and Fadrozole, a Tetrahydroimidazo[1,5-a]pyridine derivative, is a nonsteroidal aromatase inhibitor for breast cancer treatment [159–162].

Moreover, there are several reports on imidazopyridine-bearing compounds possessing antibacterial, antiviral (HIV, etc.), and antiparasitic (anti-leishmanial and anti-trypanosomal) properties [163–174]. Also, imidazopyridine derivatives are often studied as anticancer agents [175–181].

The imidazopyridine scaffold has been reported in the structures of various kinase inhibitors, such as KDR kinase, calmodulin-dependent kinase II (CaMKII), Glycogen Synthase Kinase-3, cyclin-dependent kinase (CDK), Bruton's tyrosine kinase, AKT Kinase, c-Met kinase, VEGFR2 kinase, FLT3 kinase, Pan-JAK, Aurora-A kinase, phosphatidylinositol-3-kinase (PI3K) and apoptosis signal-regulating kinase 1 (ASK1) [182–195].

In addition to these bioactivities, imidazopyridine ring isomers expressed several including positive modulation of GABA-A receptor, positive allosteric modulation of metabotropic glutamate receptor 2 (mGluR2), angiotensin II receptor antagonist, receptor-related orphan receptor gamma (RORc) inverse agonist, melanin-concentrating hormone receptor 1 (MCHR1) antagonist, anti-inflammatory, anticonvulsant, phosphodiesterase (PDE) inhibitory, platelet-activating factor antagonist, TNF-α suppressing, mammalian target of rapamycin (mTOR) inhibitory, autotaxin inhibitory, cholinesterase inhibitory, and PARP-1 inhibitory activities in the literature [196–209].

2.6 Pyrazolopyridines

The synthesis of pyrazolopyridines was reported firstly by Englert and McElvain (**Figure 6**) [210]. Shortly after the synthesis, compounds containing pyrazolopyridine moiety with anti-inflammatory, antipyretic, and analgesic activity were reported [211]. Additionally, antibacterial (against both gram-positive and gram-negative bacteria), antiviral (anti-enterovirus), and antifungal and antiparasitic (antimalarial) activity reports of pyrazolopyridine-bearing compounds were reported in the literature [212–217]. Moreover, anticancer activity of various pyrazolopyridine derivatives was investigated in many studies [218–222].

Apart from these, many kinase inhibitors, namely CDK1/CDK2, glycogen synthase kinase-3, protein kinase Cθ (PKCθ), phosphatidylinositol-3-kinases (PI3K), aurora-A kinase, pim-kinase, TYK2, ALK5 (activin receptor-like kinase 5), anaplastic lymphoma kinase (ALK), and mitogen-activated protein kinase kinase 4 (MKK4) inhibitors, have pyrazolopyridine ring in their scaffold [223–232].

Lastly, in addition to activities mentioned before, anxiolytic, adenosine A1 receptor antagonist, PDE4, PDE5, PDE9 inhibitory, mTOR inhibitory, guanylate cyclase agonist, B-Raf^{V600E} inhibitory, dopamine D3 receptor agonist, and tubulin polymerization inhibitory and cholinesterase inhibitory activity of pyrazolopyridine derivatives were reported [233–244].

2.7 Thiazolopyridines and isothiazolopyridines

Thiazolo[4,5-b]pyridine ring was synthesized by Saikachi in 1944 [245]. The first reported bioactivity of thiazolopyridine was antituberculous activity of thiazolo[4,5-c]pyridine derivatives (**Figure 7**) [246].

Antibacterial (against both gram-positive and gram-negative bacteria), antiviral, antifungal, antituberculose, and antiparasitic activity of compounds containing thiazolopyridine structure were reported [247–251]. Additionally, cytotoxic and anticancer activity of thiazolopyridine derivatives were investigated in many studies [252–255].

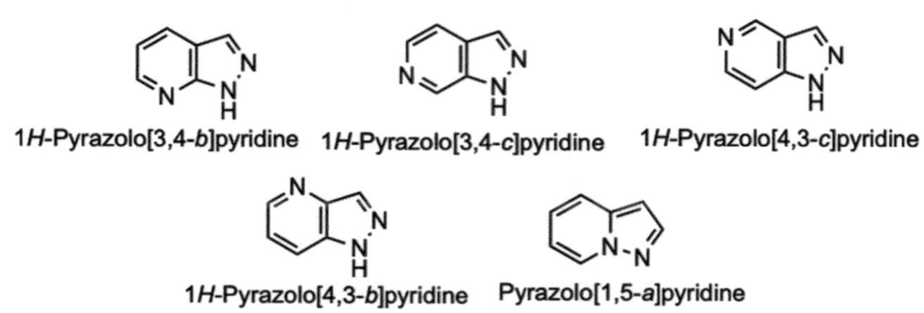

Figure 6.
Pyrazolopyridine isomeric structures.

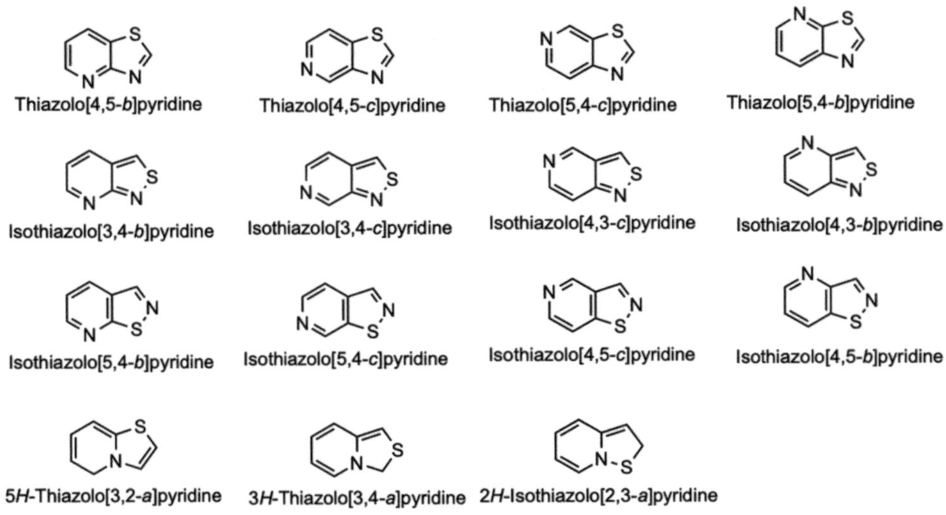

Figure 7.
Thiazolopyridine and isothiazolopyridine isomeric structures.

Moreover, there are many thiazolopyridine-bearing compounds with various bioactivity profile, such as histamine H3-receptor antagonistic activity, mGluR5—metabotropic glutamate receptor subtype 5-antagonist, sphingosine-1-phosphate (S1P) agonist, DNA Gyrase B (GyrB) ATPase inhibitor, anti-inflammatory activity, phosphoinositide 3-kinase inhibitor, and allosteric inhibitor of MALT1 [129, 256–261].

On the other hand, synthesis of isothiazolopyridines was firstly reported by Taurins and Khouw in 1997 [262]. Later, in vivo anorectic action activity of isothiazolo[5, 4-b]pyridine derivatives was reported by Malinka and Rutkowska [263]. There have been a few reports on bioactivity of isothiazolopyridine derivatives such as antitumor and radioprotective activities, in vitro antibacterial activity, analgesic activity, cyclin G-associated kinase inhibition, antiviral activity, and COX-1/2 inhibitory activity [264–269].

2.8 Triazolopyridines

Triazolopyridine scaffold is an isostere of purine ring; therefore, there are several bioactivity reports on compounds containing triazolopyridine ring.

The first report on the synthesis of (3*H*)1,2,3-triazolo[4,5-*c*]pyridine derivatives and their analeptic activity was published by Reitmann in 1936 [270].

1,2,3-Triazolo[4,5-*b*]pyridine and 1,2,3-triazolo[4,5-*c*]pyridine derivatives were reported possessing depressant, tranquilizing, anticonvulsant, and cardiovascular activities [143].

An antidepressant drug Trazodone, 1,2,4-triazolo[4,3-*a*]pyridine derivative, was first reported in 1968 and has been used commonly for the treatment of depression (**Figure 8**) [271]. In addition to its antidepressant effect, it was recently reported that trazodone inhibits tau amyloidogenesis [272].

On the other hand, several triazolopyridine-containing compounds were reported having antibacterial, antiviral, antifungal, antituberculose, and antiparasitic activity [273–278]. Additionally, triazolopyridine derivatives were investigated in many studies for their anticancer activity [279–281].

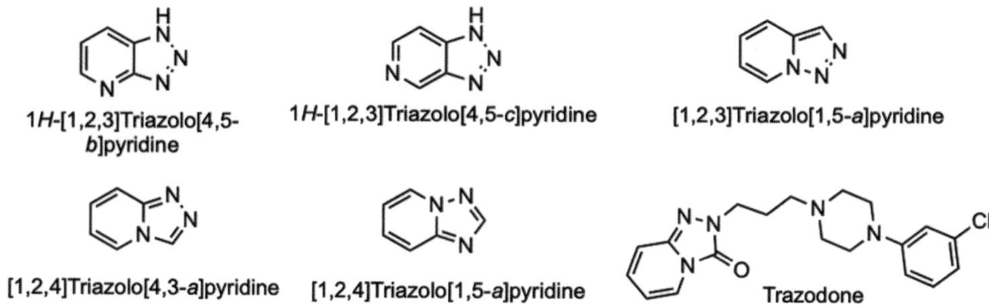

Figure 8.
Triazolopyridine isomeric structures and example drug molecule-bearing pyrrolopyridine ring.

Similar to other fused pyridine derivatives, triazolopyridine scaffold has been reported in many papers as kinase inhibitors, such as PIM kinase, JAK1, JAK2, PI3K-gama-delta, ALK-5, VEGFR2 kinase, spleen tyrosine kinase (Syk), c-met kinase, and monopolar spindle 1 (MPS1) kinase inhibitors [282–290].

Lastly, compounds containing triazolopyridine ring were evaluated for their bioactivities, such as anti-inflammatory, p38R, 11beta-hydroxysteroid dehydrogenase-type 1 (11beta-HSD-1), prolylhydroxylase domain-1 (PHD-1), myeloperoxidase, tubulin polymerization, polycomb repressive complex 2 (PRC2) inhibitory, HIV-1 allosteric inhibitor activity, mGlu receptor 2 (mGluR2) PAM, muscarinic acetylcholine receptor subtype 1 (M1) PAM, and retinoic acid receptor-related orphan nuclear receptor gama-t (RORγt) inverse agonist [174, 291–303].

2.9 The other five-membered heteroaromatic ring fused pyridine derivatives

Apart from fused pyridine derivatives mentioned before, there are several reports on five-membered heteroaromatic fused pyridine ring derivatives possessing bioactivity (**Figure 9**).

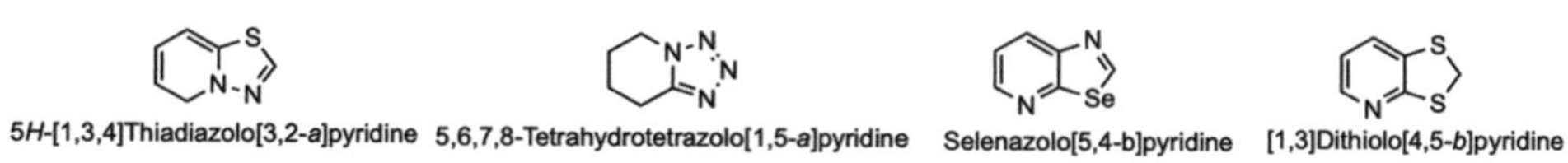

Figure 9.
Five-membered heteroaromatic ring fused pyridine derivatives.

For instance, 1,3,4-thiadiazolo[3,2-a]pyridine derivatives were reported having antimicrobial effects [304]. On the other hand, tetrahydrotetrazolopyridine scaffold was found in bovine liver-D-glucuronidase and human-alfa-L-iduronidase inhibitors [305]. Interestingly, an unusual fused pyridine derivative selenazolo[5,4-b]pyridine scaffold can highly induce apoptosis in human breast carcinoma MCF-7 cells [306]. Lastly, dithiolo[4,5-b]pyridine derivatives were reported possessing antimicrobial activity [307].

2.10 Conclusion

In conclusion, fused five-membered pyridine heteroaromatic rings are privileged scaffolds in medicinal chemistry. Therefore, selected ring systems and their bioactivities are covered in this chapter.

There are several drugs containing these heteroaromatic rings on the market, and several phase trials are ongoing on various compounds. Considering the chemical similarity between fused pyridine rings and nucleobases and amino acids, the wide variety of the bioactivity is unsurprising. The most commonly reported bioactivities of these kinds of derivatives are antimicrobial, anticancer, and kinase inhibition.

Author details

Huseyin Istanbullu[1*], Gulsah Bayraktar[2] and Merve Saylam[1]

1 Faculty of Pharmacy, Izmir Katip Celebi University, Department of Pharmaceutical Chemistry, Izmir, Turkey

2 Faculty of Pharmacy, Department of Pharmaceutical Chemistry, Ege University, Izmir, Turkey

*Address all correspondence to: huseyin.istanbullu@ikc.edu.tr

References

[1] Maffrand JP, Eloy F. Synthesis of thienopyridines and furopyridines of therapeutic interest. European Journal of Medicinal Chemistry. 1974;**9**(5):483-486

[2] Podesta M, Aubert D, Ferrand JC. Pharmacological study of thienopyridine and furopyridine analogs. European Journal of Medicinal Chemistry. 1974;**9**(5):487-490

[3] Sato Y, Shimoji Y, Fujita H, Mizuno H, et al. Synthetic studies on cardiovascular agents. IV. Synthesis of fused dihydropyridine derivatives. Yakugaku Zasshi. 1978;**98**(4):448-465. DOI: 10.1248/yakushi1947.98.4_448

[4] Garay RP, Nazaret C, Diez J, Etienne A, et al. Stimulation of potassium fluxes by diuretic drugs in human red cells. Biochemical Pharmacology. 1984;**33**(13):2013-2020. DOI: 10.1016/0006-2952(84)90567-7

[5] Schoeffter P, Ghysel-Burton J, Cabanie M, Godfraind T. Competitive and stereoselective histamine H_1-antagonistic effect of cicletanide in guinea pig isolated ileum. European Journal of Pharmacology. 1987;**136**(2):235-237. DOI: 10.1016/0014-2999(87)90716-3

[6] Auguet M, Delaflotte S, Hellegouarch A, Guillon JM, et al. In vitro cardiovascular antihistamine properties of cicletanine in comparison with diphenhydramine. Drugs under Experimental and Clinical Research. 1988;**14**(2-3):149-153

[7] Salem MS, Sakr SI, El-Senousy WM, Madkour HMF. Synthesis, antibacterial, and antiviral evaluation of new heterocycles containing the pyridine moiety. Archiv der Pharmazie. 2013;**346**:766-773. DOI: 10.1002/ardp.201300183

[8] Hung JM, Arabshahi HJ, Leung E, Reynisson J, et al. Synthesis and cytotoxicity of thieno[2,3-b]pyridine and furo[2,3-b]pyridine derivatives. European Journal of Medicinal Chemistry. 2014;**86**:420-437. DOI: 10.1016/j.ejmech.2014.09.001

[9] Naresh KR, Poornachandra Y, Nagender P, Mallareddy G, et al. Synthesis of novel trifluoromethyl substituted furo[2,3-b]pyridine and pyrido[3′,2′:4,5]furo[3,2-d]pyrimidine derivatives as potential anticancer agents. European Journal of Medicinal Chemistry. 2016;**108**:68-78. DOI: 10.1016/j.ejmech.2015.11.007

[10] Parcella K, Eastman K, Yeung K-S, Grant-Young KA, et al. Improving metabolic stability with deuterium: The discovery of BMT-052, a pan-genotypic HCV NS5B polymerase inhibitor. ACS Medicinal Chemistry Letters. 2017;**8**(7):771-774. DOI: 10.1021/acsmedchemlett.7b00211

[11] Santhosh KG, Poornachandra Y, Kumar GS, Ratnakar RK, et al. Synthesis of novel hetero ring fused pyridine derivatives; their anticancer activity, CoMFA and CoMSIA studies. Bioorganic & Medicinal Chemistry Letters. 2018;**28**(13):2328-2337. DOI: 10.1016/j.bmcl.2018.04.031

[12] Laxmi DS, Vardhini SV, Guttikonda VR, Rao MVB, et al. Synthesis of 2-substituted furo[3,2-b]pyridines under Pd/C-Cu catalysis assisted by ultrasound: Their evaluation as potential cytotoxic agents. Anti-Cancer Agents in Medicinal Chemistry.

2020;**20**(8):932-940. DOI: 10.2174/18715 20620666200311102304

[13] Said AB, Al-Refai M, Geyer A, Mansi IA, et al. Synthesis, characterization, antibacterial and cytotoxic evaluation of new 6-(chlorothiophenyl)-2-(2-oxopropoxy) pyridine-3-carbonitrile derivatives and their corresponding furo[2,3-b] pyridine derivatives. Heterocycles. 2021;**102**(11):2153-2167. DOI: 10.3987/com-21-14534

[14] Silva DG, Anna J, de Melo SMG, Fumagalli F, et al. Synthesis and structure-activity relationships of imidazopyridine/pyrimidine- and Furopyridine-based anti-infective agents against trypanosomiases. ChemMedChem. 2021;**16**(6):966-975. DOI: 10.1002/cmdc.202000616

[15] Houpis IN, Choi WB, Reider PJ, Audrey M, et al. Synthesis of functionalized furo[2,3-b]pyridines via the Pd-catalyzed coupling of acetylenes to iodopyridones. Preparation of a key intermediate to a new HIV protease inhibitor L-754,394. Tetrahedron Letters. 1994;**35**(50):9355-9358. DOI: 10.1016/s0040-4039(00)78541-8

[16] Huff JR, Vacca JP, Dorsey BD. HIV Protease Inhibitors. WO9516688 A1. 1995

[17] Sakemi S, Jon B, DeCosta DL, Dekker KA, et al. CJ-15,696 and its analogs, new furopyridine antibiotics from the fungus *Cladobotryum varium*: Fermentation, isolation, structural elucidation, biotransformation and antibacterial activities. Journal of Antibiotics. 2002;**55**(1):6-18. DOI: 10.7164/antibiotics.55.6

[18] Václav N, Michaela H, Lukáš M, Jana F, et al. Furo[3,2-b]pyridine: A privileged scaffold for highly selective kinase inhibitors and effective modulators of the hedgehog pathway. Angewandte Chemie. 2019;**58**(4):1062-1066. DOI: 10.1002/anie.201810312

[19] Nemec V, Maier L, Berger B-T, Chaikuad A, et al. Highly selective inhibitors of protein kinases CLK and HIPK with the furo[3,2-b] pyridine core. European Journal of Medicinal Chemistry. 2021;**215**:113299. DOI: 10.1016/j.ejmech.2021.113299

[20] Abdel-Rahman AA-H, Shaban AKF, Nassar IF, EL-Kady DS, et al. Discovery of new pyrazolopyridine, furopyridine, and pyridine derivatives as CDK2 inhibitors: Design, synthesis, docking studies, and anti-proliferative activity. Molecules. 2021;**26**(13):3923. DOI: 10.3390/molecules26133923

[21] Schade N, Koch P, Ansideri F, Krystof V, et al. Evaluation of novel substituted furopyridines as inhibitors of protein kinases related to tau pathology in Alzheimer's disease. Medicinal Chemistry (Sharjah, United Arab Emirates). 2021;**17**(8):844-855. DOI: 10.2174/1573406417666210601144510

[22] Guzzo P, Surman MD. 5-Furopyridinone Substituted Indazoles. WO2008086409. 2008

[23] Couhert A, Delagrange P, Caignard D-H, Chartier A, et al. Synthesis of 2-arylfuro[3,2-b]pyridines: Effect of the C2-aryl group on melatoninergic activity. European Journal of Medicinal Chemistry. 2016;**109**:268-275. DOI: 10.1016/j.ejmech.2016.01.008

[24] Miyazaki Y, Nakano M, Sato H, Truesdale AT, et al. Design and effective synthesis of novel templates, 3,7-diphenyl-4-amino-thieno and furo-[3,2-c]pyridines as protein kinase inhibitors and in vitro evaluation targeting angiogenetic kinases. Bioorganic & Medicinal Chemistry

Letters. 2007;**17**:250-254. DOI: 10.1016/j.bmcl.2006.09.050

[25] Rodriguez AL, Williams R, Zhou Y, Lindsley SR, et al. Discovery and SAR of novel mGluR5 non-competitive antagonists not based on an MPEP chemotype. Bioorganic & Medicinal Chemistry Letters. 2009;**19**:3209-3213. DOI: 10.1016/j.bmcl.2009.04.110

[26] Debenham JS, Madsen-Duggan CB, Toupence RB, Walsh TF, et al. Furo[2,3-b]pyridine-based cannabinoid-1 receptor inverse agonists: Synthesis and biological evaluation. Bioorganic & Medicinal Chemistry Letters. 2010;**20**(4):1448-1452. DOI: 10.1016/j.bmcl.2009.12.065

[27] Miyata K, Schepmann D, Wuensch B. Synthesis and σ receptor affinity of regioisomeric spirocyclic furopyridines. European Journal of Medicinal Chemistry. 2014;**83**:709-716. DOI: 10.1016/j.ejmech.2014.06.073

[28] Asagarasu A, Matsui T, Hayashi H, Tamaoki S, et al. Design and synthesis of piperazinylpyridine derivatives as novel 5-HT_{1A} agonists/5-HT_3 antagonists for the treatment of irritable bowel syndrome (IBS). Chemical & Pharmaceutical Bulletin. 2009;**57**(1):34-42. DOI: 10.1248/cpb.57.34

[29] Mathes BM, Hudziak KJ, Schaus JM, Xu Y-C, et al. Substituted furo[3,2-b] pyridines: Novel bioisosteres of 5-HT1F receptor agonists. Bioorganic & Medicinal Chemistry Letters. 2004;**14**(1):167-170. DOI: 10.1016/j.bmcl.2003.09.091

[30] Taylor DJ, Greenberg J. Experimental chemotherapy of endamoeba histolytica infections in the guinea pig. American Journal of Hygiene: Monographic Series. 1952;**56**(1):58-70. DOI: 10.1093/oxfordjournals.aje.a119541

[31] Feliste R, Delebassee D, Simon MF, Chap H, et al. Broad spectrum anti-platelet activity of ticlopidine and PCR 4099 involves the suppression of the effects of released ADP. Thrombosis Research. 1987;**48**(4):403-415. DOI: 10.1016/0049-3848(87)90398-7

[32] Sugidachi A, Asai F, Ogawa T, Inoue T, et al. The in vivo pharmacological profile of CS-747, a novel antiplatelet agent with platelet ADP receptor antagonist properties. British Journal of Pharmacology. 2000;**129**(7):1439-1446. DOI: 10.1038/sj.bjp.0703237

[33] Hongbin S, Jiaqi S, Boyu Z, Fang Y, et al. Optical-Activity 2-Hydroxytetrahydrothienopyridine Derivative, Preparation Method and Application Thereof in Pharmacy. CN102120744 A. 2011

[34] Gilis PM, Haemers A, Bollaert W. Synthesis and antibacterial evaluation of 4,7-dihydro-4-oxothieno[2,3-b]pyridine-5-carboxylic acids. European Journal of Medicinal Chemistry. 1978;**13**(3):265-269

[35] Attaby FA, Elneairy MAA, Elsayed MS. Synthesis and antimicrobial evaluation of new pyridine, thienopyridine and pyridothienopyrazole derivatives. Archives of Pharmacal Research. 1999;**22**(2):194-201. DOI: 10.1007/bf02976546

[36] Abdel-Rahman AE, Bakhite EA, Al-Taifi EA. Synthesis and antimicrobial testing of some new S-substituted-thiopyridines, thienopyridines, pyridothienopyrimidines and pyridothienotriazines. Die Pharmazie. 2003;**58**(6):372-377. DOI: 10.1002/chin.200339135

[37] Bernardino AMR, Pinheiro LCS, Rodrigues CR, Loureiro NI, et al. Design, synthesis, SAR, and biological evaluation

of new 4-(phenylamino)thieno[2,3-b]pyridine derivatives. Bioorganic & Medicinal Chemistry. 2006;**14**(16):5765-5770. DOI: 10.1016/j.bmc.2006.03.013

[38] Zeng X-X, Zheng R-L, Zhou T, He H-Y, et al. Novel thienopyridine derivatives as specific anti-hepatocellular carcinoma (HCC) agents: Synthesis, preliminary structure-activity relationships, and in vitro biological evaluation. Bioorganic & Medicinal Chemistry Letters. 2010;**20**(21):6282-6285. DOI: 10.1016/j.bmcl.2010.08.088

[39] Abreu RMV, Ferreira ICFR, Calhelha RC, Lima RT, et al. Anti-hepatocellular carcinoma activity using human HepG2 cells and hepatotoxicity of 6-substituted methyl 3-aminothieno[3,2-b]pyridine-2-carboxylate derivatives: In vitro evaluation, cell cycle analysis and QSAR studies. European Journal of Medicinal Chemistry. 2011;**46**(12):5800-5806. DOI: 10.1016/j.ejmech.2011.09.029

[40] Romagnoli R, Baraldi PG, Kimatrai Salvador MPD, Aghazadeh Tabrizi M, et al. Synthesis and biological evaluation of 2-(Alkoxycarbonyl)-3-Anilinobenzo[b]thiophenes and Thieno[2,3-b]pyridines as new potent anticancer agents. Journal of Medicinal Chemistry. 2013;**56**(6):2606-2618. DOI: 10.1021/jm400043d

[41] Nakamura RL, Burlingame MA, Yang S, Crosby DC, et al. Identification and optimization of thienopyridine carboxamides as inhibitors of HIV regulatory complexes. Antimicrobial Agents and Chemotherapy. 2017;**61**(7):e02366-16/1-e02366-16/14. DOI: 10.1128/aac.02366-16

[42] El-Deen EMM, El-Meguid EAA, Hasabelnaby S, Karam EA, et al. Synthesis, docking studies, and in vitro evaluation of some novel thienopyridines and fused thienopyridine-quinolines as antibacterial agents and DNA gyrase inhibitors. Molecules. 2019;**24**(20):3650. DOI: 10.3390/molecules24203650

[43] Sanad SMH, Mekky AEM. Novel nicotinonitriles and thieno[2,3-b]pyridines as potent biofilm and COX-2 inhibitors: Synthesis, in vitro and in silico studies. ChemistrySelect. 2020;**5**(28):8494-8503. DOI: 10.1002/slct.202001208

[44] Mugengana AK, Vita NA, Brown GA, Moran K, et al. The discovery and development of thienopyrimidines as inhibitors of helicobacter pylori that act through inhibition of the respiratory complex I. ACS Infectious Diseases. 2021;**7**(5):1044-1058. DOI: 10.1021/acsinfecdis.0c00300

[45] Elnaggar DH, Mohamed AM, Abdel Hafez NA, Azab ME, et al. Antiproliferative activity of some newly synthesized substituted nicotinamides candidates using pyridine-2(1H) thione derivatives as synthon. ACS Omega. 2022;**7**(12):10304-10316. DOI: 10.1021/acsomega.1c06951

[46] Munchhof MJ, Beebe JS, Casavant JM, Cooper BA, et al. Design and SAR of thienopyrimidine and thienopyridine inhibitors of VEGFR-2 kinase activity. Bioorganic & Medicinal Chemistry Letters. 2004;**14**:21-24. DOI: 10.1016/j.bmcl.2003.10.030

[47] Boschelli DH, Wu B, Sosa ACB, Durutlic H, et al. Synthesis and Src kinase inhibitory activity of 2-phenyl- and 2-thienyl-7-phenylaminothieno[3,2-b]pyridine-6-carbonitriles. Journal of Medicinal Chemistry. 2005;**48**(11):3891-3902. DOI: 10.1021/jm050175p

[48] Pevet I, Brule C, Tizot A, Gohier A, et al. Synthesis and pharmacological

evaluation of thieno[2,3-b]pyridine derivatives as novel c-Src inhibitors. Bioorganic & Medicinal Chemistry. 2011;**19**(8):2517-2528. DOI: 10.1016/j.bmc.2011.03.021

[49] Curtin ML, Frey RR, Heyman HR, Soni NB, et al. Thienopyridine ureas as dual inhibitors of the VEGF and Aurora kinase families. Bioorganic & Medicinal Chemistry Letters. 2012;**22**(9):3208-3212. DOI: 10.1016/j.bmcl.2012.03.035

[50] Morwick T, Berry A, Brickwood J, Cardozo M, et al. Evolution of the thienopyridine class of inhibitors of IκB kinase-β: Part I: Hit-to-lead strategies. Journal of Medicinal Chemistry. 2006;**49**(10):2898-2908. DOI: 10.1021/jm0510979

[51] Heyman HR, Frey RR, Bousquet PF, Cunha GA, et al. Thienopyridine urea inhibitors of KDR kinase. Bioorganic & Medicinal Chemistry Letters. 2007;**17**(5):1246-1249. DOI: 10.1016/j.bmcl.2006.12.015

[52] Cusack K, Allen H, Bischoff A, Clabbers A, et al. Identification of a selective thieno[2,3-c]pyridine inhibitor of COT kinase and TNF-α production. Bioorganic & Medicinal Chemistry Letters. 2009;**19**(6):1722-1725. DOI: 10.1016/j.bmcl.2009.01.088

[53] Gopalsamy A, Shi M, Hu Y, Lee F, et al. B-Raf kinase inhibitors: Hit enrichment through scaffold hopping. Bioorganic & Medicinal Chemistry Letters. 2010;**20**(8):2431-2434. DOI: 10.1016/j.bmcl.2010.03.030

[54] Yang B, Vasbinder MM, Hird AW, Su Q, et al. Adventures in scaffold morphing: Discovery of fused ring heterocyclic checkpoint kinase 1 (CHK1) inhibitors. Journal of Medicinal Chemistry. 2018;**61**(3):1061-1073. DOI: 10.1021/acs.jmedchem.7b01490

[55] Schenkel LB, Huang X, Cheng A, Deak HL, et al. Discovery of potent and highly selective thienopyridine Janus kinase 2 inhibitors. Journal of Medicinal Chemistry. 2011;**54**(24):8440-8450. DOI: 10.1021/jm200911r

[56] Naguib BH, El-Nassan HB. Synthesis of new thieno[2,3-b]pyridine derivatives as pim-1 inhibitors. Journal of Enzyme Inhibition and Medicinal Chemistry. 2016;**31**(6):1718-1725. DOI: 10.3109/14756366.2016.1158711

[57] Suzuki M, Iwasaki H, Fujikawa Y, Sakashita M, et al. Synthesis and biological evaluations of condensed pyridine and condensed pyrimidine-based HMG-CoA reductase inhibitors. Bioorganic & Medicinal Chemistry Letters. 2001;**11**:1285-1288. DOI: 10.1016/s0960-894x(01)00203-7

[58] Sasaki S, Cho N, Nara Y, Harada M, et al. Discovery of a thieno[2,3-d]pyrimidine-2,4-dione bearing a p-methoxyureidophenyl moiety at the 6-position: A highly potent and orally bioavailable non-peptide antagonist for the human luteinizing hormone-releasing hormone receptor. Journal of Medicinal Chemistry. 2003;**46**(1):113-124. DOI: 10.1021/jm020180i

[59] Horton JR, Woodcock CB, Chen Q, Liu X, et al. Structure-based engineering of irreversible inhibitors against histone lysine demethylase KDM5A. Journal of Medicinal Chemistry. 2018;**61**(23):10588-10601. DOI: 10.1021/acs.jmedchem.8b01219

[60] Mermerian AH, Case A, Stein RL, Cuny GD. Structure-activity relationship, kinetic mechanism, and selectivity for a new class of ubiquitin C-terminal hydrolase-L1 (UCH-L1) inhibitors. Bioorganic & Medicinal Chemistry Letters. 2007;**17**(13):3729-3732. DOI: 10.1016/j.bmcl.2007.04.027

[61] Saito K, Nakao A, Shinozuka T, Shimada K, et al. Discovery and structure-activity relationship of thienopyridine derivatives as bone anabolic agents. Bioorganic & Medicinal Chemistry. 2013;**21**(7):1628-1642. DOI: 10.1016/j.bmc.2013.01.071

[62] Nogradi K, Wagner G, Domany G, Bobok A, et al. Thieno[2,3-b]pyridines as negative allosteric modulators of metabotropic GluR5 receptors: Hit-to-lead optimization. Bioorganic & Medicinal Chemistry Letters. 2014;**24**(16):3845-3849. DOI: 10.1016/j.bmcl.2014.06.057

[63] Childress ES, Wieting JM, Felts AS, Breiner MM, et al. Discovery of novel central nervous system penetrant metabotropic glutamate receptor subtype 2 (mGlu2) negative allosteric modulators (NAMs) based on functionalized pyrazolo[1,5-a]pyrimidine-5-carboxamide and thieno[3,2-b]pyridine-5-carboxamide cores. Journal of Medicinal Chemistry. 2019;**62**(1):378-384. DOI: 10.1021/acs.jmedchem.8b01266

[64] Barbaro L, Rodriguez AL, Blevins AN, Dickerson JW, et al. Discovery of "molecular switches" within a series of mglu5 allosteric ligands driven by a "magic methyl" effect affording both PAMs and NAMs with in vivo activity, derived from an M1 PAM chemotype. ACS Biological and Medicinal Chemistry AU. 2021;**1**(1):21-30. DOI: 10.1021/acsbiomedchemau.1c00024

[65] Lim CJ, Oh SA, Lee BH, Oh K-S, et al. Synthesis and SAR of thieno[3,2-b]pyridinyl urea derivatives as urotensin-II receptor antagonists. Bioorganic & Medicinal Chemistry Letters. 2014;**24**(24):5832-5835. DOI: 10.1016/j.bmcl.2014.09.089

[66] Huynh T, Valant C, Crosby IT, Sexton PM, et al. Synthesis and pharmacological evaluation of M4 muscarinic receptor positive allosteric modulators derived from VU10004. ACS Chemical Neuroscience. 2015;**6**(6):838-844. DOI: 10.1021/acschemneuro.5b00035

[67] Masch A, Nasereddin A, Alder A, Bird MJ, et al. Structure-activity relationships in a series of antiplasmodial thieno[2,3-b]pyridines. Malaria Journal. 2019;**18**(1):89. DOI: 10.1186/s12936-019-2725-y

[68] Zhao Y, Li M, Li B, Zhang S, et al. Discovery and optimization of thienopyridine derivatives as novel urea transporter inhibitors. European Journal of Medicinal Chemistry. 2019;**172**:131-142. DOI: 10.1016/j.ejmech.2019.03.060

[69] Zaccariotto E, Cachon-Gonzalez MB, Wang B, Lim S, et al. A novel brain-penetrant oral UGT8 inhibitor decreases in vivo galactosphingolipid biosynthesis in murine Krabbe disease. Biomedicine & Pharmacotherapy. 2022;**149**:112808. DOI: 10.1016/j.biopha.2022.112808

[70] Hooper M, Patterson DA, Wibberley DG. Preparation and antibacterial activity of isatogens and related compounds. Journal of Pharmacy and Pharmacology. 1965;**17**(11):734-741. DOI: 10.1111/j.2042-7158.1965.tb07596.x

[71] Ibrahim P, Artis D, Bremer R, Mamo S, et al. Pyrrolo[2,3-b] Pyridine Derivatives as Protein Kinase Inhibitors. WO2007002325A1. 2007

[72] Aziz N, Moler E, Stuart D, Heise C, et al. Biomarkers of Target Modulation, Efficacy, Diagnosis and/or Prognosis for RAF Inhibitors. WO2008082730A2. 2008

[73] Foote KM, Nissink JWM, Turner P. Morpholino Pyrimidines and Their Use in Therapy. WO2011154737 A1. 2011

[74] Toja E, Tarzia G, Ferrari P, Tuan G. Pyrrolopyridine analogs of nalidixic acid. 1. Pyrrolo[2,3-b]pyridines. Journal of Heterocyclic Chemistry. 1986;**23**(5):1555-1560. DOI: 10.1002/jhet.5570230560

[75] Minakata S, Itoh S, Komatsu M, Ohshiro Y. Functionalization of 1H-pyrrolo[2,3-b]pyridine. Bulletin of the Chemical Society of Japan. 1992;**65**(11):2992-2997. DOI: 10.1246/bcsj.65.2992

[76] Paget SD, Boggs CM, Foleno BD, Goldschmidt RM, et al. Antibacterial activity of pyrrolopyridine-substituted oxazolidinones: Synthesis and in vitro SAR of various C-5 acetamide replacements. Bioorganic & Medicinal Chemistry Letters. 2006;**16**(17):4537-4542. DOI: 10.1016/j.bmcl.2006.06.023

[77] Khoje AD, Charnock C, Wan B, Franzblau S, et al. Synthesis and antimycobacterial activities of non-purine analogs of 6-aryl-9-benzylpurines: Imidazopyridines, pyrrolopyridines, benzimidazoles, and indoles. Bioorganic & Medicinal Chemistry. 2011;**19**(11):3483-3491. DOI: 10.1016/j.bmc.2011.04.023

[78] Jose G, Suresha Kumara TH, Sowmya HBV, Sriram D, et al. European Journal of Medicinal Chemistry. Synthesis, molecular docking, antimycobacterial and antimicrobial evaluation of new pyrrolo[3,2-c]pyridine Mannich bases. 2017;**131**:275-288. DOI: 10.1016/j.ejmech.2017.03.015

[79] Saigal D, Ghanem YSA, Uddin A, Khan S, et al. Synthesis, biological evaluation and docking studies of functionalized Pyrrolo[3,4-b]pyridine derivatives. ChemistrySelect. 2021;**6**(9):2323-2334. DOI: 10.1002/slct.202004781

[80] Da Settimo A, Primofiore G, Da Settimo F, Simorini F, et al. Synthesis of pyrrolo[3,4-c]pyridine derivatives possessing an acid group and their in vitro and in vivo evaluation as aldose reductase inhibitors. European Journal of Medicinal Chemistry. 1996;**31**:49-58. DOI: 10.1016/s0223-5234(96)80006-7

[81] Kulagowski JJ, Broughton HB, Curtis NR, Mawer IM, et al. 3-[[4-(4-chlorophenyl)piperazin-1-yl]methyl]-1H-pyrrolo[2,3-b]pyridine: An antagonist with high affinity and selectivity for the human dopamine D4 receptor. Journal of Medicinal Chemistry. 1996;**39**(10):1941-1942. DOI: 10.1021/jm9600712

[82] Altomare C, Summo L, Cellamare S, Varlamov AV, et al. Pyrrolo[3,2-c]pyridine derivatives as inhibitors of platelet aggregation. Bioorganic & Medicinal Chemistry Letters. 2000;**10**:581-584. DOI: 10.1016/s0960-894x(00)00052-4

[83] Guillard J, Decrop M, Gallay N, Espanel C, et al. Synthesis and biological evaluation of 7-azaindole derivatives, synthetic cytokinin analogues. Bioorganic & Medicinal Chemistry Letters. 2007;**17**:1934-1937. DOI: 10.1016/j.bmcl.2007.01.033

[84] Kim HJ, Jung M-H, Kim H, El-Gamal MI, et al. Synthesis and antiproliferative activity of pyrrolo[3,2-b]pyridine derivatives against melanoma. Bioorganic & Medicinal Chemistry Letters. 2010;**20**(1):413-417. DOI: 10.1016/j.bmcl.2009.08.005

[85] El-Gamal MI, Jung M-H, Lee WS, Sim T, et al. Design, synthesis, and antiproliferative activity of new 1H-pyrrolo[3,2-c]pyridine derivatives against melanoma cell lines. European Journal of Medicinal Chemistry. 2011;**46**(8):3218-3226

[86] Jung M-H, El-Gamal MI, Abdel-Maksoud MS, Sim T, et al. Design,

synthesis, and antiproliferative activity of new 1H-pyrrolo[3,2-c]pyridine derivatives against melanoma cell lines. Part 2. Bioorganic & Medicinal Chemistry Letters. 2012;**22**(13):4362-4367. DOI: 10.1016/j.bmcl.2012.05.004

[87] Carbone A, Parrino B, Di Vita G, Attanzio A, et al. Synthesis and antiproliferative activity of thiazolyl-bis-pyrrolo[2,3-b]pyridines and indolyl-thiazolyl-pyrrolo[2,3-c]pyridines, nortopsentin analogues. Marine Drugs. 2015;**13**(1):460-492. DOI: 10.3390/md13010460

[88] Narva S, Chitti S, Bala BR, Alvala M, et al. Synthesis and biological evaluation of pyrrolo[2,3-b]pyridine analogues as antiproliferative agents and their interaction with calf thymus DNA. European Journal of Medicinal Chemistry. 2016;**114**:220-231. DOI: 10.1016/j.ejmech.2016.02.059

[89] Tang Q, Duan Y, Wang L, Wang M, et al. Synthesis and antiproliferative activity of pyrrolo[2,3-b]pyridine derivatives bearing the 1,8-naphthyridin-2-one moiety. European Journal of Medicinal Chemistry. 2018;**143**:266-275. DOI: 10.1016/j.ejmech.2017.11.034

[90] Ullah S, El-Gamal MI, El-Gamal R, Pelletier J, et al. Synthesis, biological evaluation, and docking studies of novel pyrrolo[2,3-b]pyridine derivatives as both ectonucleotide pyrophosphatase/phosphodiesterase inhibitors and antiproliferative agents. European Journal of Medicinal Chemistry. 2021;**217**:113339. DOI: 10.1016/j.ejmech.2021.113339

[91] Zhang J, Dai J, Lan X, Zhao Y, et al. Synthesis, bioevaluation and molecular dynamics of pyrrolo-pyridine carboxamide derivatives as potential antitumor agents in vitro and in vivo. European Journal of Medicinal Chemistry. 2022;**233**:114215. DOI: 10.1016/j.ejmech.2022.114215

[92] Pin F, Buron F, Saab F, Colliandre L, et al. Synthesis and biological evaluation of 2,3-bis(het)aryl-4-azaindole derivatives as protein kinase inhibitors. MedChemComm. 2011;**2**(9):899-903. DOI: 10.1039/C1MD00141H

[93] Cai Z-W, Wei D, Schroeder GM, Cornelius LAM, et al. Discovery of orally active pyrrolopyridine- and aminopyridine-based met kinase inhibitors. Bioorganic & Medicinal Chemistry Letters. 2008;**18**(11):3224-3229. DOI: 10.1016/j.bmcl.2008.04.047

[94] Kim KS, Zhang L, Schmidt R, Cai Z-W, et al. Discovery of pyrrolopyridine-pyridone based inhibitors of met kinase: Synthesis, X-ray crystallographic analysis, and biological activities. Journal of Medicinal Chemistry. 2008;**51**(17):5330-5341. DOI: 10.1021/jm800476q

[95] Wang W, Xu S, Duan Y, Liu X, et al. Synthesis and bioevaluation and docking study of 1H-pyrrolo[2,3-b]pyridine derivatives bearing aromatic hydrazone moiety as c-Met inhibitors. European Journal of Medicinal Chemistry. 2018;**145**:315-327. DOI: 10.1016/j.ejmech.2017.12.078

[96] Patnaik S, Stevens KL, Gerding R, Deanda F, et al. Discovery of 3,5-disubstituted-1H-pyrrolo[2,3-b]pyridines as potent inhibitors of the insulin-like growth factor-1 receptor (IGF-1R) tyrosine kinase. Bioorganic & Medicinal Chemistry Letters. 2009;**19**(11):3136-3140. DOI: 10.1016/j.bmcl.2008.12.110

[97] Song P, Chen M, Ma X, Xu L, et al. Identification of novel inhibitors of Aurora A with a 3-(pyrrolopyridin-2-yl)indazole scaffold. Bioorganic &

Medicinal Chemistry. 2015;**23**(8):1858-1868. DOI: 10.1016/j.bmc.2015.02.004

[98] Park E, Lee SJ, Moon H, Park J, et al. Discovery and biological evaluation of N-methyl-pyrrolo[2,3-b]pyridine-5-carboxamide derivatives as JAK1-selective inhibitors. Journal of Medicinal Chemistry. 2021;**64**(2):958-979. DOI: 10.1021/acs.jmedchem.0c01026

[99] Weir MC, Hellwig S, Tan L, Yao L, et al. Dual inhibition of Fes and flt3 tyrosine kinases potently inhibits flt3-itd+ aml cell growth. PLoS One. 2017;**12**(7):e0181178/1-e0181178/19. DOI: 10.1371/journal.pone.0181178

[100] Thakkar M, Bhuniya D, Kaduskar R, Mengawade T, et al. Discovery and evaluation of 1H-pyrrolo[2,3-b] pyridine based selective and reversible small molecule BTK inhibitors for the treatment of rheumatoid arthritis. Bioorganic & Medicinal Chemistry Letters. 2017;**27**(8):1867-1873. DOI: 10.1016/j.bmcl.2017.02.026

[101] Halkina T, Henderson JL, Lin EY, Himmelbauer MK, et al. Discovery of potent and brain-penetrant tau tubulin kinase 1 (TTBK1) inhibitors that lower tau phosphorylation in vivo. Journal of Medicinal Chemistry. 2021;**64**(9):6358-6380. DOI: 10.1021/acs.jmedchem.1c00382

[102] Kircher T, Pantsar T, Oder A, Peter von Kries J, et al. Design and synthesis of novel fluorescently labeled analogs of vemurafenib targeting MKK4. European Journal of Medicinal Chemistry. 2021;**209**:112901. DOI: 10.1016/j.ejmech.2020.112901

[103] Yang B, Wu Q, Huan X, Wang Y, et al. Discovery of a series of 1H-pyrrolo[2,3-b]pyridine compounds as potent TNIK inhibitors. Bioorganic & Medicinal Chemistry Letters. 2021;**33**:127749. DOI: 10.1016/j.bmcl.2020.127749

[104] Koller M, Carcache DA, Orain D, Ertl P, et al. Discovery of 1H-pyrrolo[2,3-c]pyridine-7-carboxamides as novel, allosteric mGluR5 antagonists. Bioorganic & Medicinal Chemistry Letters. 2012;**22**(20):6454-6459. DOI: 10.1016/j.bmcl.2012.08.053

[105] Kim MO, Lee S, Choi K, Lee S, et al. Discovery of a novel class of diacylglycerol acyltransferase 2 inhibitors with a 1H-pyrrolo[2,3-b]pyridine core. Biological & Pharmaceutical Bulletin. 2014;**37**(10):1655-1660. DOI: 10.1248/bpb.b14-00447

[106] Bala K, Leblanc C, Sandham DA, Turner KL, et al. Organic Compounds. WO2005123731 A2. 2005

[107] Hilmy KMH, Abdul-Wahab HG, Soliman DH, Khalifa MMA, et al. Novel pyrrolo[2,3-d]pyrimidines and pyrrolo[2,3-b]pyridines: Design, synthesis, and in vivo TNF-α inhibitory activity. Medicinal Chemistry Research. 2015;**25**(4):2097-2110. DOI: 10.1007/s00044-014-1281-9

[108] Lajarin-Cuesta R, Arribas RL, Nanclares C, Garcia-Frutos EM, et al. Design and synthesis of multipotent 3-aminomethylindoles and 7-azaindoles with enhanced protein phosphatase 2A-activating profile and neuroprotection. European Journal of Medicinal Chemistry. 2018;**157**:294-309. DOI: 10.1016/j.ejmech.2018.07.030

[109] Crocetti L, Giovannoni MP, Schepetkin IA, Quinn MT, et al. 1H-pyrrolo[2,3-b]pyridine: A new scaffold for human neutrophil elastase (HNE) inhibitors. Bioorganic & Medicinal Chemistry. 2018;**26**(21):5583-5595. DOI: 10.1016/j.bmc.2018.09.034

[110] Cantini N, Khlebnikov AI, Crocetti L, Schepetkin IA, et al. Exploration of nitrogen heterocycle scaffolds for the development of potent human neutrophil elastase inhibitors. Bioorganic & Medicinal Chemistry. 2021;**29**:115836. DOI: 10.1016/j.bmc.2020.115836

[111] Schnute ME, Wennerstal M, Alley J, Bengtsson M, et al. Discovery of 3-cyano-N-(3-(1-isobutyrylpiperidin-4-yl)-1-methyl-4-(trifluoromethyl)-1H-pyrrolo[2,3-b]pyridin-5-yl) benzamide: A potent, selective, and orally bioavailable retinoic acid receptor-related orphan receptor C2 inverse agonist. Journal of Medicinal Chemistry. 2018;**61**(23):10415-10439. DOI: 10.1021/acs.jmedchem.8b00392

[112] Chrovian CC, Soyode-Johnson A, Wall JL, Rech JC, et al. 1H-Pyrrolo[3,2-b] pyridine GluN2B-selective negative allosteric modulators. ACS Medicinal Chemistry Letters. 2019;**10**(3):261-266. DOI: 10.1021/acsmedchemlett.8b00542

[113] Filla SA, Mathes BM, Johnson KW, Phebus LA, et al. Novel potent 5-HT1F receptor agonists: Structure-activity studies of a series of substituted N-[3-(1-methyl-4-piperidinyl)-1H-pyrrolo[3,2-b] pyridin-5-yl]amides. Journal of Medicinal Chemistry. 2003;**46**(14):3060-3071. DOI: 10.1021/jm030020m

[114] Bignan GC, Battista K, Connolly PJ, Orsini MJ, et al. 3-(4-piperidinyl)indoles and 3-(4-piperidinyl)pyrrolo[2,3-b] pyridines as ligands for the ORL-1 receptor. Bioorganic & Medicinal Chemistry Letters. 2006;**16**(13):3524-3528. DOI: 10.1016/j.bmcl.2006.03.094

[115] Blaazer AR, Lange JHM, van der Neut MAW, Mulder A, et al. Novel indole and azaindole (pyrrolopyridine) cannabinoid (CB) receptor agonists: Design, synthesis, structure-activity relationships, physicochemical properties and biological activity. European Journal of Medicinal Chemistry. 2011;**46**(10):5086-5098. DOI: 10.1016/j.ejmech.2011.08.021

[116] Sparkes E, Cairns EA, Kevin RC, Lai F, et al. Structure-activity relationships of valine, tert-leucine, and phenylalanine amino acid-derived synthetic cannabinoid receptor agonists related to ADB-BUTINACA, APP-BUTINACA, and ADB-P7AICA. RSC Medicinal Chemistry. 2022;**13**(2):156-174. DOI: 10.1039/D1MD00242B

[117] Fraser J, Tittensor E. Oxazolopyridines and oxazoloquinolines. Part I. 2′-Alkyl and 2′-aryl derivatives of oxazolo(4′:5′-3:4)pyridine and oxazolo(4′:5′-3:4)quinolone. Journal of the Chemical Society. 1956:1781-1784. DOI: 10.1039/JR9560001781

[118] Ruefenacht K, Kristinsson H, Mattern G. Investigations on phosphoric acid and thiophosphoric acid esters with a heterocyclic substituent. 10th and last communication. Aza analogy. II: Derivatives of oxazolo[4,5-b]pyridin-2(3H)-one, an aza analog of benzoxazol-2(3H)-one. Helvetica Chimica Acta. 1976;**59**(5):1593-1612

[119] Clark RL, Pessolano AA, Witzel B, Lanza T, et al. 2-(substituted phenyl) oxazolo[4,5-b]pyridines and 2-(substituted phenyl)oxazolo[5,4-b] pyridines as nonacidic antiinflammatory agents. Journal of Medicinal Chemistry. 1978;**21**(11):1158-1162. DOI: 10.1021/jm00209a014

[120] Yalçin İ, Ören İ, Şener E, Akin A, et al. The synthesis and the structure-activity relationships of some substituted benzoxazoles, oxazolo(4,5-b)pyridines, benzothiazoles and benzimidazoles as antimicrobial agents. European Journal of Medicinal

Chemisty. 1992;**27**(4):401-406. DOI: 10.1016/0223-5234(92)90154-S

[121] Tatipaka HB, Gillespie JR, Chatterjee AK, Norcross NR, et al. Substituted 2-phenylimidazopyridines: A new class of drug leads for human african trypanosomiasis. Journal of Medicinal Chemistry. 2014;**57**(3):828-835. DOI: 10.1021/jm401178t

[122] Reen GK, Kumar A, Sharma P. In vitro and in silico evaluation of 2-(substituted phenyl) oxazolo[4,5-b] pyridine derivatives as potential antibacterial agents. Medicinal Chemistry Research. 2017;**26**(12):3336-3344. DOI: 10.1007/s00044-017-2026-3

[123] Akbay A, Oren I, Temiz-Arpaci O, Aki-Sener E, et al. Synthesis and HIV-1 reverse transcriptase inhibitor activity of some 2,5,6-substituted benzoxazole, benzimidazole, benzothiazole and oxazolo(4,5-b)pyridine derivatives. Arzneimittel-Forschung. 2003;**53**(4):266-271. DOI: 10.1055/s-0031-1297107

[124] Sireesha R, Tej MB, Poojith N, Sreenivasulu R, et al. Synthesis of substituted aryl incorporated oxazolo[4,5-b]pyridine-triazole derivatives: Anticancer evaluation and molecular docking studies. Polycyclic Aromatic Compounds Journal. 2021. [ahead-of-print, 1-18]. DOI: 10.1080/10406638.2021.2021256

[125] Boger DL, Sato H, Lerner AE, Hedrick MP, et al. Exceptionally potent inhibitors of fatty acid amide hydrolase: The enzyme responsible for degradation of endogenous oleamide and anandamide. Proceedings of the National Academy of Sciences of the United States of America. 2000;**97**(10):5044-5049. DOI: 10.1073/pnas.97.10.5044

[126] Boger DL, Miyauchi H, Hedrick MP. α-Keto heterocycle inhibitors of fatty acid amide hydrolase: Carbonyl group modification and α-substitution. Bioorganic & Medicinal Chemistry Letters. 2001;**11**(12):1517-1520. DOI: 10.1016/s0960-894x(01)00211-6

[127] Pinar A, Yurdakul P, Yildiz I, Temiz-Arpaci O, et al. Some fused heterocyclic compounds as eukaryotic topoisomerase II inhibitors. Biochemical and Biophysical Research Communications. 2004;**317**(2):670-674. DOI: 10.1016/j.bbrc.2004.03.093

[128] Karatas E, Foto E, Ertan-Bolelli T, Yalcin-Ozkat G, et al. Discovery of 5-(or 6)-benzoxazoles and oxazolo[4,5-b] pyridines as novel candidate antitumor agents targeting hTopo Iiα. Bioorganic Chemistry. 2021;**112**:104913. DOI: 10.1016/j.bioorg.2021.104913

[129] Walczyński K, Zuiderveld OP, Timmerman H. Non-imidazole histamine H3 ligands. Part III. New 4-n-propylpiperazines as non-imidazole histamine H3-antagonists. European Journal of Medicinal Chemistry. 2005;**40**(1):15-23. DOI: 10.1016/j.ejmech.2004.09.010

[130] Vu CB, Bemis JE, Disch JS, Ng PY, et al. Discovery of ımidazo[1,2-b]thiazole derivatives as novel SIRT1 activators. Journal of Medicinal Chemistry. 2009;**52**(5):1275-1283. DOI: 10.1021/jm8012954

[131] Park HR, Kim J, Kim T, Jo S, et al. Oxazolopyridines and thiazolopyridines as monoamine oxidase B inhibitors for the treatment of Parkinson's disease. Bioorganic & Medicinal Chemistry. 2013;**21**(17):5480-5487. DOI: 10.1016/j.bmc.2013.05.066

[132] Tantray MA, Khan I, Hamid H, Alam MS, et al. Oxazolo[4,5-b]pyridine-based piperazinamides as GSK-3β

inhibitors with potential for attenuating inflammation and suppression of pro-inflammatory mediators. Archiv der Pharmazie (Weinheim, Germany). 2017;**350**(8):e1700022. DOI: 10.1002/ardp.201700022

[133] Qi X-Y, Cao Y, Li Y-L, Mo M-G, et al. Discovery of the selective sphingomyelin synthase 2 inhibitors with the novel structure of oxazolopyridine. Bioorganic & Medicinal Chemistry Letters. 2017;**27**(15):3511-3515. DOI: 10.1016/j.bmcl.2017.05.074

[134] Markillie JH. Nouveaux Isoxazoles et leurs procédés de Fabrication. FR1513038. 1968

[135] Nordmann R, Graff P, Maurer R, Gaehwiler BH. Synthesis and pharmacological evaluation of cis-2,3,3a,4,5,6,7,7a-octahydro-3-oxoisoxazolo[5,4-c]pyridine: A structural analog of the GABA agonist THIP. Journal of Medicinal Chemistry. 1985;**28**(8):1109-1111. DOI: 10.1021/jm00146a024

[136] Rajanarendar E, Raju S, Reddy MN, Krishna SR, et al. Multi-component synthesis and in vitro and in vivo anticancer activity of novel arylmethylene bis-isoxazolo[4,5-b]pyridine-N-oxides. European Journal of Medicinal Chemistry. 2012;**50**:274-279. DOI: 10.1016/j.ejmech.2012.02.004

[137] Hanan EJ, Fucini RV, Romanowski MJ, Elling RA, et al. Design and synthesis of 2-amino-isoxazolopyridines as Polo-like kinase inhibitors. Bioorganic & Medicinal Chemistry Letters. 2008;**18**(19):5186-5189. DOI: 10.1016/j.bmcl.2008.08.091

[138] Qin J, Dhondi P, Huang X, Mandal M, et al. Discovery of fused 5,6-bicyclic heterocycles as γ-secretase modulators. Bioorganic & Medicinal Chemistry Letters. 2011;**21**(2):664-669. DOI: 10.1016/j.bmcl.2010.12.012

[139] Bailey AS, Heaton MW, Murphy JI. Preparation of a nitropyrido[3,4-c] furoxan: 7-nitro[1,2,5]oxadiazolo[3,4-c] pyridine 3-oxide. Journal of the Chemical Society [Section] C: Organic. 1971:1211-1213. DOI: 10.1039/j39710001211

[140] Li Z, Huang D, Ma C, Xu X, et al. Convenient aminative ring-opening reaction of 7-amino-6-nitro-[1,2,5] oxadiazolo[3,4-b]pyridine-1-oxide and antitumor activity of corresponding products. Chinese Journal of Organic Chemistry. 2016;**36**(9):2236-2241. DOI: 10.6023/cjoc201602023

[141] Gorohmaru H, Thiemann T, Sawada T, Takahashi K, et al. Preparation of 4,7-dihetaryl-1,2,5-oxadiazolo[3,4-c] pyridines as red fluorescent materials. Heterocycles. 2002;**56**(1-2):421-431. DOI: 10.3987/COM-01-S(K)64

[142] Takahashi T, Yajima S. Synthesis of heterocyclic compounds of nitrogen. XXVI. Pyrimidazoles (imidazopyridines). Yakugaku Zasshi. 1946;**66**(2A):31

[143] Vohra MM, Pradhan SN, Jain PC, Chatterjee SK, et al. Synthesis and structure-activity relations of some amino-pyridines, imidazopyridines, and triazolopyridines. Journal of Medicinal Chemistry. 1965;**8**(3):296-304. DOI: 10.1021/jm00327a006

[144] Nakanishi M, Muro T, Nakatsu O, Nakao T, et al. Phenylalkancarbonsaeure-derivate, verfahren zu ihrer herstellung und arzneimittel. DE2432410A1. 1975

[145] Kaplan J-P, George P. Imidazo(1,2-a) pyridine Derivatives, Process for their Preparation and their Therapeutical Use. EP50563 A1. 1982

[146] Goto K, Hisadome M, Maruyama Y, Imamura H. Effects of 2-{4-(2-imidazo[1,2-a]pyridyl)phenyl} propionic acid (Y-9213) and antiinflammatory drugs on erythrocytes, polymorphonuclear leukocytes and lysosomes in vitro. Japanese Journal of Pharmacology. 1978;**28**(3):433-446. DOI: 10.1254/jjp.28.433

[147] Almirante L, Polo L, Mugnaini A, Provinciali E, et al. Derivatives of imidazole. I. Synthesis and reactions of imidazo[1,2-a]pyridines with analgesic, antiinflammatory, antipyretic, and anticonvulsant activity. Journal of Medicinal Chemistry. 1965;**8**(3):305-312. DOI: 10.1021/jm00327a007

[148] Yamanaka M, Miyake K, Suda S, Ohara H, et al. 3-Imidazo[1,2-a]pyridin-6-ylpyridine Derivatives. JP61218589 A. 1986

[149] George P, Giron C. 3-Acylaminomethylimidazo[1,2-a] pyridines and Their Therapeutical Use. EP172096 A1. 1986

[150] George P, Giron C. Preparation of 3-(acylaminomethyl)imidazo[1,2-a] pyridines and Pharmaceutical Compositions Containing Them as Anxiolytics, Sedatives, Analgesics, Anticonvulsants, and Ulcer inhibitors. US4650796 A. 1987

[151] Isomura Y, Takeuchi M, Abe T. Heterocyclic Bisphosphonic Acid Derivatives as Bone Resorption Inhibitors. EP354806 A2. 1990

[152] Bristol JA, Puchalski C. Imidazo[1,2-a]pyridines and Pharmaceutical Compositions Containing Them. EP33094 A1. 1981

[153] Diederen W, Weisenberger H. Studies on the mechanism of the positive-inotropic action of AR-L 115 BS, a new cardiotonic drug. Arzneimittel-Forschung. 1981;**31**(1A):177-182

[154] Herzig JW, Feile K, Rueegg JC. Activating effects of AR-L 115 BS on the calcium(2+) sensitive force, stiffness and unloaded shortening velocity (Vmax) in isolated contractile structures from mammalian heart muscle. Arzneimittel-Forschung. 1981;**31**(1A):188-191

[155] Daly JW, Hong O, Padgett WL, Shamim MT, et al. Non-xanthine heterocycles: Activity as antagonists of A1- and A2-adenosine receptors. Biochemical Pharmacology. 1988;**37**(4):655-664. DOI: 10.1016/0006-2952(88)90139-6

[156] Luo G, Dubowchik GM, Macor JE. Cycloheptapyridine Derivatives as CGRP Receptor Antagonists and Their Preparation and Use in the Treatment of CGRP-Related Diseases Such as Migraine. WO2011046997 A1. 2011

[157] Burgey CS, Deng ZJ, Nguyen DN, Paone DV, et al. Preparation of Piperidine Derivatives as CGRP Receptor Antagonists. WO2004092166 A2. 2004

[158] Matsuishi N, Takeda H, Iizumi K, Murakami K, et al. Preparation, Testing, and Formulation of Pyridylmethylsulfinylimidazopyridines as Ulcer Inhibitors. EP254588 A1. 1988

[159] Resurii JB. Imidazo[1,5-a]pyridines. JP59118785 A. 1984

[160] Bondy SS, Dahl TC, Oare DA, Oliyai R, et al. Novel Pyridazine-Containing Imidazopyridazine Compound and Uses Thereof. WO2008005519 A2. 2008

[161] Dowdy ED, Kent KM, Tom NJ, Zia V. Preparation of Crystalline 5-[[6-[2,4-bis(trifluoromethyl) phenyl]-3-pyridazinyl]

methyl]-2-(2-fluorophenyl)-5H-imidazo[4,5-c]pyridine for Treatment and Prophylaxis of Hepatitis C Viral Infections. WO2009009001A1. 2009

[162] Shih I, Vliegen I, Peng B, Yang H, et al. Mechanistic characterization of GS-9190 (tegobuvir), a novel nonnucleoside inhibitor of hepatitis C virus NS5B polymerase. Antimicrobial Agents and Chemotherapy. 2011;**55**(9):4196-4203. DOI: 10.1128/aac.00307-11

[163] Fisher MH, Lusi A. Imidazo[1,2-a]pyridine anthelmintic and antifungal agents. Journal of Medicinal Chemistry. 1972;**15**(9):982-985. DOI: 10.1021/jm00279a026

[164] Bochis RJ, Olen LE, Fisher MH, Reamer RA, et al. Isomeric phenylthioimidazo[1,2-a]pyridines as anthelmintics. Journal of Medicinal Chemistry. 1981;**24**(12):1483-1487. DOI: 10.1021/jm00144a022

[165] Elhakmaoui A, Gueiffier A, Milhavet J-C, Blache Y, et al. Synthesis and antiviral activity of 3-substituted imidazo[1,2-a]pyridines. Bioorganic & Medicinal Chemistry Letters. 1994;**16**(4):1937-1940

[166] Wang Q, Wolff M, Polat T, Du Y, et al. Inhibition of neuraminidase with neuraminic acid C-glycosides. Bioorganic & Medicinal Chemistry. 2002;**10**(4):941-946. DOI: 10.1016/s0960-894x(00)00132-3

[167] Castera-Ducros C, Paloque L, Verhaeghe P, Casanova M, et al. Targeting the human parasite leishmania donovani: Discovery of a new promising anti-infectious pharmacophore in 3-nitroimidazo[1,2-a]pyridine series. Bioorganic & Medicinal Chemistry. 2013;**21**(22):7155-7164. DOI: 10.1016/j.bmc.2013.09.002

[168] Jose G, Suresha Kumara TH, Nagendrappa G, Sowmya HBV, et al. New polyfunctional imidazo[4,5-C]pyridine motifs: Synthesis, crystal studies, docking studies and antimicrobial evaluation. European Journal of Medicinal Chemistry. 2014;**77**:288-297. DOI: 10.1016/j.ejmech.2014.03.019

[169] Silva DG, Gillespie JR, Ranade RM, Herbst ZM, et al. New class of Antitrypanosomal agents based on imidazopyridines. ACS Medicinal Chemistry Letters. 2017;**8**(7):766-770. DOI: 10.1021/acsmedchemlett.7b00202

[170] Vera B, Dashti HS, Gomez-Abellan P, Hernandez-Martinez AM, Esteban A, et al. Modifiable lifestyle behaviors, but not a genetic risk score, associate with metabolic syndrome in evening chronotypes. Scientific Reports. 2018;**8**(1):1-7. DOI: 10.1038/s41598-017-18268-z

[171] Zhou S, Chen G, Huang G. Design, synthesis and biological evaluation of imidazo[1,2-a]pyridine analogues or derivatives as anti-helmintic drug. Chemical Biology & Drug Design. 2019;**93**(4):503-510. DOI: 10.1111/cbdd.13441

[172] Nandikolla A, Srinivasarao S, Karan Kumar B, Murugesan S, et al. Synthesis, study of antileishmanial and antitrypanosomal activity of imidazo pyridine fused triazole analogues. RSC Advances. 2020;**10**(63):38328-38343. DOI: 10.1039/d0ra07881f

[173] Silva DG, Junker A, de Melo SMG, Fumagalli F, et al. Front cover: Synthesis and structure-activity relationships of imidazopyridine/pyrimidine- and Furopyridine-based anti-infective agents against trypanosomiases. ChemMedChem. 2021;**16**(6):898. DOI: 10.1002/cmdc.202100141

[174] Parcella K, Patel M, Tu Y, Eastman K, et al. Scaffold modifications to the 4-(4,4-dimethylpiperidinyl) 2,6-dimethylpyridinyl class of HIV-1 allosteric integrase inhibitors. Bioorganic & Medicinal Chemistry. 2022;**67**:116833. DOI: 10.1016/j.bmc.2022.116833

[175] Temple CJ, Smith BH, Elliott RD, Montgomery JA. Synthesis of potential anticancer agents. Preparation of some 1-deazapurines and pyrimidines. Journal of Medicinal Chemistry. 1973;**16**(3):292-294. DOI: 10.1021/jm00261a031

[176] Cristalli G, Franchetti P, Grifantini M, Vittori S, et al. Improved synthesis and antitumor activity of 1-deazaadenosine. Journal of Medicinal Chemistry. 1987;**30**(9):1686-1688. DOI: 10.1021/jm00392a029

[177] Ismail MA, Arafa RK, Wenzler T, Brun R, et al. Synthesis and antiprotozoal activity of novel bis-benzamidino imidazo[1,2-a]pyridines and 5,6,7,8-tetrahydro-imidazo[1,2-a] pyridines. Bioorganic & Medicinal Chemistry. 2008;**16**(2):683-691. DOI: 10.1016/j.bmc.2007.10.042

[178] Dahan-Farkas N, Langley C, Rousseau AL, Yadav DB, et al. 6-Substituted imidazo[1,2-a]pyridines: Synthesis and biological activity against colon cancer cell lines HT-29 and Caco-2. European Journal of Medicinal Chemistry. 2011;**46**(9):4573-4583. DOI: 10.1016/j.ejmech.2011.07.036

[179] Reddy Gangireddy M, Mantipally M, Gundla R, Nayak Badavath V, et al. Design and synthesis of piperazine-linked imidazo[1,2-a]pyridine derivatives as potent anticancer agents. ChemistrySelect. 2019;**4**(46):13622-13629. DOI: 10.1002/slct.201902955

[180] Rani CS, Reddy AG, Susithra E, Mak K-K, et al. Synthesis and anticancer evaluation of amide derivatives of imidazo-pyridines. Medicinal Chemistry Research. 2021;**30**(1):74-83. DOI: 10.1007/s00044-020-02638-w

[181] Mannem GR, Navudu R, Dubasi N, Mohammed MA, et al. Design, and synthesis of aryl derivatives of imidazopyridine-thiadiazoles as possible anticancer agents. ChemistrySelect. 2022;**7**(19):e202200455. DOI: 10.1002/slct.202200455

[182] Wu Z, Fraley ME, Bilodeau MT, Kaufman ML, et al. Design and synthesis of 3,7-diarylimidazopyridines as inhibitors of the VEGF-receptor KDR. Bioorganic & Medicinal Chemistry Letters. 2004;**14**(4):909-912. DOI: 10.1016/j.bmcl.2003.12.007

[183] Koltun DO, Parkhill EQ, Kalla R, Perry TD, et al. Discovery of potent and selective inhibitors of calmodulin-dependent kinase II (CaMKII). Bioorganic & Medicinal Chemistry Letters. 2018;**28**(3):541-546. DOI: 10.1016/j.bmcl.2017.10.040

[184] Engler TA, Henry JR, Malhotra S, Cunningham B, et al. Substituted 3-imidazo[1,2-a]pyridin-3-yl- 4-(1,2,3,4-tetrahydro-[1,4] diazepino- [6,7,1-hi]indol-7-yl)pyrrole-2,5-diones as highly selective and potent inhibitors of glycogen synthase kinase-3. Journal of Medicinal Chemistry. 2004;**47**(16):3934-3937. DOI: 10.1021/jm049768a

[185] Jaramillo C, De Diego JE, Hamdouchi C, Collins E, et al. Aminoimidazo[1,2-a]pyridines as a new structural class of cyclin-dependent kinase inhibitors. Part 1: Design, synthesis, and biological evaluation. Bioorganic & Medicinal Chemistry Letters. 2004;**14**(24):6095-6099. DOI: 10.1016/j.bmcl.2004.09.053

[186] Krajcovicova S, Jorda R, Vanda D, Soural M, et al. 1,4,6-trisubstituted imidazo[4,5-c]pyridines as inhibitors of Bruton's tyrosine kinase. European Journal of Medicinal Chemistry. 2021;**211**:113094. DOI: 10.1016/j.ejmech.2020.113094

[187] Rhodes N, Heerding DA, Duckett DR, Eberwein DJ, et al. Characterization of an Akt kinase ınhibitor with potent pharmacodynamic and antitumor activity. Cancer Research. 2008;**68**(7):2366-2374. DOI: 10.1158/0008-5472.can-07-5783

[188] Chen D, Wang Y, Ma Y, Xiong B, et al. Discovery of 3H-imidazo[4,5-b] pyridines as potent c-met kinase inhibitors: Design, synthesis, and biological evaluation. ChemMedChem. 2012;**7**(6):1057-1070. DOI: 10.1002/cmdc.201200120

[189] Matsumoto S, Miyamoto N, Hirayama T, Oki H, et al. Structure-based design, synthesis, and evaluation of imidazo[1,2-b]pyridazine and imidazo[1,2-a]pyridine derivatives as novel dual c-Met and VEGFR2 kinase inhibitors. Bioorganic & Medicinal Chemistry. 2013;**21**(24):7686-7698. DOI: 10.1016/j.bmc.2013.10.028

[190] Frett B, McConnell N, Smith CC, Wang Y, et al. Computer aided drug discovery of highly ligand efficient, low molecular weight imidazopyridine analogs as FLT3 inhibitors. European Journal of Medicinal Chemistry. 2015;**94**:123-131. DOI: 10.1016/j.ejmech.2015.02.052

[191] Bach J, Eastwood P, Gonzalez J, Gomez E, et al. Identification of 2-imidazopyridine and 2-aminopyridone purinones as potent pan-janus kinase (JAK) inhibitors for the inhaled treatment of respiratory diseases. Journal of Medicinal Chemistry. 2019;**62**(20):9045-9060. DOI: 10.1021/acs.jmedchem.9b00533

[192] Hayakawa M, Kaizawa H, Kawaguchi K, Ishikawa N, et al. Synthesis and biological evaluation of imidazo[1,2-a]pyridine derivatives as novel PI3 kinase p110α inhibitors. Bioorganic & Medicinal Chemistry. 2007;**15**(1):403-412. DOI: 10.1016/j.bmc.2006.09.047

[193] Bach J, Eastwood P, Gonzalez J, Gomez E, et al. Identification of 2-imidazopyridine and 2-aminopyridone purinones as potent pan-janus kinase (JAK) inhibitors for the inhaled treatment of respiratory diseases. Journal of Medicinal Chemistry. 2010;**53**(14):5213-5228. DOI: 10.1021/acs.jmedchem.9b00533

[194] Fan Y-H, Li W, Liu D-D, Bai M-X, et al. Design, synthesis, and biological evaluation of novel 3-substituted imidazo[1,2-a]pyridine and quinazolin-4(3H)-one derivatives as PI3Kα inhibitors. European Journal of Medicinal Chemistry. 2017;**139**:95-106. DOI: 10.1016/j.ejmech.2017.07.074

[195] Terao Y, Suzuki H, Yoshikawa M, Yashiro H, et al. Design and biological evaluation of imidazo[1,2-a]pyridines as novel and potent ASK1 inhibitors. Bioorganic & Medicinal Chemistry Letters. 2012;**22**(24):7326-7329. DOI: 10.1016/j.bmcl.2012.10.084

[196] Tikhonova TA, Rassokhina IV, Kondrakhin EA, Fedosov MA, et al. Development of 1,3-thiazole analogues of imidazopyridines as potent positive allosteric modulators of GABAA receptors. Bioorganic Chemistry. 2020;**94**:103334. DOI: 10.1016/j.bioorg.2019.103334

[197] Krenitsky TA, Rideout JL, Chao EY, Koszalka GW, et al. Imidazo[4,5-c]

pyridines (3-deazapurines) and their nucleosides as immunosuppressive and antiinflammatory agents. Journal of Medicinal Chemistry. 1986;**29**(1):138-143. DOI: 10.1021/jm00151a022

[198] Kelley JL, Linn JA, Rideout JL, Soroko FE. Synthesis and anticonvulsant activity of 1-benzyl-4-alkylamino-1H-imidazo[4,5-c]pyridines. Journal of Heterocyclic Chemistry. 1988;**25**(4):1255-1258. DOI: 10.1002/jhet.5570250441

[199] Mantlo NB, Chakravarty PK, Ondeyka DL, Siegl PKS, et al. Potent, orally active imidazo[4,5-b]pyridine-based angiotensin II receptor antagonists. Journal of Medicinal Chemistry. 1991;**34**(9):2919-2922. DOI: 10.1021/jm00113a035

[200] Coates WJ, Connolly B, Dhanak D, Flynn ST, et al. Cyclic nucleotide phosphodiesterase inhibition by imidazopyridines: Analogs of sulmazole and isomazole as inhibitors of the cGMP specific phosphodiesterase. Journal of Medicinal Chemistry. 1993;**36**(10):1387-1392. DOI: 10.1021/jm00062a011

[201] Carceller E, Merlos M, Giral M, Balsa D, et al. Design, synthesis, and structure-activity relationship studies of novel 1-[(1-acyl-4-piperidinyl)methyl]-1H-2-methylimidazo[4,5-c]pyridine derivatives as potent, orally active platelet-activating factor antagonists. Journal of Medicinal Chemistry. 1996;**39**(2):487-493. DOI: 10.1021/jm950555i

[202] Izumi T, Sakaguchi J, Takeshita M, Tawara H, et al. 1H-Imidazo[4,5-c]quinoline derivatives as novel potent TNF-α suppressors: Synthesis and structure-activity relationship of 1-, 2-and 4-substituted 1H-imidazo[4,5-c]quinolines or 1H-imidazo[4,5-c]pyridines. Bioorganic & Medicinal Chemistry. 2003;**11**(12):2541-2550. DOI: 10.1016/s0968-0896(03)00178-0

[203] Tresadern G, Cid JM, Macdonald GJ, Vega JA, et al. Scaffold hopping from pyridones to imidazo[1,2-a]pyridines. New positive allosteric modulators of metabotropic glutamate 2 receptor. Bioorganic & Medicinal Chemistry Letters. 2010;**20**(1):175-179. DOI: 10.1016/j.bmcl.2009.11.008

[204] Peterson EA, Boezio AA, Andrews PS, Boezio CM, et al. Discovery and optimization of potent and selective imidazopyridine and imidazopyridazine mTOR inhibitors. Bioorganic & Medicinal Chemistry Letters. 2012;**22**(15):4967-4974. DOI: 10.1016/j.bmcl.2012.06.033

[205] Hintermann S, Guntermann C, Mattes H, Carcache DA, et al. Synthesis and biological evaluation of new triazolo- and imidazolopyridine RORγt inverse agonists. ChemMedChem. 2016;**11**(24):2640-2648. DOI: 10.1002/cmdc.201600500

[206] Igawa H, Takahashi M, Kakegawa K, As K, et al. Melanin-concentrating hormone receptor 1 antagonists lacking an aliphatic amine: Synthesis and structure-activity relationships of novel 1-(imidazo[1,2-a]pyridin-6-yl)pyridin-2(1H)-one derivatives. Journal of Medicinal Chemistry. 2016;**59**(3):1116-1139. DOI: 10.1021/acs.jmedchem.5b01704

[207] Joncour A, Desroy N, Housseman C, Bock X, et al. Discovery, structure-activity relationship, and binding mode of an imidazo[1,2-a]pyridine series of autotaxin inhibitors. Journal of Medicinal Chemistry. 2017;**60**(17):7371-7392. DOI: 10.1021/acs.jmedchem.7b00647

[208] Lee JM, Choi HS, Kim ES, Keum B, et al. Characterization of irreversible

electroporation on the stomach: A feasibility study in rats. Scientific Reports. 2019;**9**(1):1-15. DOI: 10.1038/s41598-019-54656-3

[209] Shen H, Ge Y, Wang J, Li H, et al. Design, synthesis and biological evaluation of novel molecules as potent PARP-1 inhibitors. Bioorganic & Medicinal Chemistry Letters. 2021;**47**:128169. DOI: 10.1016/j.bmcl.2021.128169

[210] Englert SME, McElvain SM. Pyrazolones derived from the carbethoxypiperidones. Journal of the American Chemical Society. 1934;**56**:700-702. DOI: 10.1021/ja01318a051

[211] Schmidt P, Eichenberger K, Rossi A, Wilhelm M. Preparation of Pyrazolopyridines. CH416659. 1967

[212] Sekikawa I, Nishie J, Tonooka S, Tanaka Y, et al. Antituberculous compounds. XXVIII. Synthesis of pyrazolopyridines. Journal of Heterocyclic Chemistry. 1973;**10**(6):931-932. DOI: 10.1002/jhet.5570100607

[213] Gudmundsson KS, Johns BA, Wang Z, Turner EM, et al. Synthesis of novel substituted 2-phenylpyrazolopyridines with potent activity against herpesviruses. Bioorganic & Medicinal Chemistry. 2005;**13**(18):5346-5361. DOI: 10.1016/j.bmc.2005.05.043

[214] El-borai MA, Rizk HF, Abd-Aal MF, El-Deeb IY. Synthesis of pyrazolo[3,4-b]pyridines under microwave irradiation in multi-component reactions and their antitumor and antimicrobial activities. Part 1. European Journal of Medicinal Chemistry. 2012;**48**:92-96. DOI: 10.1016/j.ejmech.2011.11.038

[215] Quiroga J, Villarreal Y, Galvez J, Ortiz A, et al. Synthesis and antifungal in vitro evaluation of pyrazolo[3,4-b]pyridines derivatives obtained by aza-Diels-Alder reaction and microwave irradiation. Chemical & Pharmaceutical Bulletin. 2017;**65**(2):143-150. DOI: 10.1248/cpb.c16-00652

[216] Hu Y, Kitamura N, Musharrafieh R, Wang J. Discovery of potent and broad-spectrum pyrazolopyridine-containing antivirals against enteroviruses D68, A71, and coxsackievirus B3 by targeting the viral 2C protein. Journal of Medicinal Chemistry. 2021;**64**(12):8755-8774. DOI: 10.1021/acs.jmedchem.1c00758

[217] Niemand J, van Biljon R, van der Watt M, van Heerden A, et al. Chemogenomic fingerprints associated with stage-specific gametocytocidal compound action against human malaria parasites. ACS Infectious Diseases. 2021;**7**(10):2904-2916. DOI: 10.1021/acsinfecdis.1c00373

[218] Sanghvi YS, Larson SB, Willis RC, Robins RK, et al. Synthesis and biological evaluation of certain C-4 substituted pyrazolo[3,4-b]pyridine nucleosides. Journal of Medicinal Chemistry. 1989;**32**(5):945-951. DOI: 10.1021/jm00125a004

[219] Mohamed AM, El-Sayed WA, Alsharari MA, Al-Qalawi HRM, et al. Anticancer activities of some newly synthesized pyrazole and pyrimidine derivatives. Archives of Pharmacal Research. 2013;**36**(9):1055-1065. DOI: 10.1007/s12272-013-0163-x

[220] Milisiunaite V, Arbaciauskiene E, Reznickova E, Jorda R, et al. Synthesis and anti-mitotic activity of 2,4- or 2,6-disubstituted- and 2,4,6-trisubstituted-2H-pyrazolo[4,3-c]pyridines. European Journal of Medicinal Chemistry. 2018;**150**:908-919. DOI: 10.1016/j.ejmech.2018.03.037

[221] Abozeid MA, El-Sawi AA, Abdelmoteleb M, Awad H, et al. Synthesis of novel naphthalene-heterocycle hybrids with potent antitumor, anti-inflammatory and antituberculosis activities. RSC Advances. 2020;**10**(70):42998-43009. DOI: 10.1039/d0ra08526j

[222] Papastathopoulos A, Lougiakis N, Kostakis IK, Marakos P, et al. New bioactive 5-arylcarboximidamidopyrazolo[3,4-c] pyridines: Synthesis, cytotoxic activity, mechanistic investigation and structure-activity relationships. European Journal of Medicinal Chemistry. 2021;**218**:113387. DOI: 10.1016/j.ejmech.2021.113387

[223] Misra RN, Rawlins DB, Xiao H, Shan W, et al. 1H-pyrazolo[3,4-b] pyridine inhibitors of cyclin-dependent kinases. Bioorganic & Medicinal Chemistry Letters. 2003;**13**(6):1133-1136. DOI: 10.1016/s0960-894x(03)00034-9

[224] Witherington J, Bordas V, Garland SL, Hickey DMB, et al. 5-aryl-pyrazolo[3,4-b]pyridines: Potent inhibitors of glycogen synthase kinase-3 (GSK-3). Bioorganic & Medicinal Chemistry Letters. 2003;**13**(9):1577-1580. DOI: 10.1016/S0960-894X(03)00134-3

[225] Collier PN, Twin HC, Knegtel RMA, Boyall D, et al. Discovery of selective, orally bioavailable pyrazolopyridine inhibitors of protein kinase Cθ (PKCθ) that ameliorate symptoms of experimental autoimmune encephalomyelitis. ACS Medicinal Chemistry Letters. 2019;**10**(8):1134-1139. DOI: 10.1021/acsmedchemlett.9b00134

[226] Gilbert AM, Nowak P, Brooijmans N, Bursavich MG, et al. Novel purine and pyrazolo[3,4-d]pyrimidine inhibitors of PI3 kinase-α: Hit to lead studies. Bioorganic & Medicinal Chemistry Letters. 2010;**20**(2):636-639. DOI: 10.1016/j.bmcl.2009.11.051

[227] Shi J, Xu G, Zhu W, Ye H, et al. Design and synthesis of 1,4,5,6-tetrahydropyrrolo[3,4-c] pyrazoles and pyrazolo[3,4-b]pyridines for Aurora-a kinase inhibitors. Bioorganic & Medicinal Chemistry Letters. 2010;**20**(14):4273-4278. DOI: 10.1016/j.bmcl.2010.04.083

[228] Nishiguchi GA, Atallah G, Bellamacina C, Burger MT, et al. Discovery of novel 3,5-disubstituted indole derivatives as potent inhibitors of Pim-1, Pim-2, and Pim-3 protein kinases. Bioorganic & Medicinal Chemistry Letters. 2011;**21**(21):6366-6369. DOI: 10.1016/j.bmcl.2011.08.105

[229] Yogo T, Nagamiya H, Seto M, Sasaki S, et al. Structure-based design and synthesis of 3-amino-1,5-dihydro-4H-pyrazolopyridin-4-one derivatives as tyrosine kinase 2 inhibitors. Journal of Medicinal Chemistry. 2016;**59**(2):733-749. DOI: 10.1021/acs.jmedchem.5b01857

[230] Sabat M, Wang H, Scorah N, Lawson JD, et al. Design, synthesis and optimization of 7-substituted-pyrazolo[4,3-b]pyridine ALK5 (activin receptor-like kinase 5) inhibitors. Bioorganic & Medicinal Chemistry Letters. 2017;**27**(9):1955-1961. DOI: 10.1016/j.bmcl.2017.03.026

[231] Nam Y, Hwang D, Kim N, Seo H-S, et al. Identification of 1H-pyrazolo[3,4-b] pyridine derivatives as potent ALK-L1196M inhibitors. Journal of Enzyme Inhibition and Medicinal Chemistry. 2019;**34**(1):1426-1438. DOI: 10.1080/14756366.2019.1639694

[232] Pfaffenrot B, Kloevekorn P, Juchum M, Selig R, et al. Design and synthesis of 1H-pyrazolo[3,4-b]pyridines targeting mitogen-activated protein kinase kinase 4 (MKK4)—A promising target for liver regeneration. European Journal of Medicinal Chemistry.

2021;**218**:113371. DOI: 10.1016/j.ejmech.2021.113371

[233] Bare TM, McLaren CD, Campbell JB, Firor JW, et al. Synthesis and structure-activity relationships of a series of anxioselective pyrazolopyridine ester and amide anxiolytic agents. Journal of Medicinal Chemistry. 1989;**32**(12):2561-2573. DOI: 10.1021/jm00132a011

[234] Akahane A, Katayama H, Mitsunaga T, Kita Y, et al. Discovery of FK453, a novel non-xanthine adenosine A1 receptor antagonist. Bioorganic & Medicinal Chemistry Letters. 1996;**6**(17):2059-2062. DOI: 10.1016/0960-894x(96)00368-x

[235] Ochiai H, Ishida A, Ohtani T, Kusumi K, et al. New orally active PDE4 inhibitors with therapeutic potential. Bioorganic & Medicinal Chemistry Letters. 2004;**14**(1):29-32. DOI: 10.1016/j.bmcl.2003.10.025

[236] Yu G, Mason HJ, Wu X, Wang J, et al. Substituted pyrazolopyridines as potent and selective PDE5 inhibitors: Potential agents for treatment of erectile dysfunction. Journal of Medicinal Chemistry. 2001;**44**(7):1025-1027. DOI: 10.1021/jm0155042

[237] Wu Y, Zhou Q, Zhang T, Li Z, et al. Discovery of potent, selective, and orally bioavailable inhibitors against phosphodiesterase-9, a novel target for the treatment of vascular dementia. Journal of Medicinal Chemistry. 2019;**62**(8):4218-4224. DOI: 10.1021/acs.jmedchem.8b01041

[238] Kaplan J, Verheijen JC, Brooijmans N, Toral-Barza L, et al. Discovery of 3,6-dihydro-2H-pyran as a morpholine replacement in 6-aryl-1H-pyrazolo[3,4-d]pyrimidines and 2-arylthieno[3,2-d]pyrimidines: ATP-competitive inhibitors of the mammalian target of rapamycin (mTOR). Bioorganic & Medicinal Chemistry Letters. 2010;**20**(2):640-643. DOI: 10.1016/j.bmcl.2009.11.050

[239] Griebenow N, Schirok H, Mittendorf J, Alexander S, et al. Identification of acidic heterocycle-substituted 1H-pyrazolo[3,4-b]pyridines as soluble guanylate cyclase stimulators. Bioorganic & Medicinal Chemistry Letters. 2013;**23**(5):1197-1200. DOI: 10.1016/j.bmcl.2013.01.028

[240] Wenglowsky S, Ahrendt KA, Buckmelter AJ, Feng B, et al. Pyrazolopyridine inhibitors of B-RafV600E. Part 2: Structure-activity relationships. Bioorganic & Medicinal Chemistry Letters. 2011;**21**(18):5533-5537. DOI: 10.1016/j.bmcl.2011.06.097

[241] Wenglowsky S, Li R, Ahrendt KA, Laird ER, et al. Pyrazolopyridine inhibitors of B-RafV600E. Part 1: The development of selective, orally bioavailable, and efficacious inhibitors. ACS Medicinal Chemistry Letters. 2011;**2**(5):342-347. DOI: 10.1021/ml200025q

[242] Elsner J, Boeckler F, Heinemann FW, Huebner H, et al. Pharmacophore-guided drug discovery investigations leading to bioactive 5-aminotetrahydropyrazolopyridines. Implications for the binding mode of heterocyclic dopamine D3 receptor agonists. Journal of Medicinal Chemistry. 2005;**48**(18):5771-5779. DOI: 10.1021/jm0503805

[243] Zhai M, Liu S, Gao M, Wang L, et al. 3,5-Diaryl-1H-pyrazolo[3,4-b]pyridines as potent tubulin polymerization inhibitors: Rational design, synthesis and biological evaluation. European Journal of Medicinal Chemistry. 2019;**168**:426-435. DOI: 10.1016/j.ejmech.2018.12.053

[244] Pan T, Xie S, Zhou Y, Hu J, et al. Dual functional cholinesterase and PDE4D inhibitors for the treatment of Alzheimer's disease: Design, synthesis and evaluation of tacrine-pyrazolo[3,4-b]pyridine hybrids. Bioorganic & Medicinal Chemistry Letters. 2019;**29**(16):2150-2152. DOI: 10.1016/j.bmcl.2019.06.056

[245] Saikachi H. Pyridine derivatives containing sulfur. VII. Syntheses of pyridoxazoles and pyridothiazoles. Yakugaku Zasshi. 1944;**64**:201-202

[246] Ozawa S. Antituberculous activity of heterocyclic compounds. IV. Pyridine, pyridothiazole, phenylpyridyl ether, nicotinoyl hydrazide, and isonicotinoyl hydrazide derivatives. Kyoto Daigaku Kekkaku Kenkyusho Nempo. 1956;**4**:284-294

[247] Leysen DC, Haemers A, BollaertW. Thiazolopyridine analogs of nalidixic acid. 1. Thiazolo[5,4-b]pyridines. Journal of Heterocyclic Chemistry. 1984;**21**(2):401-406. DOI: 10.1002/jhet.5570210226

[248] El-Hag Ali GAM, Khalil A, Ahmed AHA, El-Gaby MSA. Studies on thiazolopyridines. Part 2. Synthesis and antimicrobial activity of novel thiazolo[3,2-a]pyridine and thiazolo[3,2-a] [1,8]naphthyridine derivatives having two different aryl moieties. Acta Chimica Slovenica. 2002;**49**(2):365-376

[249] Chaban T, Klenina O, Drapak I, Ogurtsov V, et al. Synthesis of some novel thiazolo[4,5-b]pyridines and their tuberculostatic activity evaluation. Chemistry & Chemical Technology. 2014;**8**(3):287-292. DOI: 10.23939/chcht08.03.287

[250] El-Mawgoud HKA. Synthesis, in-vitro cytotoxicity and antimicrobial evaluations of some novel thiazole based heterocycles. Chemical & Pharmaceutical Bulletin. 2019;**67**(12):1314-1323. DOI: 10.1248/cpb.c19-00681

[251] Othman IMM, Gad-Elkareem MAM, Radwan HA, Badraoui R, et al. Synthesis, structure-activity relationship and in silico studies of novel pyrazolothiazole and thiazolopyridine derivatives as prospective antimicrobial and anticancer agents. ChemistrySelect. 2021;**6**(31):7860-7872. DOI: 10.1002/slct.202101622

[252] Shi F, Li C, Xia M, Miao K, et al. Green chemoselective synthesis of thiazolo[3,2-a]pyridine derivatives and evaluation of their antioxidant and cytotoxic activities. Bioorganic & Medicinal Chemistry Letters. 2009;**19**(19):5565-5568. DOI: 10.1016/j.bmcl.2009.08.046

[253] Lozynskyi A, Zimenkovsky B, Ivasechko I, Senkiv J, et al. Synthesis and cytotoxicity of new 2-oxo-7-phenyl-2,3-dihydrothiazolo[4,5-b]pyridine-5-carboxylic acid amides. Phosphorus, Sulfur and Silicon and the Related Elements. 2019;**194**(12):1149-1157. DOI: 10.1080/10426507.2019.1633318

[254] Yahia HB, Sabri S, Essehli R, Kasak P, et al. Crystal growth, single crystal structure, and biological activity of thiazolo-pyridine dicarboxylic acid derivatives. ACS Omega. 2020;**5**(43):27756-27765. DOI: 10.1021/acsomega.0c01769

[255] Raslan RR, Hessein SA, Fouad SA, Shmiess NAM. Synthesis and antitumor evaluation of some new thiazolopyridine, nicotinonitrile, pyrazolopyridine, and polyhydroquinoline derivatives using ceric ammonium nitrate as a green catalyst. Journal of Heterocyclic

Chemistry. 2022;**59**(5):832-846. DOI: 10.1002/jhet.4423

[256] Kulkarni SS, Newman AH. Discovery of heterobicyclic templates for novel metabotropic glutamate receptor subtype 5 antagonists. Bioorganic & Medicinal Chemistry Letters. 2007;**17**(11):2987-2991. DOI: 10.1016/j.bmcl.2007.03.066

[257] Cee VJ, Lin J, Yu XY, Zhang Z. S1p1 Receptor Agonists and Use Thereof. WO2009154775 A1. 2012

[258] Kale MG, Raichurkar A, Shahul HP, Waterson D, et al. Thiazolopyridine ureas as novel antitubercular agents acting through inhibition of DNA gyrase B. Journal of Medicinal Chemistry. 2013;**56**(21):8834-8848. DOI: 10.1021/jm401268f

[259] Chaban TI, Ogurtsov VV, Matiychuk VS, Chaban IG, et al. Synthesis, anti-inflammatory and antioxidant activities of novel 3H-thiazolo[4,5-b]pyridines. Acta Chimica Slovenica. 2019;**66**(1):103-111. DOI: 10.17344/acsi.2018.4570

[260] Zhu J, Li K, Xu L, Jin J. Insight into the selective mechanism of phosphoinositide 3-kinase γ with benzothiazole and thiazolopiperidine γ-specific inhibitors by in silico approaches. Chemical Biology & Drug Design. 2019;**93**(5):818-831. DOI: 10.1111/cbdd.13469

[261] Scott DA, Hatcher JM, Liu H, Fu M, et al. Quinoline and thiazolopyridine allosteric inhibitors of MALT1. Bioorganic & Medicinal Chemistry Letters. 2019;**29**(14):1694-1698. DOI: 10.1016/j.bmcl.2019.05.040

[262] Taurins A, Khouw VT, Isothiazolopyridines I. Synthesis and spectra of isothiazolo[3,4-b]-, 3-aminoisothiazolo[4,3-b]-, isothiazolo[5,4-b]-, and 3-methylisothiazolo[5,4-c]pyridines. Preparation and spectra of some 2,3 and 3,4-disubstituted pyridines. Canadian Journal of Chemistry. 1973;**51**(11):1741-1748. DOI: 10.1139/v73-262

[263] Malinka W, Rutkowska M. Synthesis and anorectic activity of 2H-4,6-dimethyl-2-[(4-phenylpiperazin-1-yl)methyl]-3-oxo-2,3-dihydroisothiazolo[5,4-b]pyridine. Farmaco. 1997;**52**(10):595-601

[264] Ghorab M-M, Hassan A-Y, Nassar O-M. Synthesis of novel heterocyclic compounds for antitumor and radioprotective activities. Phosphorus, Sulfur and Silicon and the Related Elements. 1998;**134**(135):447-462. DOI: 10.1080/10426509808545486

[265] Malinka W, Sieklucka-Dziuba M, Rajtar G, Zgodzinski W, et al. Synthesis and preliminary screening of derivatives of 2-(4-arylpiperazin-1-ylalkyl)-3-oxoisothiazolo[5,4-b]pyridines as CNS and antimycobacterial agents. Die Pharmazie. 2000;**55**(6):416-425

[266] Malinka W, Swiatek P, Filipek B, Sapa J, et al. Synthesis, analgesic activity and computational study of new isothiazolopyridines of Mannich base type. Farmaco. 2005;**60**(11-12):961-968. DOI: 10.1016/j.farmac.2005.08.005

[267] Li J, Kovackova S, Pu S, Rozenski J, et al. Isothiazolo[4,3-b] pyridines as inhibitors of cyclin G associated kinase: Synthesis, structure-activity relationship studies and antiviral activity. MedChemComm. 2015;**6**(9):1666-1672. DOI: 10.1039/C5MD00229J

[268] Martinez-Gualda B, Saul S, Froeyen M, Schols D, et al. Discovery of 3-phenyl- and 3-N-piperidinyl-isothiazolo[4,3-b]pyridines as highly

potent inhibitors of cyclin G-associated kinase. European Journal of Medicinal Chemistry. 2021;**213**:113158. DOI: 10.1016/j.ejmech.2021.113158

[269] Swiatek P, Strzelecka M, Urniaz R, Gebczak K, et al. Synthesis, COX-1/2 inhibition activities and molecular docking study of isothiazolopyridine derivatives. Bioorganic & Medicinal Chemistry. 2016;**25**(1):316-326. DOI: 10.1016/j.bmc.2016.10.036

[270] Reitmann J. Pyridine compounds with analeptic action. Medizin und Chemie. Abhandlungen aus den Medizinisch-chemischen Forschungsstätten der I.G. Farbenindustrie Aktiengesellschaft. 1936;**3**:399-402

[271] Palazzo G, Silvestrini B. S-Triazolo[4,3-a]pyridines. US3381009 A. 1968

[272] Akbari V, Ghobadi S, Mohammadi S, Khodarahmi R. The antidepressant drug; trazodone inhibits tau amyloidogenesis: Prospects for prophylaxis and treatment of AD. Archives of Biochemistry and Biophysics. 2020;**679**:108218. DOI: 10.1016/j.abb.2019.108218

[273] Sadana AK, Mirza Y, Aneja KR, Prakash O. Hypervalent iodine mediated synthesis of 1-aryl/hetaryl-1,2,4-triazolo[4,3-a]pyridines and 1-aryl/hetaryl-5-methyl-1,2,4-triazolo[4,3-a] quinolines as antibacterial agents. European Journal of Medicinal Chemistry. 2003;**38**(5):533-536. DOI: 10.1016/S0223-5234(03)00061-8

[274] East SP, White CB, Barker O, Barker S, et al. DNA gyrase (GyrB)/topoisomerase IV (ParE) inhibitors: Synthesis and antibacterial activity. Bioorganic & Medicinal Chemistry Letters. 2009;**19**(3):894-899. DOI: 10.1016/j.bmcl.2008.11.102

[275] Liu X-H, Sun Z-H, Yang M-Y, Tan C-X, et al. Microwave assistant one pot synthesis, crystal structure, antifungal activities and 3D-QSAR of novel 1,2,4-triazolo[4,3-a]pyridines. Chemical Biology & Drug Design. 2014;**84**(3):342-347. DOI: 10.1111/cbdd.12323

[276] Hudson NO, Buck-Koehntop BA. Zinc finger readers of methylated DNA. Molecules. 2018;**23**(10):1-15. DOI: 10.3390/molecules23102555

[277] Hartwich A, Zdzienicka N, Schols D, Andrei G, et al. Design, synthesis and antiviral evaluation of novel acyclic phosphonate nucleotide analogs with triazolo[4,5-b] pyridine, imidazo[4,5-b]pyridine and imidazo[4,5-b]pyridin-2(3H)-one systems. Nucleosides, Nucleotides & Nucleic Acids. 2020;**39**(4):542-591. DOI: 10.1080/15257770.2019.1669046

[278] Butani SC, Vekariya MK, Dholaria PV, Kapadiya KM, et al. Copper(I)-catalyzed click chemistry-based synthesis and antimicrobial evaluation of triazolopyridine-triazole congeners. Russian Journal of Organic Chemistry. 2022;**58**(3):405-411. DOI: 10.1134/s1070428022030204

[279] Zhang G, Hu Y. Synthesis and antitumor activities of 2-(substituted) phenyl-1,2,4-triazolo[1,5-a]pyridines. Journal of Heterocyclic Chemistry. 2007;**44**(4):919-922. DOI: 10.1002/jhet.5570440428

[280] Sachdeva T, Low ML, Mai C-W, Cheong SL, et al. Design, synthesis and characterisation of novel phenothiazine-based triazolopyridine derivatives: Evaluation of anti-breast cancer activity on human breast carcinoma. ChemistrySelect. 2019;**4**(43):12701-12707. DOI: 10.1002/slct.201903203

[281] Tian N, Wu H, Zhang H, Yang D, et al. Discovery of [1,2,4]triazolo[4,3-a]

pyridines as potent smoothened inhibitors targeting the hedgehog pathway with improved antitumor activity in vivo. Bioorganic & Medicinal Chemistry. 2020;**28**(16):115584. DOI: 10.1016/j.bmc.2020.115584

[282] Pastor J, Oyarzabal J, Saluste G, Alvarez RM, et al. Hit to lead evaluation of 1,2,3-triazolo[4,5-b]pyridines as PIM kinase inhibitors. Bioorganic & Medicinal Chemistry Letters. 2012;**22**(4):1591-1597. DOI: 10.1016/j.bmcl.2011.12.130

[283] Menet CJ, Fletcher SR, Van Lommen G, Geney R, et al. Triazolopyridines as selective JAK1 inhibitors: From hit ıdentification to GLPG0634. Journal of Medicinal Chemistry. 2014;**57**(22):9323-9342. DOI: 10.1021/jm501262q

[284] Dugan BJ, Gingrich DE, Mesaros EF, Milkiewicz KL, et al. A selective, orally bioavailable 1,2,4-triazolo[1,5-a]pyridine-based inhibitor of janus kinase 2 for use in anticancer therapy: Discovery of CEP-33779. Journal of Medicinal Chemistry. 2012;**55**(11):5243-5254. DOI: 10.1021/jm300248q

[285] Ellard K, Sunose M, Bell K, Ramsden N, et al. Discovery of novel PI3Kγ/δ inhibitors as potential agents for inflammation. Bioorganic & Medicinal Chemistry Letters. 2012;**22**(14):4546-4549. DOI: 10.1016/j.bmcl.2012.05.121

[286] Krishnaiah M, Jin CH, Sreenu D, Subrahmanyam VB, et al. Synthesis and biological evaluation of 2-benzylamino-4(5)-(6-methylpyridin-2-yl)-5(4)-([1,2,4]triazolo[1,5-a]pyridin-6-yl)thiazoles as transforming growth factor-β type 1 receptor kinase inhibitors. European Journal of Medicinal Chemistry. 2012;**57**:74-84. DOI: 10.1016/j.ejmech.2012.09.011

[287] Oguro Y, Cary DR, Miyamoto N, Tawada M, et al. Design, synthesis, and evaluation of novel VEGFR2 kinase inhibitors: Discovery of [1,2,4]triazolo[1,5-a]pyridine derivatives with slow dissociation kinetics. Bioorganic & Medicinal Chemistry. 2013;**21**(15):4714-4729. DOI: 10.1016/j.bmc.2013.04.042

[288] Ferguson GD, Delgado M, Plantevin-Krenitsky V, Jensen-Pergakes K, et al. A novel triazolopyridine-based spleen tyrosine kinase inhibitor that arrests joint inflammation. PLoS One. 2016;**11**(1):e0145705/1-e0145705/24. DOI: 10.1371/journal.pone.0145705

[289] Zhao J, Fang L, Zhang X, Liang Y, et al. Synthesis and biological evaluation of new [1,2,4]triazolo[4,3-a]pyridine derivatives as potential c-Met inhibitors. Bioorganic & Medicinal Chemistry. 2016;**24**(16):3483-3493. DOI: 10.1016/j.bmc.2016.05.057

[290] Schulze VK, Klar U, Kosemund D, Wengner AM, et al. Treating cancer by spindle assembly checkpoint abrogation: Discovery of two clinical candidates, BAY 1161909 and BAY 1217389, targeting MPS1 kinase. Journal of Medicinal Chemistry. 2020;**63**(15):8025-8042. DOI: 10.1021/acs.jmedchem.9b02035

[291] Nettekoven M. Combinatorial synthesis of 5-aryl-[1,2,4]-triazolo-[1,5-a]-pyridine derivatives as potential inhibitors of the adenosine 2a receptor. Synlett. 2001;**12**:1917-1920. DOI: 10.1055/s-2001-18758

[292] Shaik K, Deb PK, Mailavaram RP, Chandrasekaran B, et al. 7-Amino-2-aryl/hetero-aryl-5-oxo-5,8-dihydro[1,2,4]triazolo[1,5-a]pyridine-6-carbonitriles: Synthesis and adenosine receptor binding studies. Chemical Biology & Drug Design. 2019;**94**(2):1568-1573. DOI: 10.1111/cbdd.13528

[293] Dombroski MA, Duplantier AJ, Laird ER, Letavic MA, et al. Preparation of 6-(phenylheterocyclyl)-[1,2,4]triazolo[4,3-a]pyridines as Anti-Inflammatory Agents. WO2002072579 A1. 2002

[294] McClure KF, Abramov YA, Laird ER, Barberia JT, et al. Theoretical and experimental design of atypical kinase inhibitors: Application to p38 MAP kinase. Journal of Medicinal Chemistry. 2005;**48**(18):5728-5737. DOI: 10.1021/jm050346q

[295] Wang H, Robl JA, Hamann LG, Simpkins L, et al. Generation of 3,8-substituted 1,2,4-triazolopyridines as potent inhibitors of human 11β-hydroxysteroid dehydrogenase type 1 (11β-HSD-1). Bioorganic & Medicinal Chemistry Letters. 2011;**21**(14):4146-4149. DOI: 10.1016/j.bmcl.2011.05.101

[296] Andres J-I, Alcazar J, Cid JM, De Angelis M, et al. Synthesis, evaluation, and radiolabeling of new potent positive allosteric modulators of the metabotropic glutamate receptor 2 as potential tracers for positron emission tomography imaging. Journal of Medicinal Chemistry. 2012;**55**(20):8685-8699. DOI: 10.1021/jm300912k

[297] Engers JL, Bender AM, Kalbfleisch JJ, Cho HP, et al. Discovery of tricyclic triazolo- and ımidazopyridine lactams as M1 positive allosteric modulators. ACS Chemical Neuroscience. 2019;**10**(3):1035-1042. DOI: 10.1021/acschemneuro.8b00311

[298] Ahmed S, Ayscough A, Barker GR, Canning HE, et al. 1,2,4-triazolo-[1,5-a]pyridine HIF prolylhydroxylase domain-1 (PHD-1) ınhibitors with a novel monodentate binding ınteraction. Journal of Medicinal Chemistry. 2017;**60**(13):5663-5672. DOI: 10.1021/acs.jmedchem.7b00352

[299] Wurtz NR, Viet A, Shaw SA, Dilger A, et al. Potent triazolopyridine myeloperoxidase inhibitors. ACS Medicinal Chemistry Letters. 2018;**9**(12):1175-1180. DOI: 10.1021/acsmedchemlett.8b00308

[300] Nakajima R, Oono H, Sugiyama S, Matsueda Y, et al. Discovery of [1,2,4]triazolo[1,5-a]pyridine derivatives as potent and orally bioavailable RORγt inverse agonists. ACS Medicinal Chemistry Letters. 2020;**11**(4):528-534. DOI: 10.1021/acsmedchemlett.9b00649

[301] Yang F, Jian X-E, Diao P-C, Huo X-S, et al. Synthesis, and biological evaluation of 3,6-diaryl-[1,2,4]triazolo[4,3-a]pyridine analogues as new potent tubulin polymerization inhibitors. European Journal of Medicinal Chemistry. 2020;**204**:112625. DOI: 10.1016/j.ejmech.2020.112625

[302] Huang Y, Sendzik M, Zhang J, Gao Z, et al. Discovery of the clinical candidate MAK683: An EED-directed, allosteric, and selective PRC2 inhibitor for the treatment of advanced malignancies. Journal of Medicinal Chemistry. 2022;**65**(7):5317-5333. DOI: 10.1021/acs.jmedchem.1c02148

[303] Tian C, Zhang G, Xia Z, Chen N, et al. Identification of triazolopyridine derivatives as a new class of AhR agonists and evaluation of anti-psoriasis effect in a mouse model. European Journal of Medicinal Chemistry. 2022;**213**:114122. DOI: 10.1016/j.ejmech.2022.114122

[304] Mahmoud MR, El-Shahawi MM, Abu El-Azm FS, Abdeen MJ. Synthesis and antimicrobial activity of polyfunctionally substituted heterocyclic compounds derived from 5-cinnamoylamino-2-cyanomethyl-1,3,4-thiadiazole. Journal of Heterocyclic Chemistry. 2017;**54**(4):2352-2359. DOI: 10.1002/jhet.2824

[305] Pabba J, Rempel BP, Withers SG, Vasella A. Synthesis of glycaro-1,5-lactams and tetrahydrotetrazolopyridine-5-carboxylates: Inhibitors of β-D-glucuronidase and α-L-iduronidase. Helvetica Chimica Acta. 2006;**89**(4):635-666. DOI: 10.1002/hlca.200690066

[306] Zhou M, Ji S, Wu Z, Li Y, et al. Synthesis of selenazolopyridine derivatives with capability to induce apoptosis in human breast carcinoma MCF-7 cells through scavenge of intracellular ROS. European Journal of Medicinal Chemistry. 2015;**96**:92-97. DOI: 10.1016/j.ejmech.2015.03.069

[307] Tong YC. Preparation of 1,3-Dithiolo- and 1,4-Dithiinopyridines as Industrial Antimicrobials. US5171743 A. 1992

Chapter 8

Advances in Pyridyl-Based Fluorophores for Sensing Applications

Andreia Leite, Carla Queirós and Ana M.G. Silva

Abstract

Fluorescence sensing plays an important role in high sensitivity, selectivity, and real-time monitoring of biological and environmentally relevant species. Several classes of fluorescent dyes (fluorophores) including rhodamine, BODIPY, 1,8-naphthalimide, and coumarin-among others—when conveniently functionalized with reactive pyridyl receptors, have emerged as effective sensors to detect and quantify chemical species with high accuracy through fluorescent imaging and spectroscopy. Among the sensing targets, monitoring of harmful chemical species, e.g., metal ions (zinc, copper, iron, mercury, cadmium, lead, etc.) and anions (chloride, fluoride, sulfide, thiocyanate, etc.) can be used to understand their physiological and pathological role in live-cells and tissues, as well as to protect human health. This chapter focuses on recent advances in the molecular design of pyridyl-substituted fluorophores, their photophysical properties, and sensing applications.

Keywords: molecular design, fluorescent dyes, pyridyl receptors, photophysical properties, sensing behavior

1. Introduction

Fluorescence detection techniques have become of paramount importance for monitoring biochemical and biological processes, allowing the detection and quantification of levels of chemical species in the human body and in the surrounding environment. Indeed, fluorescence sensing is a highly sensitive technique having numerous parameters that can serve as analytical information, including decay time, energy transfer, and quenching efficiency, in addition to the more conventional measurement of fluorescence intensity or polarization. Through the design of fluorescent dyes (fluorophores), it is possible to obtain molecules and materials that respond to the presence of a target analyte through changes in its physicochemical properties, presenting typically high sensibility and selectivity, quick response time and simplicity of measurement, and quantification of the analyte [1].

When combined with specific receptor units, fluorescent dyes can be extremely useful in several applications such as detection and quantification of chemical species, as well as in understanding their physiological and pathological role in cells and tissues. Receptors based on the pyridyl group are of major importance in ligand design for many of the above applications. The pyridine ring possesses a dipole moment found to be 2.22 D; therefore, it exhibits greater electronegativity as compared with the phenyl ring [2]. The pyridyl groups, such as di-(2-picolyl)amine (DPA), are excellent metal ion binding sites for the construction of fluorescent probes and can be attached to specific fluorophores or integrated into the fluorophore as part of the metal binding group, as found in quinolines. The principal fluorescence mechanisms involved in the design of the chemosensors are schematized in **Figure 1** and include:

i. Photoinduced electron transfer (PET, **Figure 1a**): Originally proposed by A. Prasanna de Silva and coworkers [3], PET involves the use of fluorophore-spacer-receptor-type structures. The spacer is used to separate the fluorophore

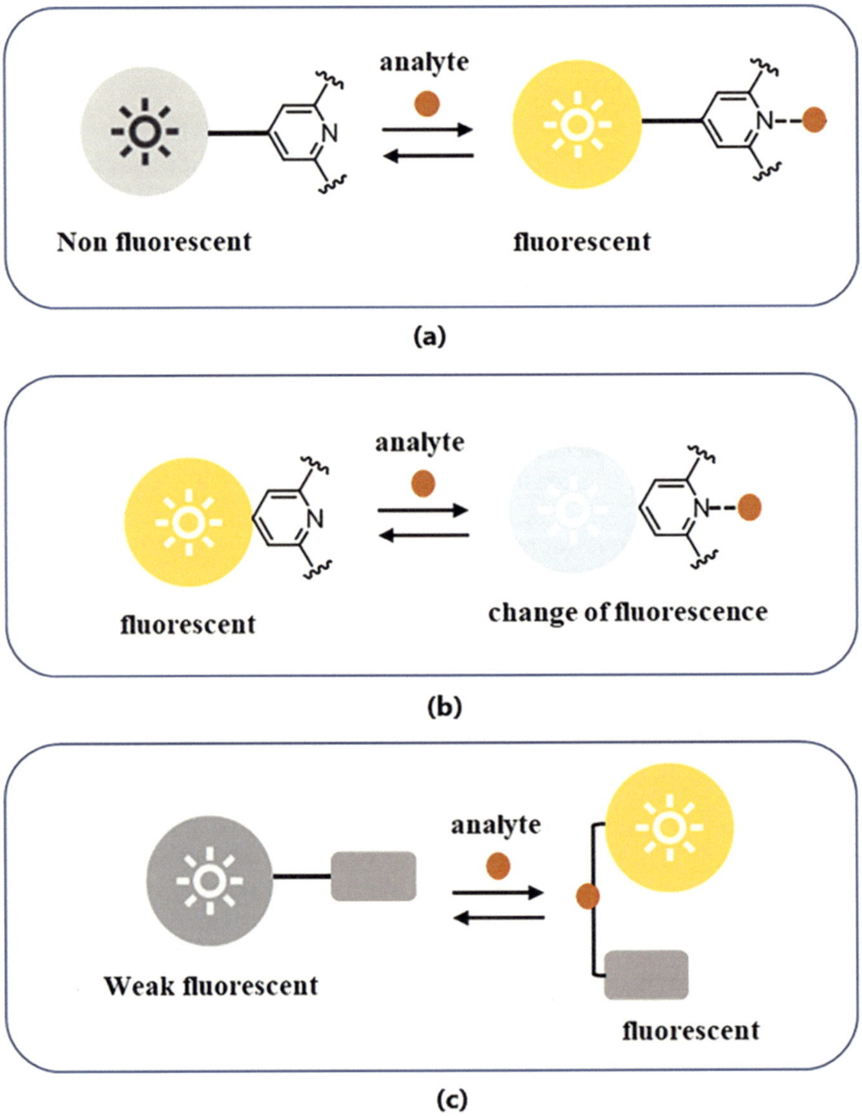

Figure 1.
Schematic representation of the main fluorescence mechanisms.

from the receptor at a certain distance while allowing the intramolecular electron transfer causes the interruption of the fluorophore's fluorescence. The interaction of the analyte with the receptor causes a change in the redox potential of the receptor and the electron transfer became energetically unfavorable, which leads to the re-establishment of fluorophore's fluorescence;

ii. Intramolecular charge transfer (ICT, **Figure 1b**): In ICT, the fluorophore can integrate the receptor unit and is characterized by a donor and an electron acceptor group, forming a push–pull system. When the analyte, in particular charged species, interacts with the receptor, causes the strengthening or weakening of the push–pull character, leading to a change in the emission band. This is a characteristic process of ratiometric sensors [4];

iii. Resonance energy transfer (FRET, **Figure 1c**). This mechanism involves the energy transfer from the excited state of a "donor" fluorophore to an "acceptor" fluorophore. In most cases, FRET occurs between two distinct fluorophores with overlapped emission spectrum of the "donor" and the absorption spectrum of the "acceptor" [4].

Such fluorescence mechanisms have inspired the development of new fluorescent structures and materials for the preparation of optical sensors for analyte detection in real scenarios. This chapter will focus precisely on recent advances in the molecular design of pyridyl substituted fluorophores, their photophysical properties, and sensing applications.

2. Pyridyl groups in fluorescent dyes

2.1 Rhodamine dyes

2.1.1 Molecular design

Xanthene is a heterocyclic organic compound with yellow coloration that contains two benzene rings connected through an oxygen atom and a methylene group (**Figure 2**). This class of dyes comprises fluorescein, rhodamine, and rhodol derivatives. Rhodamines were first produced in the late nineteenth century. They can be distinguished from other dyes by the presence of *N*-atoms at positions 3 and 6 of the xanthene core and they are one of the most widely used organic dyes with application in areas such as bioimaging, chemosensing, cosmetics, inks, and textiles. Rhodamine's photophysical properties are highly dependent on the structural features and substituent groups. The periphery of the xanthene ring can be modified using several strategies and it affects the selectivity in their metal ion-induced signaling pattern. The most common derivatizations are:

i. The derivatization reaction of the carboxylic group at position 2′ of the xanthene leads to spirocyclic derivatives (closed form);

ii. Modification at positions 3, 4, 5, and 6. In some cases, the alkylation at positions 3 and 6 can promote a bathochromic shift, which increases with the increase in the degree of alkylation, while in other cases, the functionalization of the amino groups of xanthene moiety can cause the total loss of fluorescence [5];

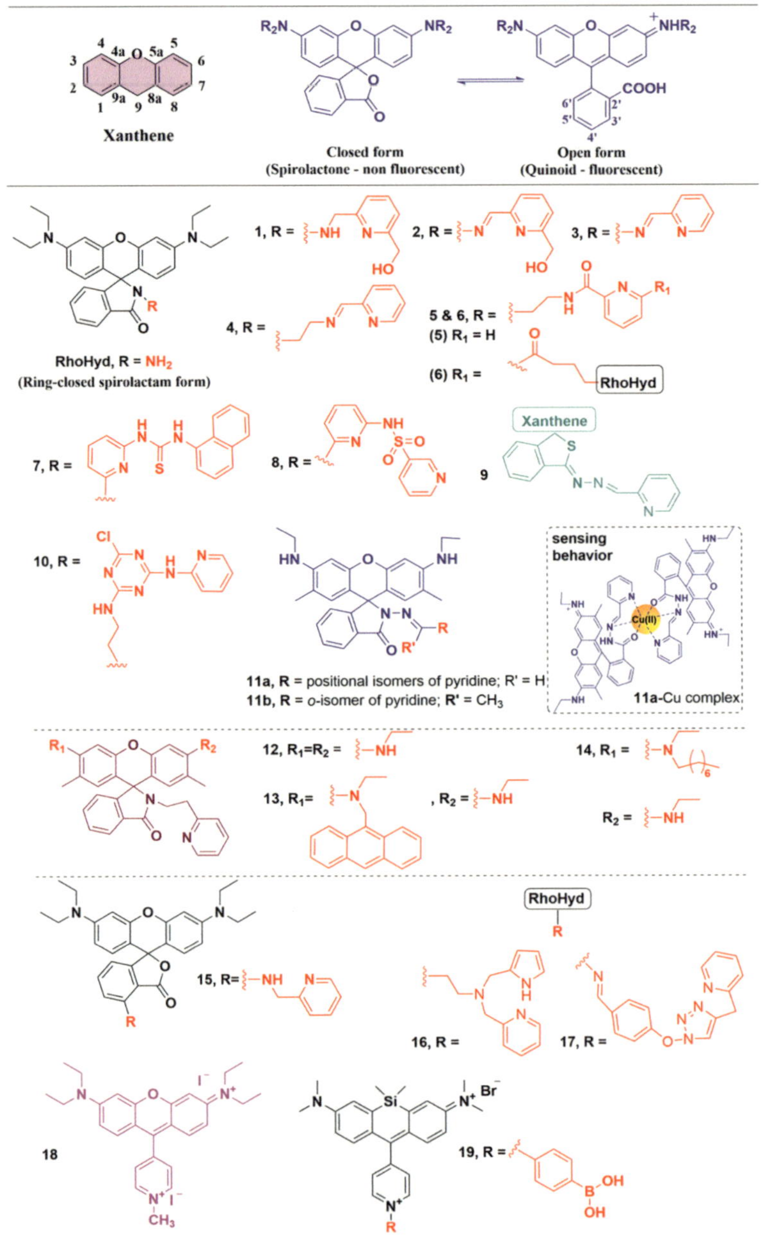

Figure 2.
Representative examples of rhodamine dyes functionalized with pyridyl groups.

iii. Modifications in the periphery of the phenyl ring at positions 4′ and/or 5′ are difficult to perform, especially when aiming to prepare isomerically pure derivatives from the sequential Friedel–Crafts reaction of an aminophenol with an asymmetric anhydride. This reaction usually led to a mixture of two isomers often difficult to separate and purify. Some of these derivatives are used for labeling molecules of interest [5];

iv. Modifications at position 9 are used for the synthesis of dihydro derivatives;

v. Substitution of the xanthene heteroatom (O), for example, by Si can potentiate the absorption and emission capacity in the near-infrared region, fluorescence quantum yield, or fluorescence intensity [6].

Some studies also focus on the influence of the positional isomers of the pyridine's nitrogen (*ortho*-, *meta*-, and *para*-) in the sensitivity and selectivity toward analytes, such as metal ions.

2.1.2 Photophysical properties

The excellence of the photophysical properties of rhodamines is one of the main reasons for their success and wide application in several areas. Rhodamines possess high molar absorptivity coefficient (ε), long absorption and emission wavelengths (>500 nm), high fluorescence quantum yield, photostability, and good water solubility [7]. These properties are directly associated with the extensive π-conjugated systems, molecular rigidity, and presence of functional groups.

One of the most interesting features of rhodamine derivatives is the existence of two isomeric forms-spirolactone (closed form) and quinoid (opened form) (**Figure 2**)-with very different optical properties. The spirolactone form is colorless and nonfluorescent, while the open form is highly fluorescent and has a pink coloration. The open form owes its properties to its extended π-conjugation and the interconversion from the closed to open form allows the rhodamine derivatives to possess an *off–on* (*turn-on*) characteristic fluorescence, usually promoted by acid or specific metal ions interactions [8].

2.1.3 Sensing applications

Rhodamines are frequently used in the preparation of highly selective, fast response, and sensitive sensing tools, employed in the detection of contaminants and environmental parameters in air, water, and waste [9]. **Figure 2** shows a series of selected examples of rhodamine derivatives/probes, those structural and photophysical features and sensing behaviors will be discussed in the next paragraphs.

One of the most explored rhodamine-based dyes for conjugation with pyridyl derivatives is rhodamine B hydrazide (**RhoHyd**, **Figure 2**), being the condensation product between the two moieties involving the terminal NH_2 of **RhoHyd**. This condensation can be achieved by attaching directly the pyridyl derivatives, through single or double bonds, or by using spacers. Uvdal and co-workers have reported probe **1,** which is prepared by appending a hydroxymethyl-pyridine group to **RhoHyd** [10]. This probe presented specific Hg^{2+}-induced color change and fluorescent enhancement in aqueous systems based on a metal binding induced ring-opening process of the spirolactam form. Probe **1** presented a limit of detection (LOD) of 15.7×10^{-9} mol dm^{-3}, and the **1**-Hg complex, with 1:1 stoichiometry, was formed by the coordinating atoms •O–N–N–O• from hydroxymethyl-pyridine and **RhoHyd** - with an association constant of 0.70×10^{5} mol^{-1} dm^{3}. The results also revealed good cell-membrane permeability and applicability of probe **1** for the detection of intracellular Hg^{2+} in living cells with almost no cytotoxicity. The simple change of linking the -NH group of **RhoHyd** to the hydroxyl-pyridyl derivative using a double bond, allowed the synthesis of probe **2** with even a higher association constant (1.27×10^7 mol^{-1} dm^3), suitable for a pH range from 5 to 9 and capable to detect basal levels of Fe^{3+}, as well as the metal ion dynamic changes in live cells at subcellular resolution [11]. The confocal laser scanning microscopy experiments showed two Fe^{3+} pools in mitochondria and endosomes/lysosomes for the first time.

In 2012, a study related to the influence of the number, nature, and size of coordinating entities was reported [12]. The synthesized probe **3** has been reported several times in literature and can be prepared from a condensation reaction between **RhoHyd** and 2-pyridinecarboxaldehyde [12–14]. In all cases, the probe was isolated in the ring-closed spirolactam form. Chereddy and co-workers [12] reported that in the presence of 50×10^{-3} mol dm^{-3} concentration of Cu^{2+} or Fe^{3+}, a clear pink color solution (0.01 mol dm^{-3} Tris HCl:CH_3CN solvent mixture, pH 7.4) was observed with the concomitant appearance of a new peak at 555 nm in the absorption spectra-ring opening mechanism: 57-fold for Fe^{3+} and 53-fold for Cu^{2+}. This lack of selectivity of probe **3** was overcome by using a CH_3CN/H_2O binary solution (7:3 v/v) [13]. A 1:1 stoichiometry of the **3**-Cu complex was estimated with a binding constant of 2.5×10^4 mol^{-1} dm^3. The UV–vis and fluorescence spectra showed an increase in the absorption maximum band and the depletion of fluorescence intensity, respectively. Besides, this complex proved to be reversible in the presence of KI. In 2017, Stalin and co-workers selected a solvent mixture of CH_3CN/H_2O (2,8, v/v) buffered with 2-[4-(2-hydroxyethyl)piperazin-1-yl]ethanesulfonic acid (HEPES), pH = 7.2, to perform their studies [14]. In this case, the probe revealed sensitivity (binding constant of 4.25×10^4 mol^{-1} dm^3 and LOD of 0.10 mol dm^{-3}) and selectivity toward Cd^{2+} by an intramolecular FRET process induced by the binding to Cd^{2+} ion and significant spectral overlap between the absorption spectrum of **3** with the emission spectrum of the pyridine fragment. Using a hand-held UV lamp, naked-eye detection of Cd^{2+} presence was possible by observing the color change from deep magenta to bright orange. On the other hand, the *in situ* generated **3**–Cd^{2+} complex was able to selectively sense S^{2-}, with the remarkable recovery of fluorescence and UV-vis absorption spectra, by means of a displacement approach-formation of a CdS complex. This same chemosensor was later explored by our research group as a probe that could allow discrimination of light-up effects induced by metal ion chelation and variation of pH [15]. The probe synthesis was optimized using a solvent-free approach under microwave irradiation and a crystal suitable for single-crystal X-ray diffraction (SCXRD) proved the isolation of the probe in the expected spirolactam form. The fluorescence properties of the probe were studied and determined that: i) the probe was fluorescent in the pH range 2–4 (max. Value at pH 3; $pK_{a1} = 2.98$ and $pK_{a2} = 2.89$); ii) the presence of Fe^{3+} triggered the opening of the spirolactam ring with the formation of a new and intense fluorescence band at 586 nm (dimethylsulfoxide (DMSO) and DMSO:H_2O (9,1, v/v); and iii) the determined 1:2 (metal: dye) was consistent with the formation of the Fe^{3+} complex with the tridentate probe.

A similar dye with a longer spacer (**4**) was synthesized via one-pot Schiff base reaction of rhodamine B, ethylenediamine, and isonicotinaldehyde and was characterized by SCXRD, where suitable orange-brown crystals were obtained by slow solvent evaporation methods [16]. The Fe^{3+} recognition mechanism, established by density-functional theory (DFT), involved a PET mechanism between the rhodamine core and pyridine and proved to be reversible by UV/Photoluminescence (PL) and time-resolved photoluminescence (TRPL) in the presence of EDTA (ethylenediaminetetraacetic acid tetrasodium salt). A LOD estimated value of 102.3×10^{-9} mol dm^{-3} was reported as well as the probe sensitivity in the pH range from 3 to 10 and cellular imaging studies revealed real applicability of the probe in Fe^{3+} detection. Another work showed its selectivity toward SCN^- in human embryonic kidney cells, including fluorescence and "naked-eye" detection of nanomolar concentration of the analyte [17]. DFT calculations suggested the existence of non-covalent interactions and long-range electrostatic forces between the analyte and the probe, and a comparison using a fluorescein derivative as a model compound allowed to establish a "lock" and "key" mechanism for the analyte sensing. The probe was used successfully in the quantification of SCN^- in real samples such as sheep blood serum and cow milk.

Kan and co-workers reported two probes (**5** and **6**) prepared by a two-step approach: i) reaction of rhodamine B with ethylenediamine followed by ii) reaction with 2-picolinic acid and pyridine-2,6-dicarbonyl dichloride, respectively [18]. Both probes exhibited excellent selectivity and sensitivity for Fe^{3+} in EtOH/H_2O solution (3:1, v/v, HEPES, 0.5×10^{-3} mol dm^{-3}, pH = 7.33) and living human breast adenocarcinoma (MCF-7) cells. Probe **5** presented a 1:1 binding stoichiometry and a lower LOD (0.067×10^{-6} mol dm^{-3}) than probe **6**, which presented a 1:2 binding stoichiometry. Both were successfully applied in the detection of trace amounts of Fe^{3+} up to 200×10^{-6} mol dm^{-3}- in tap water and real mud water with good recovery efficiency, and once again the *turn-on* mechanism was observed. Probe **6**, reported by Li and co-workers [19], operates under two different Fe^{3+} recognition mechanisms based on the solvent used: i) in acetonitrile (CH_3CN), a Fe^{3+} complex is formed causing the quenching of fluorescence, and ii) in phosphate-buffered saline (PBS), hydrolysis occurred leading to the ring opening and a 75-fold increase in fluorescence intensity, with the formation of dipicolinic acid - a result supported by mass spectrometry (MS). The fluorescent imaging of living cell revealed low cytotoxicity, cell viability, and that the probe could penetrate cell membranes.

Some probes are designed to incorporate selected receptor groups, such as sulfur derivatives-thiourea, sulfonyl, or thiol groups. Sarkar and co-workers [20] prepared a rhodamine-linked pyridyl thiourea probe (**7**) with distinct cation and anion binding sites. The probe was capable of selectively detecting different analytes: i) in CH_3CN, fluoride was detected by changes in the emission at 518 nm; ii) Al^{3+} detection occurred at concentrations of approximately 10^{-5} mol dm^{-3} by colorimetric and ratiometric responses; iii) in aqueous CH_3CN mixture, the probe was capable to distinguish between Al^{3+} and Cu^{2+}-possessing higher sensitivity and selectivity toward Al^{3+} at the emission wavelength 558 nm; and iv) the probe could also detect Ag^+ through an increase in the emission intensity at 416 nm, with a LOD of 2.09×10^{-4} mol dm^{-3}. In 2020, probe **8** based on the linkage of rhodamine B and pyridine-3-sulfonyl chloride was reported [21]. This dye resulted from the combination of an electron-donor group (amino group) for fluorescence and sensitivity enhancement and a recognition group with good ion coordination ability (pyridine-3-sulfonyl chloride). **8** presented fast (280 s) and dual response-absorption and fluorescence-upon addition of Al^{3+}, with a LOD of 14.23×10^{-9} mol dm^{-3}. The **8**-Al complex could further be used as a sensor for fluoride by fluorescence intensity decrease. The probe was used successfully in the detection of low Al^{3+} concentrations in natural water, living cells, zebrafish, and plant tissues. Other derivatizations can also be used, for example, Duan and co-workers reported a rhodamine-thiospirolactam probe prepared from the reaction of thiooxorhodamine B hydrazide and 2-pyridinecarboxaldehyde (**9**) [22]. In this case, the *N*-atom of the spirolactam was replaced by an S-atom, while the carbonyl was converted into a hydrazone linked to the pyridine derivative. The probe presented a color change from colorless to pink and a fluorescence intensity enhancement in the presence of Hg^{2+} even at the ppb level. The thioether probe was compared to its thioamide congener and revealed higher selectivity for Hg^{2+}, which was related to its poorer coordination affinity for other interference metal ions.

In 2015, Fu and co-workers prepared three novel rhodamine-triazine aminopyridine derivatives, in which the *N*-atom of the aminopyridine ring was placed in *ortho*-, *meta*-, or *para*- position [23]. These probes' design took into account the aminopyridine water solubility and the excellent reactivity properties of cyanuric chloride as the connecting bridge. The *ortho*-derivative (**10**) presented higher selectivity for Fe^{3+} in water-over other metal ions and amino acids-due to its more suitable space coordination sphere. In the presence of Fe^{3+}, a new absorption band appeared (562 nm) and the emission intensity at 582 nm increased up to 35-fold, along with the change in color from colorless to pink. This probe possessed a LOD for Fe^{3+} of 4.1×10^{-8} mol dm^{-3} and an association constant of 1.49×10^{6} mol^{-1} dm^{3}, being a 1:1 stoichiometric structure supported by Job's plot and MS. Furthermore, the probe revealed to be: i) capable to detect Fe^{3+} in environmental samples using paper-made test kits impregnated with the probe; ii) capable to detect up to 0.3 mol dm^{-3} Fe^{3+} in tap water, and iii) suitable for imaging intracellular Fe^{3+} in HL-7702 cells. Another example was the report from Bhattacharya and co-workers where the three isomers of the pyridine's nitrogen were compared toward Cu^{2+} and Hg^{2+} sensitivity using probe **11a** as the common point [24]. The dye with the pyridine nitrogen at *ortho*-position was the only isomer that presented selective colorimetric detection of Cu^{2+}-in water (pH 7.4), in a medium containing bovine serum albumin and blood serum. The detection mechanism was based on the formation of the Cu^{2+} complex (2:1 stoichiometry) involving the carbonyl oxygen, amido nitrogen, and pyridine nitrogen (see **Figure 2**). The analytes were detected in different water sources at the ppb level, and the probes could be used for rapid *on-site* detection by the preparation of portable test strips. A very similar probe to the *ortho*-derivative **11a**, with a methyl substituent in the *N*-atom attached to the pyridine (**11b**) was prepared through the condensation of **RhoHyd** and 2-acetylpyridine and applied in the selective detection of Cu^{2+}, again by a *turn-on* process due to spirolactam ring opening [25]. The probe was suitable for Cu^{2+} detection within a concentration range from 2.0 to 20.0×10^{-6} mol dm^{-3} and presented a LOD of 0.21×10^{-6} mol dm^{-3} - a value lower than the maximum concentration established by the World Health Organization (WHO).

In 2014, a study based on the influence of different substituents attached to the *N*-atom of the xanthene at positions 3 and 6 was reported [26]. The probes were prepared from the condensation of rhodamine 6G with 2-aminoethylpyridine (**12**), followed by a subsequent nucleophilic substitution (SN_2) reaction with 9-bromomethyl anthracene (**13**) or with 1-bromo-octane (**14**). All the probes revealed chromogenic and fluorogenic *turn-on* spectral responses in the presence of Pb(II) ions and **13** also presented the lowest LOD and reversibility due to the perturbation of the combined PET inhibition and FRET processes associated with its bifluorophoric nature. In the same report, a derivative with two ethyl-substituents at both *N*-atoms attached to the xanthene core is presented as a selective sensor of Hg^{2+} with a dual mode spectral amplification. The authors have concluded that changes in selectivity and signaling pattern are associated with induced amine rigidity in xanthene. Other positions of the rhodamine dye can also be used for structural modifications. For example, probe **15** based on the modification of the 3′-position of the benzolate in the rhodamine with an amino pyridine substituent was prepared [27]. This probe exhibited high selectivity and sensitivity toward Ni^{2+}, possessed a LOD down to 4.6 ppb, and the chelation of the metal ion involved the carboxylate group of the rhodamine moiety and the *N*-atom of the pyridine moiety.

Another strategy for the design of rhodamine-pyridine probes is by conjugation with other dyes or aromatic rings. In 2016, a rhodamine derivative incorporating a 2-[(1H-pyrrol-2-ylmethyl)-(2-pyridinyl-methyl) amino]- tripodal receptor was reported (**16**) and used as a sensor for the detection of accumulated Co(II) in *Hybanthus enneaspermus* plant [28]. The addition of Co(II) to a solution of **16** in THF/H_2O (8:2 v/v, 0.01 mol dm^{-3} HEPES, pH 7.4) promoted the spirolactam ring opening with the formation of a 2:1 complex (probe:Co) with LOD of 4.3×10^{-9} mol dm^{-3}. The complex was reversible in the presence of EDTA and the probe proved to be suitable for *in-situ* detection of Co(II) in a pH range from 5 to 10. Xu and co-workers designed a multidentate dye **17** with rhodamine-triazole-pyridine units for the detection of Sn^{2+} [29]. In the presence of Sn^{2+} the probe, in CH_3CN:H_2O (99:1, v/v), showed changes in color, from colorless to orange, and in the absorption and fluorescence spectra—appearance of a new band at 560 nm and intensity enhancement at 587 nm, respectively. The recognition mechanism was studied by several techniques and confirmed the formation of stable 5-member or 6-member rings between Sn^{2+} and **17** (1:1 complex).

In 2019, our work group designed a series of pyridyl analogs of rosamines (rhodamine derivatives lacking the carboxylic group at position 2′ of the benzenic ring) and studied the influence of solvent and charge on their photophysical properties [30]. It was found that the structural variation involving the position of the *N*-atom in the pyridine did not influence the absorption and fluorescence properties of dyes, the same could not be said about the charge - the introduction of a positive charge at the *N*-atom (**18**) in the pyridinium analog promoted a significant bathochromic shift in the absorption and fluorescence quenching, both effects associated to *d*-PET mechanism. Probe **18** showed extinction of color and fluorescence in the presence of EtOH, the same being true for the uncharged derivative. The detection of EtOH was more pronounced for **18** and resulted from the nucleophilic addition of the ethoxide ion to the central 9-position of the xanthene core, the process was reversible with the addition of a weak acid (trifluoroacetic acid, TFA). Two years later, Xie and co-workers reported a pyridine-Si-rhodamine-based probe (**19**) that could be used as a lysosomal-targeted near-infrared (NIR) fluorescent probe for reactive oxygen species (ROS) [31]. Probe **19** possessed a pyridine-Si-rhodamine moiety as a fluorescent reporter and a borophenylic acid moiety as a reacting group. The probe exhibited good water solubility and, in the presence of hydrogen peroxide (H_2O_2), revealed a significant enhancement in the fluorescence intensity at 680 nm, which could be attributed to the solvent effect and ICT. The response toward other ROS was also evaluated and revealed that the fluorescence enhancement would occur in the order: hypochlorous acid (HClO, 5-fold) < H_2O_2 (14-fold) < hydroxyl radical (•OH, 16-fold). The recognition mechanism, proved by high-resolution MS, indicated the oxidation of phenylboronic acid and was similar with that of phenylboronic acid-based ROS probes. **19** revealed sensitivity to detect ROS in cancer cells and in tumor-bearing mouse xenograft models, being indicative of the probe's applicability to the study of lysosomal cell death.

Many other examples of rhodamine-pyridyl derivatives can be found in literature for selectively sensing several analytes, such as picric acid [32].

2.2 BODIPY dyes

2.2.1 Molecular design

The 4,4-difluoro-4-bora-3a,4a-diaza-s-indacene, also known as boron dipyrrin or boron dipyrromethene (BODIPY), is one of the most popular families of organic fluorophores that have found numerous practical applications as fluorescence probes for bioimaging and sensing, laser dyes, and as bright pigments in various fields of technology, *e.g.* in solar fuel generation, in photovoltaic devices, in light-harvesting arrays for antenna systems, and in photocatalysis, among others. Their main credits are due to the excellent photophysical and spectral properties they possess, including insensitivity to solvent polarity and pH, high photostability with high absorption coefficients, and high fluorescence quantum yield, allowing them to be excited at rather long wavelength (~500 nm) [33, 34]. When compared with rhodamine and fluorescein dyes, the BODIPY fluorophore is smaller and more insensitive to environmental conditions, while Förster radius R_0 has about the same value [35].

From the molecular design point of view, the BODIPY dye (**Figure 3**) can be functionalized at the pyrrolic ring, at the central *meso*-position, and at the boron atom [36]. By introducing substituent groups into the different positions of the BODIPY scaffold, as well as by varying the conjugation length with appropriate spacer or π-linker, the spectroscopic, (photo)physical, and chemical characteristics of the final molecule can be fine-tuned according to the intended application.

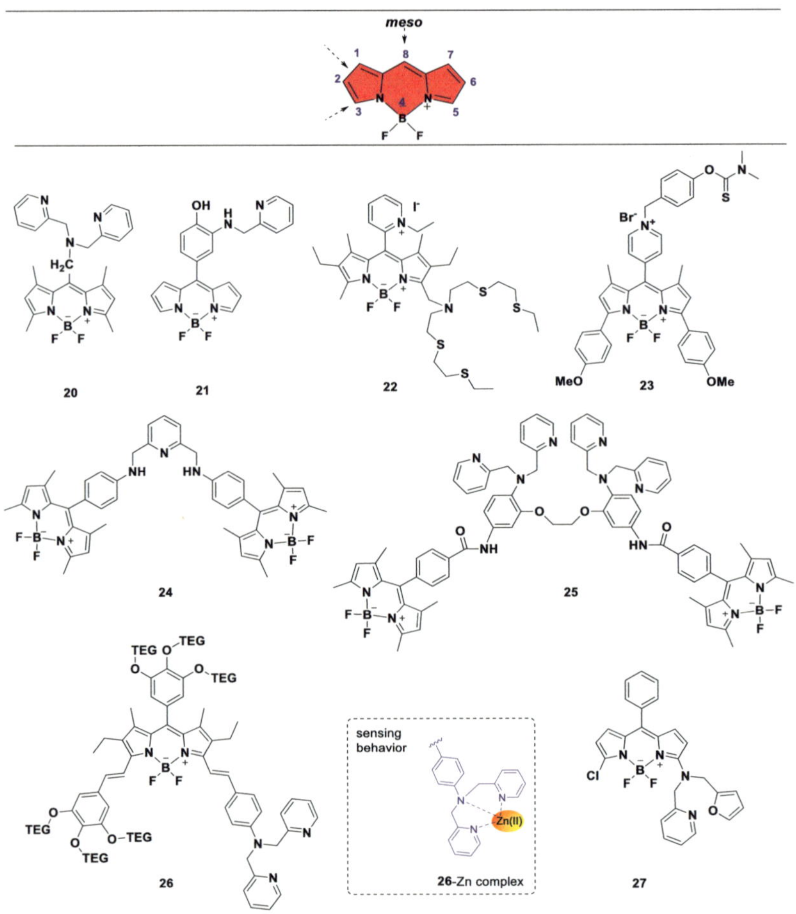

Figure 3.
Representative examples of BODIPY dyes functionalized with pyridyl and polypyridyl groups.

2.2.2 Photophysical properties

The BODIPY typically exhibits a weak absorption band in 350–450 nm region and a strong absorption band in the 450–580 nm, corresponding to $\pi \rightarrow \pi^*$ transitions. It shows a strong fluorescence spectrum in the visible region and the fluorescence quantum yields are typically higher than 0.8. The BODIPY dye shows fluorescence lifetime (s) on nanosecond scale, which is independent of the excitation and emission wavelengths, suggesting simple emission from the locally excited state [35].

2.2.3 Sensing applications

Several BODIPY derivatives having very attractive photophysical properties and photochemical stability have found very useful applications as fluorescent platforms for sensing applications. The introduction of the pyridyl or polypyridyl groups at the periphery of the BODIPY core can lead to a large variety of chemosensors for detecting anions, cations, amino acids, etc. **Figure 3** shows a series of selected examples of these BOPIDY derivatives with different sensing behaviors.

Y. Wu and co-workers [37] reported one of the most notable examples of *meso*-functionalized BODIPY for detection of Zn^{2+} by preparing the probe (**20**) through the combination of 1,3,5,7-tetramethyl-boron dipyrromethene with the DPA receptor. This probe works in an aqueous solution, it exhibits $\lambda_{abs}/\lambda_{em}$ = 491/509 nm with Φ_F = 0.077 and has the advantage of being very selective for Zn^{2+}, with the fluorescence emission of zinc-binding being pH independent in the range of pH 3–10.

Another example of a *meso*-functionalized BODIPY comes from the X. You's group [38], through the preparation of the BODIPY derivative (**21**) containing a tridentate 2-*N*-(2-pyridylmethyl)amino-phenol ligand. The probe, which is almost nonfluorescent because of the PET quenching process from the *meso*-electron-donating substituent to the excited BODIPY unit, upon addition of Hg^{2+}, the fluorescence intensity increased remarkably, showing a very high sensitivity (detection limit ≤2 ppb), a rapid response time (≤5 seconds), and high selectivity for Hg^{2+} over other metal cations.

Developed by G. T. Sfrazzetto and co-workers [39], probe **22** contains a tetratia-aza-crown receptor and an alkyl-pyridinium moiety to get water solubility and selectivity for target mitochondria. The probe was found to be highly selective to detect Cu^+ in solution and in living cells through an emission quenching response, which is attributed to the PET process between the BODIPY core and the Cu^+ chelated tetrathia-aza crown receptor.

Through a benzyl pyridinium cleavable unit at *meso* position of BODIPY, probe **23** was developed for detection of HOCl. Upon addition of HOCl, it exhibits a fast-responsive rate and a dramatic red fluorescence increase (λ_{em} = 614 nm, 170-fold) with high selectivity and sensitivity (LOD = 60 × 10^{-9} mol dm^{-3}) [40].

The dyad **24** featuring two BODIPY fluorophores linked by a *N,N′*-(pyridine-2,6-diylbis(methylene))-dianiline substituent showed a highly selective fluorescent *turn-on* response in the presence of Hg^{2+} [41]. Through theoretical calculations, it was possible to predict the photophysical properties of the **24**-Hg^{2+} complex, both the reductive and oxidative PETs are prohibited, thus justifying its strong fluorescence emission observed experimentally.

In a similar approach, dyad **25** consisting of a 2,2′-(ethane-1,2-diylbis(oxy)) bis(*N,N*-bis(pyridine-2-ylmethyl)-aniline receptor, which was covalently connected through aromatic amides with two BODIPY fluorophores, was found to selectively detect both Hg^{2+} and Cd^{2+} ions [42]. In this case, the receptor has been designed to effectively wrap around a metal ion and, at the same time, make the dye water-soluble for its operation in aqueous environment. This probe exhibited LOD values of 38×10^{-9} mol dm^{-3} for aqueous Hg^{2+} and a 77×10^{-9} mol dm^{-3} for aqueous Cd^{2+}.

The functionalization of the BODIPY dye at the δ-position has been also highly explored and one of the most representative examples is the distyryl-substituted BODIPY dye **26** developed by E. U. Akkaya's group [43]. This compound contains the DPA receptor combined with six triethylene glycol (TEG) groups to provide water solubility. It presents $\lambda_{abs}/\lambda_{em}$ = 680/726 nm and the gradual addition of Zn^{2+} ions to this compound results in a blueshift to 625 nm with a concomitant increase in emission intensity, in aqueous solutions, resulting from the coordination of Zn(II) ions to the DPA receptor (see **Figure 3**). Other similar BODIPY probes for Zn^{2+} include: (i) the BODIPY functionalized with a *N,N*-di-(2-picolyl)ethylenediamine (DPEN) receptor [44], which can detect Zn^{2+} cation through fluorescence enhancement and also detect pyrophosphate anion through a fluorescence quenching and (ii) the BODIPY featuring a DPEN and a methyl acetate group for monitoring and quantifying levels of Zn^{2+} in living cells and detecting intracellular Zn^{2+} released from intracellular metalloproteins [45].

Probe **27** is another interesting example of a BODIPY functionalized at δ-position with a 1-(furan-2-yl)-*N*-((pyridin-2-yl)methyl)methanamine group [46]. The probe is almost nonfluorescent, but upon addition of Cu^{2+}, a large bathochromic shift in the absorption and fluorescence spectra and induced fluorescence amplification at ~600 nm was observed, showing great potential for imaging and sensing of Cu^{2+} in living cells. On the other hand, by modifying BODIPY with a 4-aminostyryl group [47], a probe for Cu^{2+} with a large Stokes shift, high photostability, and high quantum yield was obtained for monitoring *in vivo* Cu^{2+} imaging in live mice.

2.3 1,8-Naphthalimide dyes

2.3.1 Molecular design

1,8-Naphthalimide (**NI**) core is considered as one of the most versatile fluorophore units due to its synthetic versatility and unique photophysical properties. The aromatic **NI** core, an electron acceptor, along with the *N*-imide site can be easily modified (**Figure 4**), which allows the introduction of an enormous variety of structural units and functional groups in the main core. Regarding their photophysical properties, naphthalimide structures are strongly influenced by the nature of the substituent. The functionalization at C-4 with donor moieties, such as amine or hydroxyl groups, induces a red-shifted ICT band with marked solvatochromic effect. These characteristics encourage the use of **NI** as probes as the changes in spectroscopic properties, such as absorption, dichroism, and fluorescence can all be used to monitor their binding to different analytes. The **NI** and its derivatives have immense potential in the development of new fluorescent probes, laser dyes, optoelectronic materials, and bioimaging but also present high antitumor and antiviral activities [48].

2.3.2 Photophysical properties

The spectroscopic properties of 1,8-naphthalimides are strongly dependent on the C-4 substituent group. To increase the fluorescent quantum yield, the substituent

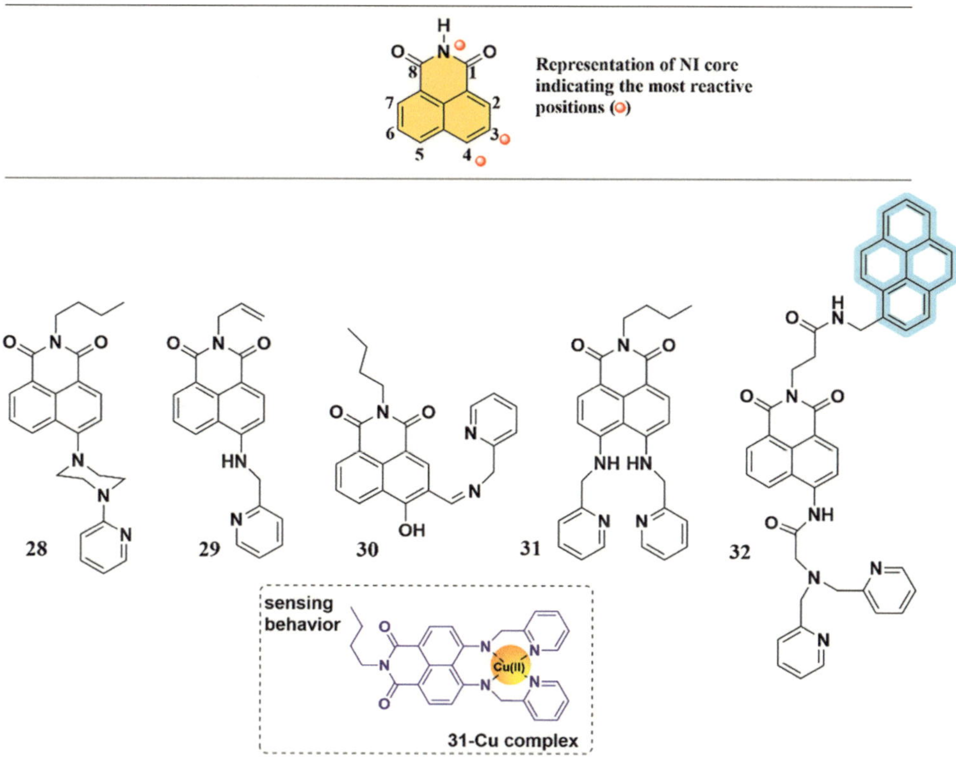

Figure 4.
Representative examples of 1,8-naphthalimide dyes functionalized with pyridyl groups.

group at the 4-position should be an electron-donating group. Other features that contribute for 1,8-naphthalimides extensive use are related with their extraordinary thermal and chemical stability.

2.3.3 Sensing applications

Several **NI** derivatives have been used as fluorescent platforms in distinct sensing applications. The introduction of one or more pyridyl units in the periphery of the **NI** core led to a large variety of chemosensors for different analytes. In **Figure 4**, a series of **NI** derivatives having different structural characteristics and sensing behaviors are shown.

The first example, probe **28**, uses a 1,8-naphthalimide unit as a receptor, and 1-(2-pyridyl)piperazine as a receptor to design a *turn-on* fluorescent probe for Fe^{3+} [46]. In this example, the sensor was achieved by mild reaction and simple post process and found to have excellent selectivity and sensitivity to Fe^{3+}. The chelation with Fe^{3+} over other cations caused a 15.8-fold fluorescence enhancement, which could be explained by the fact that the *N*-atoms in pyridine and piperazine moieties provided the binding sites for Fe^{3+} and enhancing the fluorescence by blocking the PET process. The maximal fluorescence intensity was linearly proportional to the Fe^{3+} concentration (60–140 × 10^{-6} mol dm^{-3}), a LOD of 81 × 10^{-9} mol dm^{-3} and the probe worked in a pH range of 5.0–8.0. A 1:1 complex was formed reversibly between the probe and Fe^{3+}. Moreover, tests were performed with other metal cations and it was verified a negligible influence on the fluorescence spectrum of probe **28**/Fe^{3+}.

This result indicated that probe **28** had good anti-interference ability and was a reliable high sensitivity fluorescent probe for Fe^{3+}.

In the next example, a fluorescent ion-imprinted probe (**FIIS**) for rapid and convenient detection of Cu^{2+} ions was fabricated. Probe **29** represents a fluorescent polymerizable ligand, 4-(2-aminomethyl)pyridine-*N*-allylnaphthalimide [49]. The design of this probe took into consideration to increase the fluorescence quantum yield of 1,8-naphthalimides and at the same time introduced a chelating unit, the substituent group at the 4-position should be an electron-donating group. Taking these considerations into account, 2-aminomethyl pyridine was chosen as the C-4 substituent group in the synthesis of this fluorescent functional monomer (F). The **FIIS** was prepared by surface functionalization of PVDF membrane with a thin layer of Cu^{2+} ion-imprinted polymer using the synthesized ligand as the fluorescent functional monomer. The intensity of fluorescence emission of **FIIS** decreased linearly with the increase of Cu^{2+} ions concentration in the range of 0-70.0 × 10^{-6} mol dm^{-3}. The results of selectivity tests indicated that FIIS had a high specific recognition ability for Cu^{2+} and its application in the determination of Cu^{2+} in real water samples revealed a LOD for Cu^{2+} ions in the range of 0.11–0.14 × 10^{-6} mol dm^{-3}.

The third example, probe **30**, was published by Wu and co-workers and presented the successful design and synthesis of a simple fluorescent and colorimetric probe [50]. The design involved the functionalization in the *N*-imide site but also in positions 3 and 4 of the **NI** core. This probe exhibited an excellent selective fluorescence response for the simultaneous detection of Zn^{2+} and Al^{3+} with a single excitation wavelength in the same solvent system. The LOD of probe **30** for Zn^{2+} and Al^{3+} were 14.4 × 10^{-6} and 74.0 × 10^{-6} mol dm^{-3}, respectively. In addition, the solution of probe **30** with Zn^{2+} exhibited a dramatic color change from bright green to bright blue, light, and dark blue with Al^{3+}, which could be easily detected by *naked eye* under UV.

The fourth example represents a ratiometric and selective fluorescent probe (**31**) for Cu^{2+}. This probe was easily synthesized by conjugating 2-(aminomethyl)pyridine and *N*-butyl-4-bromo-5-nitro-1,8-naphthalimide [51]. The design and synthesis took into consideration was the mechanism of ICT since this mechanism had been widely exploited for cation sensing. Another aspect that was taken into consideration was the use of a tetradentate receptor site with nitrogen and pyridyl donors since there were strong pieces of evidence that these receptors were very useful for binding Cu^{2+} ions [52]. The capture of Cu^{2+} by the receptor resulted in the reduction of the electron-donating ability of the two amino groups of the naphthalene ring; thus, the receptor showed a 50 nm blue shift of fluorescence emission and provided high selectivity for Cu^{2+} over other heavy and first transition metal ions. The fluorescence of the probe at 525 nm remains unaffected between pH 4.7–13. This probe presents high sensitivity and selectivity toward Cu^{2+} ions, allows the detection of Cu^{2+} ratiometrically, and forms a 1:1 complex (see **Figure 4**) [51].

In the last example, Lee and co-workers designed a pyrene-appended naphthalimide, probe **32**, as a ratiometric fluorescence probe that can detect Zn^{2+} ion in physiological conditions [53]. In this approach, the pyrene unit acts as a reference fluorophore emitting an unaffected fluorescence intensity for Zn^{2+} and the naphthalimide-dipicolylamine moiety acts as a Zn^{2+} sensing unit providing a fluorescence change based on a PET mechanism. This probe displayed a ratiometric change in the fluorescent intensities at 385 and 530 nm, which corresponds to the emissions of pyrene and naphthalimide units, for Zn^{2+} allowing for a precise quantitative analysis. This ratiometric change could be also visualized by a fluorescent color change from blue to green. The probe presented a rapid detection of Zn^{2+} ions in a 1:1 ratio with high sensitivity, even in the presence of other competitive metal ions, and with a LOD of 10.5 × 10^{-9} mol dm^{-3}. Moreover, this probe was able to detect Zn^{2+} ions in the pH range of 4–11 and it could be efficiently recycled by treating it with EDTA.

2.4 Coumarin dyes

2.4.1 Molecular design

Coumarins are a large family of compounds containing the 2*H*-chromen-2-one motif. This platform has been widely used in the design of fluorescent chemosensors because of its small size, excellent biocompatibility, strong and stable fluorescence emission, and good structural flexibility. Hence, this scaffold is an important unit in the development of fluorescent chemosensors with different applications in different fields, such as molecular recognition, molecular imaging, bioorganic, analytical, and materials chemistry, as well as in biology and medical science. Most coumarins were synthesized or designed *de novo* rather than *via* post-functionalization of the coumarin skeleton. The synthetic transformation of coumarins into other heterocyclic compounds and larger fused heterocycles with a coumarin moiety has also been developed [54]. In addition, the benzene subunit of the coumarin ring system is not as reactive as the unsubstituted benzene ring, while the 3 and 4 positions are highly reactive.

2.4.2 Photophysical properties

Although the coumarin unit exhibits a very weak fluorescence, the introduction of proper substituents originates new coumarin derivatives with significant fluorescence in the visible light range. Hundreds of coumarin dyes have been developed as active components due to their improved quantum yields, tunable emission wavelengths, and the fact that they are very responsive to the polarity of their microenvironments. The previously published results on the photophysical properties of fluorescent coumarins have revealed important structure-property relationships, which have also been important to guide the design of fluorescent chemosensors.

2.4.3 Sensing applications

A vast variety of coumarin-derived fluorescent chemosensors were built by combining the coumarin moiety with other functional receptors. Herein we present a series of coumarin derivatives in which the receptor is a pyridyl moiety (**Figure 5**).

L. Wang and co-workers reported two ratiometric probes, **33a** and **33b**, to be employed in the quantitative determination of pH value in acidic pH zone. The development of such ratiometric probes, employing the ratio of two emissions at different wavelengths as the detecting signal, allows for more accurate analysis [55]. The reported probes were strategically designed with a 7-diethylamino-coumarin moiety as the fluorophore and pyridine as the receptor. Both probes exhibited a fluorescence ratiometric response to acidic pH. For probe **33a**, upon decreasing the pH from 8.35 to 2.36, the fluorescence emission spectra exhibited a large red shift from 529 to 616 nm, and the emission ratio changed dramatically from 8.58 to 0.09. The emission ratio also displayed good linearity with the pH in the range of 4.0 to 6.5, which is valuable for the quantitative determination of pH values in this acidic pH window. Similar behavior was observed for probe **33b**. By performing some NMR experiments and

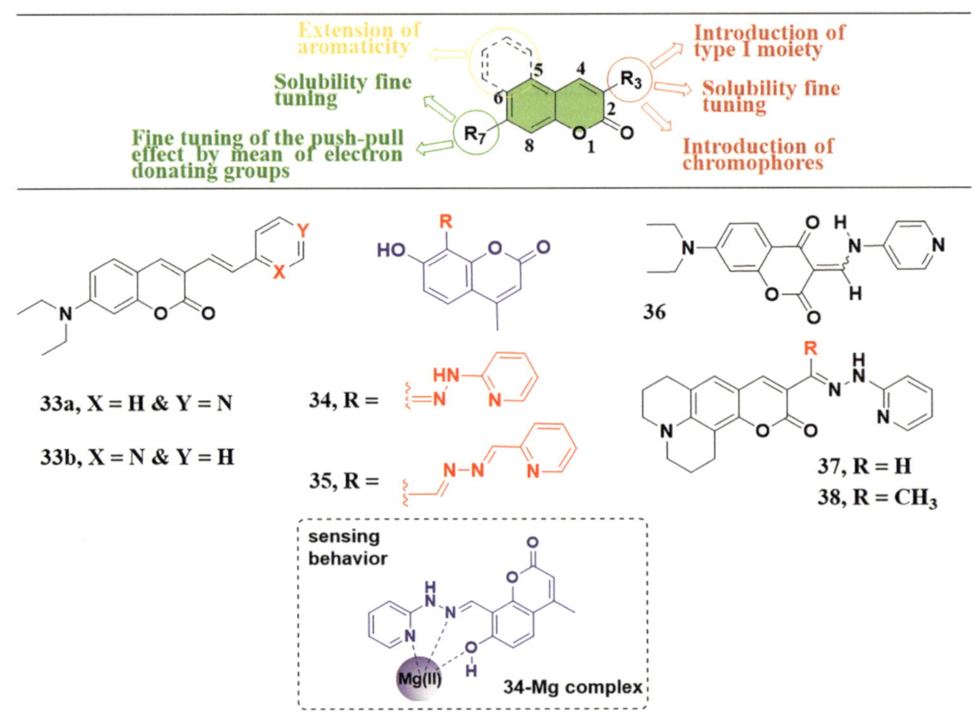

Figure 5.
Representative examples of coumarin dyes functionalized with pyridyl groups.

theoretical calculations they conclude that the ratiometric response of the probes to acidic pH was due to H^+ binding with the nitrogen of the pyridine receptor and the induced enhancement of the ICT process.

J. Portilla and co-workers reported the coumarin probe (**34**) bearing a 7-hydroxy −4-methylcoumarin unit for selective detection of Mg^{2+} [56]. The design also includes the 2-pyridylhydrazone substituent as a chelating unit as well as a phenolic hydroxyl group in the fluorophore unit. In addition, the 2-pyridylhydrazone substituent has the C=N donor system that can quench the fluorescence of the fluorophore by PET process and C=N isomerization. The coordination of probe **34** to Mg^{2+} probably disrupts these processes and increases its structural rigidity producing a fluorescence enhancement. The binding mode of the complex probe **34**- Mg^{2+} was studied by several spectroscopic methods and revealed the formation of a 1:1 complex (see **Figure 5**). The probe showed good binding ability toward Mg^{2+}, low interference from $Ca^{2+,}$ and a LOD of 105×10^{-9} mol dm^{-3} in ethanol-water solution.

Another example of a simple coumarin-pyridyl probe was presented by K. Xu and co-workers. The study presents two probes, but we will focus on probe **35** [57]. This probe contains C=N bond to enhance the ability of binding metal ions and contribute to extending the system conjugation. As a result, free probe displays a weak fluorescence due to C=N isomerization, but when a metal ion binds to the chelating unit, the isomerization process is disrupted and there is a fluorescence enhancement. The probe was synthesized for the sequential detection of Zn^{2+} ion and phosphate anion (PA) in DMF (dimethylformamide)/HEPES buffer medium. The binding of Zn^{2+} resulted in a pronounced fluorescence enhancement, accompanied by a noticeable color change in the *naked eye*. The detection limits of probe **35** toward Zn^{2+} was 1.03×10^{-7} mol dm^{-3}. Probe **35**-Zn^{2+} complex was then used as a probe for detecting phosphate anion, showing an *off–on–off* fluorescence switching response with Zn^{2+} and phosphate anion.

Probe **36** was published by K. J. Wallace, with the intention of synthesizing a planar molecule with a high degree of conjugation, which could be easily perturbed to produce a spectroscopic response, taking advantage of intramolecular hydrogen bonding [58]. The design also took into consideration that electron-withdrawing functional groups attached to a carbonyl moiety will pull electron density away from the carbon atom, consequently making this region more electrophilic and susceptible to rapid nucleophilic attack. This probe can undergo the Michael addition of cyanide at the α,β-unsaturated carbonyl, and demonstrated its selectivity for CN^- over 12 common anions with LOD of approximately 4 ppb [59].

The next examples were designed by F. Yu and co-workers and represent two-photon fluorescence probes, probes **37** and **38**, possessing coumarin derivatives, for selective and sensitive detection of Zn^{2+} [60]. Both probes exhibited excellent analytical properties for Zn^{2+} detection including rapid response, high sensitivity, and good selectivity. In each probe, the coumarin moiety acts as a fluorophore and 2-hydrazino-pyridine unit as a metal ion coordination site. Upon addition of Zn^{2+}, solutions of the weakly emissive probes **37** or **38** become strongly fluorescent with emission at 543 nm (probe **37** *ca.* seven times and probe **38** *ca.* four times) in HEPES buffer. In addition, the two-photon properties of these coumarin derivatives make them applicable to detect Zn^{2+} in biological systems.

2.5 Other pyridyl-based fluorophores

In quinoline dyes, for example, the pyridyl group is part of the fluorophore, as well as an integral part of the metal-binding group. This fluorophore can be conveniently functionalized with several substituent groups for sensing essentially Zn^{2+} including 6-methoxy-(8-*p*-toluenesulfonamido) group [61], DPA group [62], among others. Other pyridyl-based fluorophores include: (i) the triazole-pyridine system featuring two pyridine receptors, which behave as an interesting ICT chemosensor for cations and anions [63]; (ii) the 7-nitrobenzo-2-oxo-1,3-diazole dye comprising two pyridines for Zn(II) detection [64] and (iii) the carbazole derivative integrating pyridine units exhibiting fluorescence switching by acid/base exposing [65].

3. Conclusions

The optical properties of dyes as well as their sensitivity and selectivity toward analytes are highly dependent not only on the fluorophore backbone but also on its substituents and the solvent in which the detection occurs.

Throughout the chapter, several classes of fluorescent dyes-rhodamines, BODIPY's, 1,8-naphthalimides, and coumarins-functionalized with reactive pyridyl receptors were examined. The presented examples explored the strategies used for structural optimization to improve sensing abilities using the principal fluorescence sensing mechanisms. In coming years, new developments are expected toward better sensitivity and selectivity of the probes, to improve their application in the detection and quantification of important analytes in the fields of health and environment.

Acknowledgements

This work received financial support from PT national funds (FCT/MCTES, Fundação para a Ciência e a Tecnologia, and Ministério da Ciência, Tecnologia e Ensino Superior) through the project UIDB/50006/2020, UIDP/50006/2020, PTDC/QUI-QIN/28142/2017, EXPL/QUI-OUT/1554/2021 and PARSUK for the Portugal-UK Bilateral Research Fund (BRF 2022). A. M. G. Silva and A. Leite thank FCT for funding through program DL 57/2016 - Norma transitória.

Conflict of interest

The authors declare no conflict of interest.

Author details

Andreia Leite, Carla Queirós and Ana M.G. Silva*
LAQV/REQUIMTE, Chemistry and Biochemistry Department, Faculty of Sciences, University of Porto, Porto, Portugal

*Address all correspondence to: ana.silva@fc.up.pt

References

[1] Callan JF, de Silva AP, Magri DC. Luminescent sensors and switches in the Early 21st century. Tetrahedron. 2005;**61**:8551-8588. DOI: 10.1016/j.tet.2005.05.043

[2] Eicher PDT, Hauptmann PDS, P. D. A. Speicher six-membered heterocycles: Sections 6.10-6.14. The Chemistry of Heterocycles. 2003:257-310

[3] de Silva AP, Rupasinghe RADD. A new class of fluorescent pH indicators based on photo-induced electron transfer. Journal of the Chemical Society, Chemical Communications. 1985;(23):1669-1670. DOI: 10.1039/C39850001669

[4] Pal A, Karmakar M, Bhatta SR, Thakur A. A detailed insight into anion sensing based on intramolecular charge transfer (ICT) mechanism: A comprehensive review of the years 2016 to 2021. Coordination Chemistry Reviews. 2021;**448**:214167. DOI: 10.1016/j.ccr.2021.214167

[5] Beija M, Afonso CAM, Martinho JMG. Synthesis and applications of rhodamine derivatives as fluorescent probes. Chemical Society Reviews. 2009;**38**:2410-2433. DOI: 10.1039/B901612K

[6] Deng F, Xu Z. Heteroatom-substituted rhodamine dyes: Structure and spectroscopic properties. Chinese Chemical Letters. 2019;**30**:1667-1681. DOI: 10.1016/j.cclet.2018.12.012

[7] Zhang Q, Wong KMC. Photophysical, ion-sensing and biological properties of rhodamine-containing transition metal complexes. Coordination Chemistry Reviews. 2020;**416**:213336. DOI: 10.1016/j.ccr.2020.213336

[8] Wang C, Wong KM-C. Selective Hg^{2+} sensing behaviors of rhodamine derivatives with extended conjugation based on two successive ring-opening processes. Inorganic Chemistry. 2013;**52**:13432-13441. DOI: 10.1021/ic401810x

[9] Jeong Y, Yoon J. Recent progress on fluorescent chemosensors for metal ions. Inorganica Chimica Acta. 2012;**381**:2-14

[10] Hu J, Hu Z, Cui Y, Zhang X, Gao HW, Uvdal K. A rhodamine-based fluorescent probe for Hg^{2+} and its application for biological visualization. Sensors and Actuators, B: Chemical. 2014;**203**:452-458. DOI: 10.1016/j.snb.2014.06.104

[11] Ozdemir M, Zhang Y, Guo M. A highly selective "off-on" fluorescent sensor for subcellular visualization of labile iron(III) in living cells. Inorganic Chemistry Communications. 2018;**90**:73-77. DOI: 10.1016/j.inoche.2018.02.015

[12] Chereddy NR, Suman K, Korrapati PS, Thennarasu S, Mandal AB. Design and synthesis of rhodamine based chemosensors for the detection of Fe^{3+} ions. Dyes and Pigments. 2012;**95**:606-613. DOI: 10.1016/j.dyepig.2012.05.025

[13] Sikdar A, Roy S, Haldar K, Sarkar S, Panja SS. Rhodamine-based Cu^{2+}-selective fluorosensor: Synthesis, mechanism, and application in living cells. Journal of Fluorescence. 2013;**23**:495-501. DOI: 10.1007/s10895-013-1169-y

[14] Maniyazagan M, Mariadasse R, Jeyakanthan J, Lokanath NK, Naveen S, Premkumar K, et al. Rhodamine based "Turn–on" molecular switch FRET–sensor for cadmium and sulfide ions and live cell imaging study. Sensors and

Actuators, B: Chemical. 2017;**238**:565-577. DOI: 10.1016/j.snb.2016.07.102

[15] Leite A, Silva AMG, Cunha-Silva L, de Castro B, Gameiro P, Rangel M. Discrimination of fluorescence light-up effects induced by pH and metal ion chelation on a spirocyclic derivative of rhodamine B. Dalton Transactions. 2013;**42**:6110-6118. DOI: 10.1039/c2dt32198j

[16] Shellaiah M, Thirumalaivasan N, Aazaad B, Awasthi K, Sun KW, Wu SP, et al. Novel rhodamine probe for colorimetric and fluorescent detection of Fe^{3+} ions in aqueous media with cellular imaging. Spectrochimica Acta - Part A: Molecular and Biomolecular Spectroscopy. 2020;**242**. DOI: 10.1016/j.saa.2020.118757

[17] Mandal S, Sahana A, Banerjee A, Safin DA, Babashkina MG, Robeyns K, et al. A smart rhodamine-pyridine conjugate for bioimaging of thiocyanate in living cells. RSC Advances. 2015;**5**:103350-103357. DOI: 10.1039/c5ra21838a

[18] Song F, Yang C, Liu H, Gao Z, Zhu J, Bao X, et al. Dual-binding pyridine and rhodamine B conjugate derivatives as fluorescent chemosensors for ferric ions in aqueous media and living cells. Analyst. 2019;**144**:3094-3102. DOI: 10.1039/c8an01915k

[19] Wang X, Li T. A novel "off-on" rhodamine-based colorimetric and fluorescent chemosensor based on hydrolysis driven by aqueous medium for the detection of Fe^{3+}. Spectrochimica Acta - Part A: Molecular and Biomolecular Spectroscopy. 2020;**229**. DOI: 10.1016/j.saa.2019.117951

[20] Ghosh K, Panja S, Sarkar T. Rhodamine-linked pyridyl thiourea as a receptor for selective recognition of F^-, Al^{3+} and Ag^+ under different conditions. Supramolecular Chemistry. 2015;**27**:490-500. DOI: 10.1080/10610278.2014.998673

[21] Kan C, Shao X, Wu L, Zhang Y, Bao X, Zhu J. A novel "OFF–ON–OFF" fluorescence chemosensor for hypersensitive detection and bioimaging of Al(III) in living organisms and natural water environment. Journal of Photochemistry and Photobiology A: Chemistry. 2020;**398**. DOI: 10.1016/j.jphotochem.2020.112618

[22] Huang W, Song C, He C, Lv G, Hu X, Zhu X, et al. Recognition preference of rhodamine-thiospirolactams for mercury(II) in aqueous solution. Inorganic Chemistry. 2009;**48**:5061-5072. DOI: 10.1021/ic8015657

[23] Yan F, Zheng T, Shi D, Zou Y, Wang Y, Fu M, et al. Rhodamine-aminopyridine based fluorescent sensors for Fe^{3+} in water: Synthesis, quantum chemical interpretation and living cell application. Sensors and Actuators, B: Chemical. 2015;**215**:598-606. DOI: 10.1016/j.snb.2015.03.096

[24] Kumari N, Dey N, Bhattacharya S. Remarkable role of positional isomers in the design of sensors for the ratiometric detection of copper and mercury ions in water. RSC Advances. 2014;**4**:4230-4238. DOI: 10.1039/c3ra45054f

[25] Liu K, Xu S, Guo P, Liu L, Shi X, Zhu B. A novel fluoro-chromogenic Cu^{2+} probe for living-cell imaging based on rhodamine 6G-pyridine conjugation. Analytical and Bioanalytical Chemistry. 2019;**411**:3021-3028. DOI: 10.1007/s00216-019-01748-8

[26] Biswal B, Bag B. Switching selectivity between Pb^{2+} and Hg^{2+} ions through variation of substituents at xanthene end; "turn-on" signalling responses by FRET modulation. RSC Advances.

2014;**4**:33062-33073. DOI: 10.1039/c4ra04152f

[27] Zhang I, Wang Y, Wan C, Xing Z, Li W, Li M, et al. A new rhodamine based chemodosimeter for Ni^{2+} with high sensitivity and selectivity. RSC Advances. 2015;**5**:66416-66419. DOI: 10.1039/c5ra11737b

[28] Biswal B, Mallick D, Thirunavoukkarasu M, Mohanty R, Bag B. A pyridine and pyrrole coupled rhodamine derivative for Co(II) ion detection and its imaging application in plant tissues. Sensors and Actuators, B: Chemical. 2016;**232**:410-419. DOI: 10.1016/j.snb.2016.03.160

[29] Yan Z, Wei G, Guang S, Xu M, Ren X, Wu R, et al. A multidentate ligand chromophore with rhodamine-triazole-pyridine units and its acting mechanism for dual-mode visual sensing trace Sn^{2+}. Dyes and Pigments. 2018;**159**:542-550. DOI: 10.1016/j.dyepig.2018.07.028

[30] Leite A, Cunha-Silva L, Silva D, Lobo Ferreira AIMC, Santos LMNBF, Cardoso ICS, et al. Synthesis of pyridyl and N-methylpyridinium analogues of rosamines: Relevance of solvent and charge on their photophysical properties. Chemistry - A European Journal. 2019;**25**:15073-15082. DOI: 10.1002/chem.201903313

[31] Zhou Z, Yuan X, Long D, Liu M, Li K, Xie Y. A pyridine-Si-rhodamine-based near-infrared fluorescent probe for visualizing reactive oxygen species in living cells. Spectrochimica Acta - Part A: Molecular and Biomolecular Spectroscopy. 2021;**246**:118927. DOI: 10.1016/j.saa.2020.118927

[32] Sakthivel P, Sekar K, Singaravadivel S, Sivaraman G. Rhodamine-isonicotinic hydrazide analogue: A selective fluorescent chemosensor for the nanomolar detection of picric acid in aqueous media. Chemistry Select. 2019;**4**:3817-3822. DOI: 10.1002/slct.201804032

[33] Kaur P, Singh K. Recent advances in the application of BODIPY in bioimaging and chemosensing. Journal of Materials Chemistry C. 2019;**7**:11361-11405. DOI: 10.1039/C9TC03719E

[34] Poddar M, Misra R. Recent advances of BODIPY based derivatives for optoelectronic applications. Coordination Chemistry Reviews. 2020;**421**:213462. DOI: 10.1016/j.ccr.2020.213462

[35] Karolin J, Johansson LB-A, Strandberg L, Ny T. Fluorescence and absorption spectroscopic properties of dipyrrometheneboron difluoride (BODIPY) derivatives in liquids, lipid membranes, and proteins. Journal of the American Chemical Society. 1994;**116**:7801-7806. DOI: 10.1021/ja00096a042

[36] Boens N, Verbelen B, Dehaen W. Postfunctionalization of the BODIPY core: Synthesis and spectroscopy. European Journal of Organic Chemistry. 2015;**2015**:6577-6595. DOI: 10.1002/ejoc.201500682

[37] Wu Y, Peng X, Guo B, Fan J, Zhang Z, Wang J, et al. Boron dipyrromethene fluorophore based fluorescence sensor for the selective imaging of Zn(II) in living cells. Organic & Biomolecular Chemistry. **2005**;**3**:1387. DOI: 10.1039/b501795e

[38] Lu H, Xiong L, Liu H, Yu M, Shen Z, Li F, et al. A highly selective and sensitive fluorescent turn-on sensor for Hg^{2+} and its application in live cell imaging. Organic & Biomolecular Chemistry. 2009;**7**:2554. DOI: 10.1039/b902912e

[39] Giuffrida ML, Rizzarelli E, Tomaselli GA, Satriano C, Trusso Sfrazzetto G. A novel fully water-soluble Cu^{2+} probe for fluorescence live cell imaging. Chemical Communications. 2014;**50**:9835. DOI: 10.1039/C4CC02147A

[40] Wang X, Tao Y, Zhang J, Chen M, Wang N, Ji X, et al. Selective detection and visualization of exogenous/endogenous hypochlorous acid in living cells using a BODIPY-based red-emitting fluorescent probe. Chemistry – An Asian Journal. 2020;**15**:770-774. DOI: 10.1002/asia.201901709

[41] Lu H, Zhang S, Liu H, Wang Y, Shen Z, Liu C, et al. Experimentation and theoretic calculation of a BODIPY sensor based on photoinduced electron transfer for ions detection. The Journal of Physical Chemistry A. 2009;**113**:14081-14086. DOI: 10.1021/jp907331q

[42] Maity SB, Banerjee S, Sunwoo K, Kim JS, Bharadwaj PK. A fluorescent chemosensor for Hg^{2+} and Cd^{2+} ions in aqueous medium under physiological pH and its applications in imaging living cells. Inorganic Chemistry. 2015;**54**:3929-3936. DOI: 10.1021/acs.inorgchem.5b00106

[43] Atilgan S, Ozdemir T, Akkaya EU. A sensitive and selective ratiometric near IR fluorescent probe for zinc ions based on the distyryl–bodipy fluorophore. Organic Letters. 2008;**10**:4065-4067. DOI: 10.1021/ol801554t

[44] Lin J-R, Chu C-J, Venkatesan P, Wu S-P. Zinc(II) and pyrophosphate selective fluorescence probe and its application to living cell imaging. Sensors and Actuators B: Chemical. 2015;**207**:563-570. DOI: 10.1016/j.snb.2014.10.109

[45] Xia S, Shen J, Wang J, Wang H, Fang M, Zhou H, et al. Ratiometric fluorescent and colorimetric BODIPY-based sensor for zinc ions in solution and living cells. Sensors and Actuators B: Chemical. 2018;**258**:1279-1286. DOI: 10.1016/j.snb.2017.11.129

[46] Yang C, Gong D, Wang X, Iqbal A, Deng M, Guo Y, et al. A new highly copper-selective fluorescence enhancement chemosensor based on BODIPY excitable with visible light and its imaging in living cells. Sensors and Actuators B: Chemical. 2016;**224**:110-117. DOI: 10.1016/j.snb.2015.10.037

[47] Xue X, Fang H, Chen H, Zhang C, Zhu C, Bai Y, et al. In vivo fluorescence imaging for Cu^{2+} in live mice by a new NIR fluorescent sensor. Dyes and Pigments. 2016;**130**:116-121. DOI: 10.1016/j.dyepig.2016.03.017

[48] Gopikrishna P, Meher N, Iyer PK. Functional 1,8-Naphthalimide AIE/AIEEgens: Recent advances and prospects. ACS Applied Materials and Interfaces. 2018;**10**:12081-12111. DOI: 10.1021/acsami.7b14473

[49] Xu Z, Deng P, Li J, Tang S. Fluorescent ion-imprinted sensor for selective and sensitive detection of copper (II) ions. Sensors and Actuators, B: Chemical. 2018;**255**:2095-2104. DOI: 10.1016/j.snb.2017.09.007

[50] Chiou YR, Chen JH, Hu CH, Wu AT. A new fluorescence turn-on sensor for the distinct detection of Zn^{2+} and Al^{3+}: Experimental and density functional theory study. Inorganica Chimica Acta. 2020;**502**. DOI: 10.1016/j.ica.2019.119295

[51] Xu Z, Xiao Y, Qian X, Cui J, Cui D. Ratiometric and selective fluorescent sensor for Cu(II) based on internal charge transfer (ICT). Organic Letters. 2005;7:889-892. DOI: 10.1021/ol0473445

[52] Benshafrut R, Haran A, Shvarts D, Schneider B. Synthesis and characterization of a new tetradentate ligand for Cu(II) metal ions. Journal of Organic Chemistry. 2002;**67**:4040-4044. DOI: 10.1021/jo010661u

[53] Yoon SA, Lee J, Lee MH. A ratiometric fluorescent probe for Zn^{2+} based on pyrene-appended naphthalimide-dipicolylamine. Sensors and Actuators, B: Chemical. 2018;**258**:50-55. DOI: 10.1016/j.snb.2017.11.126

[54] Govori S. Convenient methods for the synthesis of pentacyclic fused heterocycles with coumarin moiety. Synthetic Communications. 2016;**46**: 569-580. DOI: 10.1080/00397911. 2016.1161802

[55] Long L, Li X, Zhang D, Meng S, Zhang J, Sun X, et al. Amino-coumarin based fluorescence ratiometric sensors for acidic ph and their application for living cells imaging. RSC Advances. 2013;**3**:12204-12209. DOI: 10.1039/ C3RA41329B

[56] Orrego-Hernández J, Nuñez-Dallos N, Portilla J. Recognition of Mg^{2+} by a new fluorescent "Turn-on" chemosensor based on pyridyl-hydrazono-coumarin. Talanta. 2016;**152**:432-437. DOI: 10.1016/j. talanta.2016.02.020

[57] Fu J, Yao K, Li B, Mei H, Chang Y, Xu K. Coumarin-based colorimetric-fluorescent sensors for the sequential detection of Zn^{2+} ion and phosphate anions and applications in cell imaging. Spectrochimica Acta - Part A: Molecular and Biomolecular Spectroscopy. 2020;**228**. DOI: 10.1016/j. saa.2019.117790

[58] Gilli G, Bellucci F, Ferretti V, Bertolasi V. Evidence for resonance-assisted hydrogen bonding from crystal-structure correlations on the enol form of the beta-diketone fragment. Journal of the American Chemical Society. 1989;**111**:1023-1028. DOI: 10.1021/ja00185a035

[59] Davis AB, Lambert RE, Fronczek FR, Cragg PJ, Wallace KJ. An activated coumarin-enamine michael acceptor for CN^-. New Journal of Chemistry. 2014;**38**:4678-4683. DOI: 10.1039/ C4NJ00862F

[60] Li M, Xing Y, Zou Y, Chen G, You J, Yu F. Imaging of the mutual regulation between zinc cation and nitrosyl via two-photon fluorescent probes in cells and in vivo. Sensors and Actuators, B: Chemical. 2020;**309**. DOI: 10.1016/j. snb.2020.127772

[61] Fahrni CJ, O'Halloran TV. Aqueous coordination chemistry of quinoline-based fluorescence probes for the biological chemistry of Zinc. Journal of the American Chemical Society. 1999;**121**:11448-11458. DOI: 10.1021/ ja992709f

[62] Lee HG, Lee JH, Jang SP, Hwang IH, Kim S-J, Kim Y, et al. Zinc selective chemosensors based on the flexible dipicolylamine and quinoline. Inorganica Chimica Acta. 2013;**394**:542-551. DOI: 10.1016/j.ica.2012.09.009

[63] Chadlaoui M, Abarca B, Ballesteros R, de Arellano CR, Aguilar J, Aucejo R, et al. Properties of a triazolopyridine system as a molecular chemosensor for metal ions, anions, and amino acids. Journal of Organic Chemistry. 2006;**71**:9030-9034. DOI: 10.1021/jo061326e

[64] Jiang W, Fu Q, Fan H, Wang W. An NBD fluorophore-based sensitive and selective fluorescent probe for zinc ion. Chemical Communications. 2008:259-261. DOI: 10.1039/B712377A

[65] Gayathri P, Pannipara M, Al-Sehemi AG, Moon D, Anthony SP. Molecular structure controlled self-assembly of pyridine appended fluorophores: Multi-stimuli fluorescence responses and fabricating rewritable/self-erasable fluorescent platforms. Materials Advances. 2021;**2**(3):996-1005. DOI: 10.1039/d0ma00736f

Chapter 9

Chemistry with Schiff Bases of Pyridine Derivatives: Their Potential as Bioactive Ligands and Chemosensors

Kaushal K. Joshi

Abstract

Pyridine is a valuable nitrogen based heterocyclic compound which is present not only in large number of naturally occurring bioactive compounds, but widely used in drug designing and development in pharmaceuticals as well as a precursor to agrochemicals and chemical-based industries. Pyridine derivatives bearing either formyl or amino group undergo Schiff base condensation reaction with appropriate substrate and under optimum conditions resulting in Schiff base as product which behave as a flexible and multidentate bioactive ligand. These Schiff bases are of great interest in medicinal chemistry as they can exhibit physiological effects similar to pyridoxal-amino acid systems which are considered to be very important in numerous metabolic reactions. They possess an interesting range of bioactivities including antibacterial, antiviral, antitubercular, antifungal, antioxidant, anticonvulsants, antidepressant, anti-inflammatory, antihypertensive, anticancer activity etc. and considered as a versatile pharmacophore group. Further, several pyridine-based Schiff bases show very strong binding abilities towards the various cations and anions with unique photophysical properties which can be used in ion recognition and they are extensively used in development of chemosensors for qualitative and quantitative detection of selective or specific ions in various kinds of environmental and biological media. These chapter insights the bioactivity and ion recognition ability of Schiff bases derived from pyridine derivatives.

Keywords: pyridine derivatives, Schiff bases, bioactive ligands, pharmacophore, chemosensors, ion recognition

1. Introduction

Nitrogen based heterocyclic compounds are well dispersed in nature and present in large number of alkaloids, vitamins, essential oils, amino acids, metabolites etc. all of them are essential for various biochemical processes and cellular life. Pyridine is considered among the most important nitrogen based heterocyclic compounds which is present in numerous bioactive compounds. Pyridine acts as a versatile solvent and gives different types of reactions including nucleophilic substitution, electrophilic substitution, N-protonation easily. It also possesses some unique optical properties. Due to its important physical, chemical and biological properties, pyridine forms large number of derivatives which are found to be less toxic, but possess much enhanced chemical and biological activities as compared to parent compound. These pyridine derivatives are frequently used in various chemical-based industries like paints and adhesives, dyes and textiles, flavors and perfumes, disinfectants and explosives and so on. They are also used in large scale as a precursor for production of various agro-chemicals like herbicides, insecticides, fungicides etc. Pyridine moieties or scaffold are also present in large number of lifesaving drugs and dietary supplements. Pyridine has capability to bind with number of transition metal ions and form innumerable metal complexes. Some of them are widely used as organometallic catalysts in chemical reactions whereas some others possess unique photophysical and luminescence properties and can be used as electrochemical or colorimetric sensors. The most important applications of pyridine and its derivatives are found in pharmaceutical field due to their significant biological activities. Pyridine nucleus is found to be basic skeleton of large number of bioactive molecules which ranges from Antitubercular, Antibacterial, Antiviral, Antianginal, Antihistaminic, Antiulcer, Antitumor drugs etc. Such bioactive pyridine derivatives bearing excellent coordination and strong binding ability can act as important bioactive ligand and can effectively bind with important biomolecules such as proteins, DNA, coenzymes, amino acids and other metabolites by reflecting their pharmacological potential. Thus, pyridine derivatives or scaffolds form the basis of a potent pharmacophore group having biological significance with important therapeutic applications.

Pyridine derivatives bearing either formyl or amino group readily undergo Schiff base condensation reaction with appropriate substrate and optimum conditions. Schiff bases are the condensation products of primary amines and carbonyl compounds and considered as sub-class of imines. They act as an effective organic ligand due to the presence of imine nitrogen which is basic in nature and exhibits π-acceptor properties. Further, if some other hetero atoms like nitrogen, oxygen or sulfur of a specific functional group is present in vicinity of azomethine group, the schiff base act as multidentate ligands with flexibility in structure. Thus, Schiff bases of pyridine can be regarded as much better ligand as compared to pyridine itself in terms of strong binding ability, flexibility in structure and greater bioactivity. Schiff bases derived from pyridine derivatives are of great interest in medicinal chemistry due to their role of bioactive ligand as these can exhibit physiological effects similar to pyridoxal-amino acid systems which are considered to be very important in numerous metabolic reactions. They possess a wide variety of biological activities that include antibacterial, antiviral, antitubercular, antifungal, antioxidant, anti-inflammatory, anticonvulsants, antidepressant, antihypertensive, anticancer activity and so on. Due to their vast pharmacological activities, they are considered as a versatile pharmacophore. Further, pyridine-based Schiff bases also play important role in analytical chemistry. As Schiff bases show very strong binding abilities towards the various cations and anions, flexibility in their structure and unique photophysical properties, they can be used in ion recognition and therefore they are extensively used in development of different types of chemosensors for selective detection of specific ions in various kinds of environmental and biological media as well as in industrial and agricultural fields.

In the view of the versatile pharmacological properties as possessed by Schiff bases derived from pyridine derivatives, it is expected that they have high potential in the field of various biological activities that are still unexplored and can be used effectively in drug

DOI: http://dx.doi.org/10.5772/intechopen.106749

discovery. Further, the designing of specific sensor for the recognition of various ions is one of the most demanding areas of chemical research due to their significant contribution in analytical, industrial, agricultural, environmental and biological fields and there is an urgent need to explore the chemistry of pyridine-based Schiff bases to find out their applications as chemosensors for ions recognition studies. This chapter throws some light on chemistry and biological significance of pyridine derivatives, reviews the recent work done on Schiff bases derived from pyridine derivatives and their potential as effective bioactive ligands as well as efficient chemosensors.

2. Pyridine derivatives: chemistry and biological significance

2.1 Pyridine

2.1.1 A valuable N-based heterocyclic compound

Heterocyclic compounds are widely distributed in nature and they are found to be essential for various biochemical processes. They also play a vital role in the metabolism of all living cells as well as in the composition of genetic material of the cells. Many of them are pharmacologically active and are in clinical usage. Among these heterocyclic compounds, those based on nitrogen are of great importance as they are widely spread in nature, possess more therapeutic values and less toxicity as compared to other heterocycles based on oxygen or sulfur. Moreover, their structure can be subtly manipulated to achieve a required modification in function. Such nitrogen based heterocyclic compounds represent important building blocks in both natural and synthetic bioactive compounds. Among these, pyridine is the simplest monoazine compound but considered as one of the most valuable N-based heterocyclic. An important property of pyridine is that it's a polar solvent but aprotic in nature. Thus, it can be easily mixed with polar as well as with many non-polar organic solvents which makes it a versatile solvent. Further, the derivatives of pyridine are found to be less toxic, but possess much more important chemical and biological properties as compared to parent pyridine and therefore, they are frequently used as precursors for many important chemicals, agrochemicals and pharmaceuticals. Owing to their important chemical properties and biological significance, pyridine derivatives find applications in variety of fields as shown in **Figure 1**. These properties of

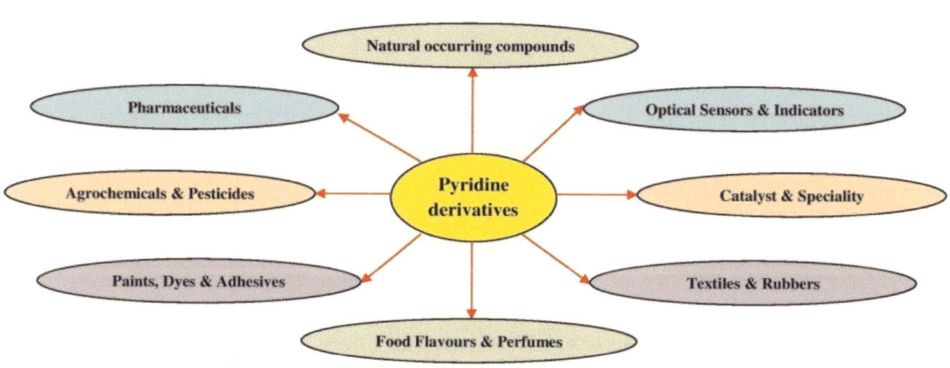

Figure 1.
Applications of pyridine derivatives in variety of fields.

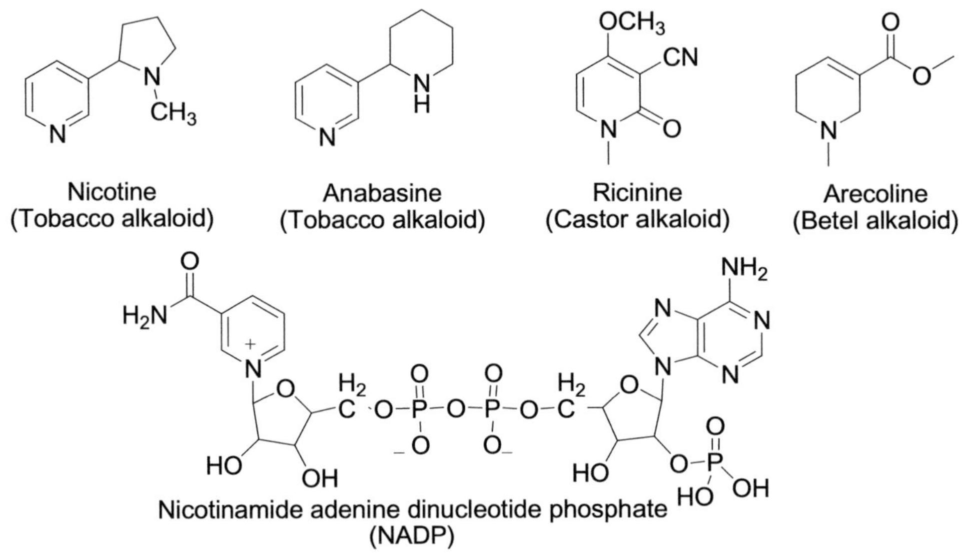

Figure 2.
Naturally occurring compounds containing pyridine ring.

pyridine and its derivatives make them useful in synthesis of innumerable products such as medicines, agrochemicals, catalysts, optical sensors, food flavorings, perfumes, dyestuffs, paints, adhesives, rubber products, textile fabrics etc. [1–5].

2.1.2 Naturally occurring compounds

Pyridine derivatives are the fundamentally important nitrogen-based heterocycles which are present in large number of naturally occurring compounds. They are often present as a partial structure in many plant-based alkaloids. For example: Nicotine and Anabasine are found in tobacco whereas Ricinine is present in castor oil and Arecoline is present in betelnut. Nicotinamide adenine dinucleotide phosphate is a cofactor used in anabolic reactions and nucleic acid syntheses which is used by all forms of cellular life (**Figure 2**) [6].

2.1.3 Vitamins and dietary supplements

Some essential B group vitamins such as Niacin (Vitamin B_3) and Pyridoxine (Vitamin B_6) are simply the derivatives of pyridine. Chromium picolinate and Zinc picolinate are used as dietary supplements (**Figure 3**) [6, 7].

2.1.4 Pharmaceutical compounds

Pyridine moieties are present in large number of bioactive compounds and form the basis of pharmacophore group. They can be used as prodrugs or drug molecules themselves which possess wide range of medicinal applications including Antitubercular, Antibacterial, Anticholinesterase, Antihistamine, Antiulcer, Antianginal etc. (**Figure 4**). Further detailed studies on bioactivity of pyridine derivatives are given in Section 2.2 [8, 9].

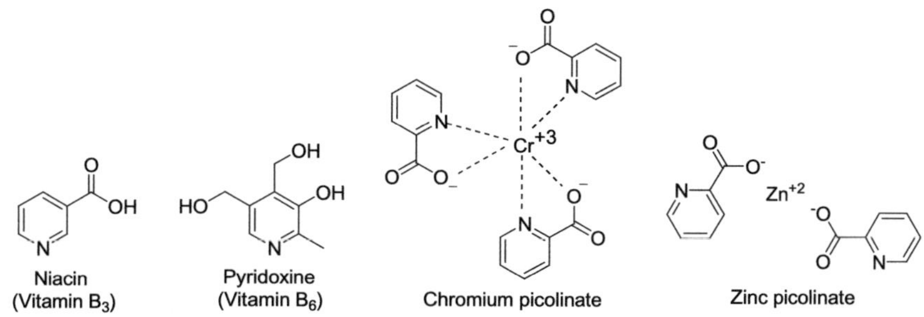

Figure 3.
Vitamins and dietary supplements based on pyridine derivatives.

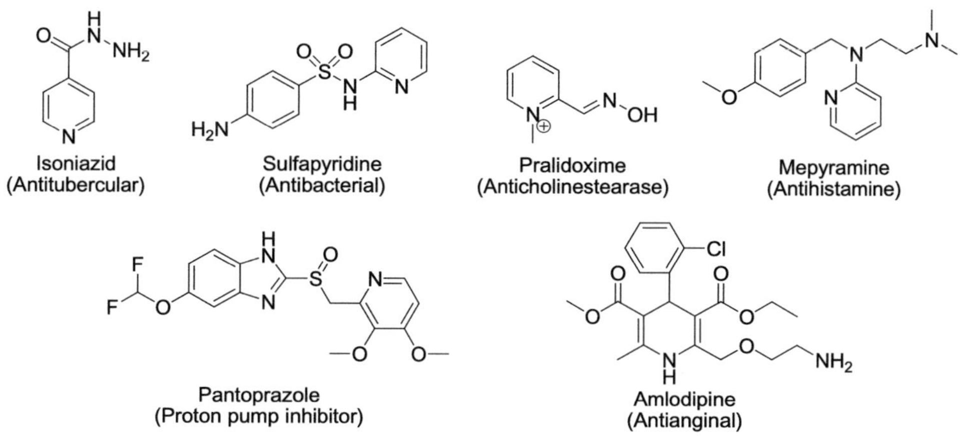

Figure 4.
Pharmaceutical compounds having pyridine moiety.

2.1.5 Agrochemicals

Pyridine or its derivatives are used as starting materials for synthesis of many agrochemicals or pesticides. They act as the precursor or intermediates for many important herbicides, fungicides and insecticides (**Figure 5**) [10, 11].

2.1.6 Catalysts

Several pyridinium salts are used as catalyst in many organic reactions. For example: Collins reagent is used to convert primary alcohols into aldehydes; Cornforth reagent is used for oxidation of primary and secondary alcohols into carbonyls whereas PCC is used primarily for selective oxidation of alcohols into carbonyls. Pyridineborane is used as a reducing agent with improved stability and solubility over $NaBH_4$ (**Figure 6**). Crabtree catalyst and Milstein catalyst are well known organometallic catalyst used for hydrogenation and dehydrocoupling of alcohols respectively [12, 13].

2.1.7 Optical sensors

Several bipyridine or terpyridine based metal complexes exhibit intense luminescence and can be used as fluorescent chemosensors. For example: $[Ru(bipy)_3]^{+2}$ is used as a luminophore whereas $[Fe(bipy)_3]^{+2}$ is used in redox titrations and colorimetric analysis. The complex $[Fe(phen)_3]^{2+}$ is widely used as Ferroin indicator in redox titrations and for the photometric determination of Fe (II) (**Figure 7**) [14, 15].

2.1.8 Chemical based industries

Pyridine derivatives are also used on large scale in many chemical-based industries. Pyridone based azo disperse dyes are widely used for making dyestuffs. Pyridine derivative ADP is applied to improve network capacity of cotton in textile industries. Polyvinyl pyridines are used as copolymer with styrene for making adhesives and install water proofing properties in paint industries. Several alkyl or acyl derivatives of pyridines are the main source of flavors and essential oils which are widely used in food industries and cosmetic industries (**Figure 8**) [16–19].

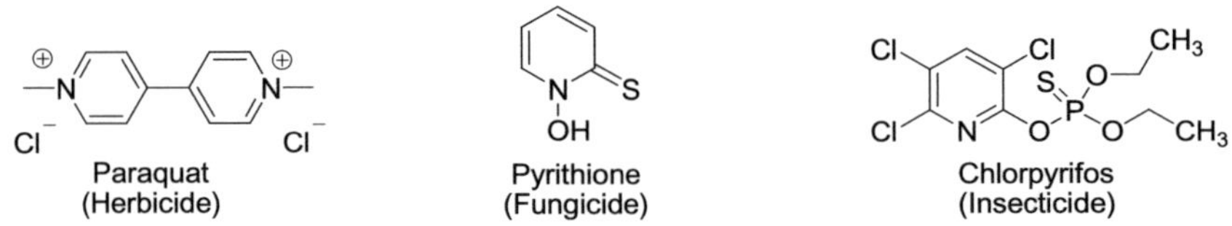

Figure 5.
Agrochemicals based on pyridine derivatives.

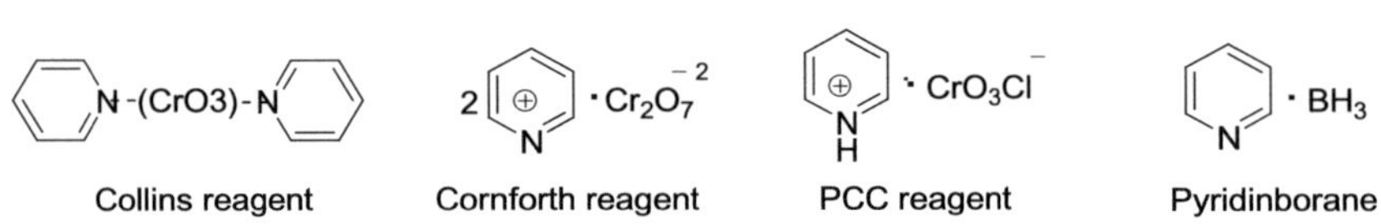

Figure 6.
Catalysts based on pyridinium salts.

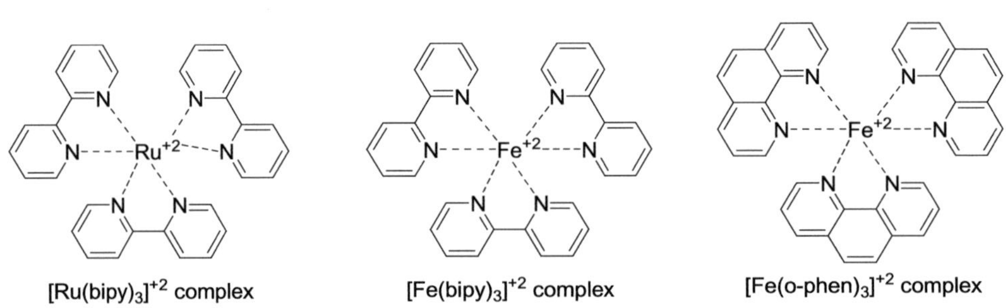

Figure 7.
Optical sensors based on bipyridine and o-phenanthroline metal complex.

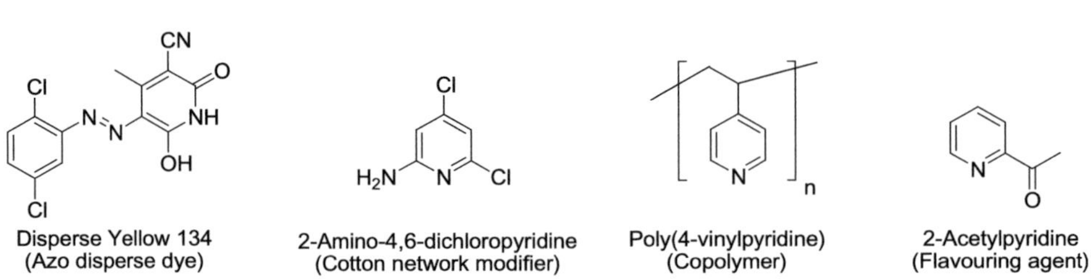

Figure 8.
Pyridine derivatives used in chemical-based industries.

DOI: http://dx.doi.org/10.5772/intechopen.106749

2.1.9 Speciality reagents

Many specialty reagents used in chemical lab are based on pyridine. Pyridine is often used as a reaction solvent for many organic reactions because of its polar nature, low reactivity and miscibility with wide range of solvents. For example: pyridine is an important constitutes of Karl Fischer reagent for determining traces of water in pharmaceuticals, deuterated pyridine is used as common solvent in ^{1}H-NMR spectroscopy, and pyridine is also used as denaturant for making anti freezing mixtures of ethyl alcohol.

Hence, pyridine and its derivatives have significant applications in various fields, especially in the medicinal and agrochemicals. Due to such wide range of applications and extremely usage in industries, pyridine and its derivatives are considered among the most important and valuable N-based heterocyclic compounds which is also evident from the current annual worldwide production of pyridine which is approximately 20,000 ton per year.

2.2 Biological importance of pyridine derivatives

Pyridine is one of the most important nitrogen-based heterocyclic compounds which is present in large number of naturally occurring compounds. It is widely used as a precursor to agrochemicals and pharmaceuticals. Pyridine moieties are present in large number of drug molecules as well as in essential dietary supplements. This indicates that pyridine compounds can be used as precursor of drugs and with their proper structural modification or derivatization they can be led to important prodrugs or drugs itself of therapeutic value. Pyridine is an important heterocyclic organic compound. Pyridine and their heterocyclic annulated derivatives are of great interest due to the wide variety of biological activities as observed in these compounds. Pyridine nucleus is found to be basic skeleton of large number of bioactive molecules which ranges from Antitubercular, Antibacterial, Antiviral, Antiseptic, Antihistaminic, Antianginal, Anticholinesterase, Anti-inflammatory, Antiulcer, Anticancer etc.

2.2.1 B-group vitamins

Pyridine ring is present as basic nucleus in various B group vitamins such as Nicotinamide, Nicotinic Acid and Pyridoxine which are used as essential dietary supplements and for therapeutic effect (**Figure 9**).

Figure 9.
B-group vitamins based on pyridine derivatives.

2.2.2 Antituberculars

These drugs are medications used to treat bacterial infection caused by *Mycobacterium tuberculosis*. Pyridine nucleus is found to be basic skeleton of major antitubercular drugs such as Isoniazid, Ethionamide and Prothionamide which are used in treatment of tuberculosis (**Figure 10**).

2.2.3 Antibacterials

These drugs are a principal type of antimicrobial agent or antibiotic which are used to either kill or inhibit the growth of certain bacteria. Sulfapyridine and Sulfasalazine are sulpha drugs containing pyridine nuclei which act as antibacterial agents used to inhibit bacterial infection (**Figure 11**).

2.2.4 Antihistamines

These drugs are used to oppose the activity of histamine receptors in human body so that to treat different allergic conditions like allergic rhinitis, common cold, influenza etc. Betahistine, Chlorpheniramine, Dexchlorpheniramine, Mepyramine, Pheniramine and Triprolidine are Histamine H1-receptor antagonist and used as antihistaminic drugs for allergic disorders. All of them contain the pyridine ring as an important part of their structure (**Figure 12**).

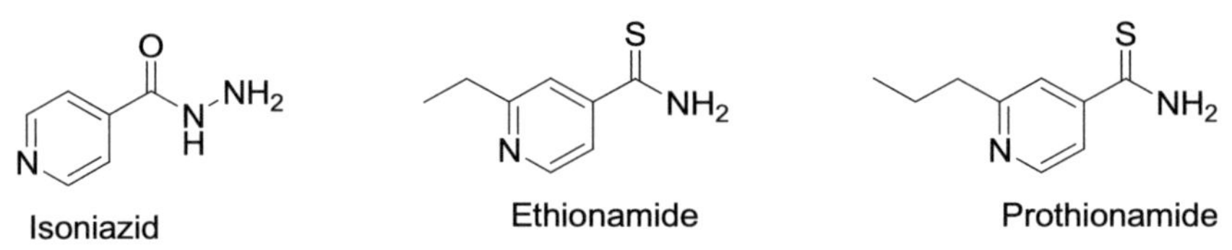

Figure 10.
Antitubercular drugs containing pyridine as basic skeleton.

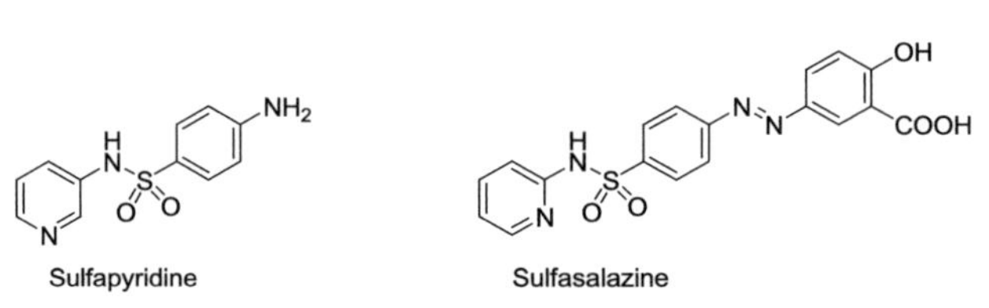

Figure 11.
Antibacterial drugs containing pyridine nuclei.

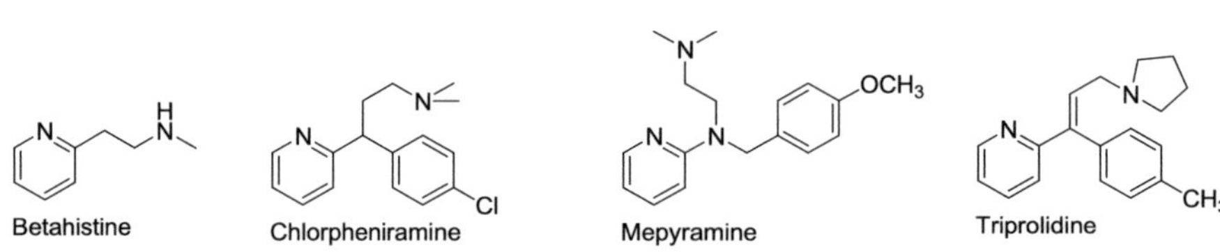

Figure 12.
Antihistamine drugs having pyridine nucleus.

2.2.5 Antianginals & antihypertensive drugs

Antianginal drugs are used in treatment of angina pectoris, a type of heart disease. They are also classified as calcium channel blockers or beta blockers. Antihypertensive drugs are used to prevent conditions of high blood pressure, stroke and myocardial infarction. Amlodipine, Azelnidipine, Clinidipine, Felodipine, Lacidipine, Nicardipine and Nifedipine are some Antianginal/Antihypertensive drugs which contain the pyridine as core structure (**Figure 13**).

2.2.6 Anticholinesterase drugs

These drugs act as antidote for cholinesterase inhibitors and prevent the breakdown of neurotransmitter acetylcholine. Examples are Pralidoxime and Pyridostigmine which are simply the pyridinium salt derivatives (**Figure 14**).

2.2.7 Analgesic and anti-inflammatory drugs

These drugs are used to reduce pain, decreases inflammation and also reduce fever. Etoricoxib, Phenyramidol, and Piroxicam are used as analgesic and ant-inflammatory drugs that contain the pyridine scaffold (**Figure 15**).

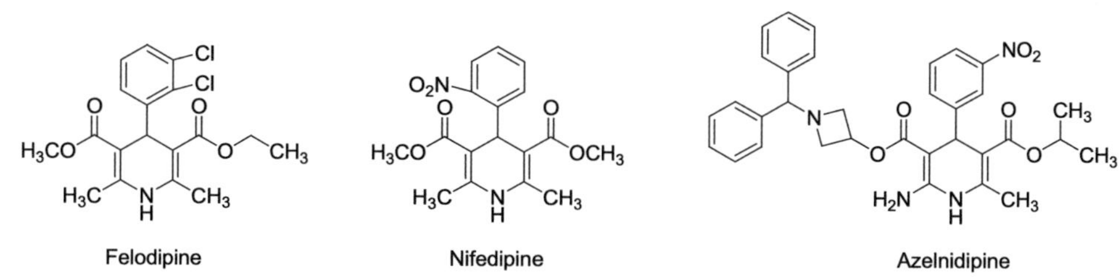

Figure 13.
Antianginal drugs bearing pyridine as core structure.

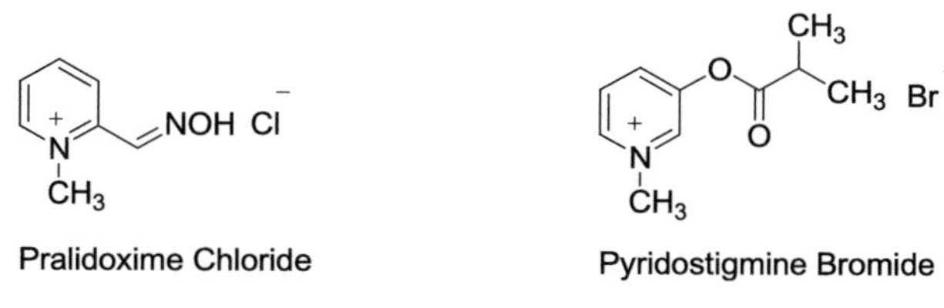

Figure 14.
Anticholinestearase drugs based on pyridinium salts.

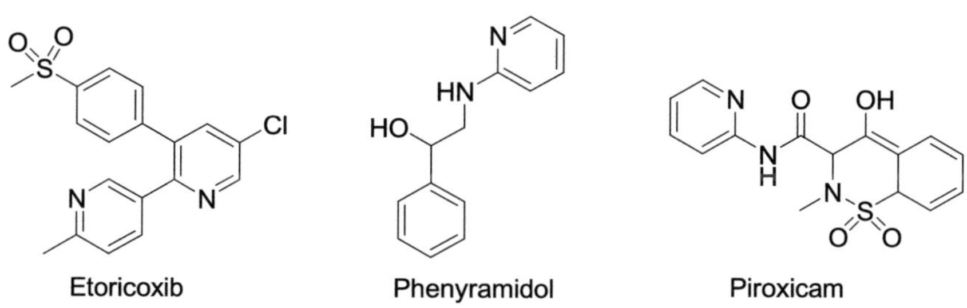

Figure 15.
Analgesic/anti-inflammatory drugs having pyridine scaffold.

2.2.8 Antiulcer drugs

These are class of drugs used to treat peptic ulcer or gastrointestinal tract infections. They also include the class proton pump inhibitor that is used in reduction of gastric acid production. Lansoprazole, Omeprazole, Pantoprazole and Rabeprazole are proton pump inhibitor and used as antiulcer drugs. All of them contain pyridine nucleus as an important part of their structure (**Figure 16**).

2.2.9 Anticancer drugs

These drugs are effective in the treatment of malignant or cancerous disease by inhibiting the cell division and proliferation. Abiraterone, Imatinib and Sorafenib are used as anticancer drugs that consist of pyridine ring (**Figure 17**).

2.2.10 Antivirals

These drugs are used in treatment of viral infections. They do not destroy the target pathogen but inhibit its growth. Atazanavir and Indinavir are antiretroviral drugs that are used in treatment of HIV/AIDS. Both of them have pyridine nuclei as a part of their structure (**Figure 18**).

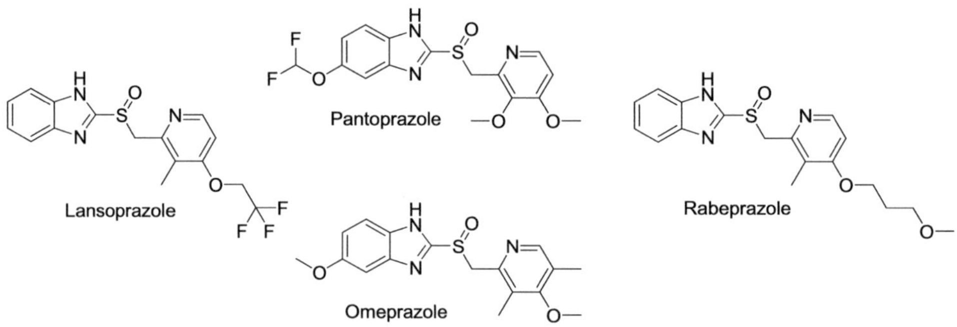

Figure 16.
Antiulcer drugs containing pyridine nuclei.

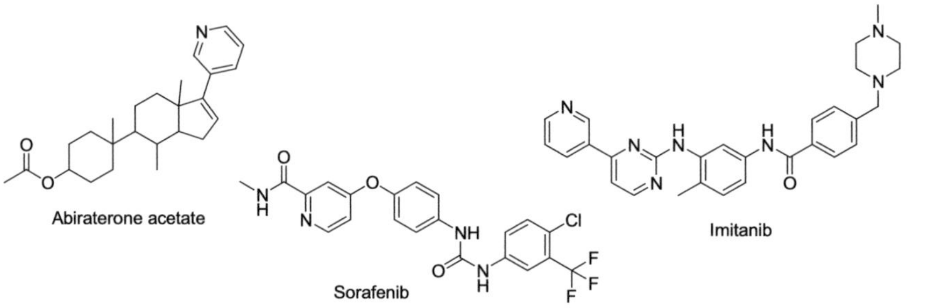

Figure 17.
Anticancer drugs bearing pyridine ring as part of their structure.

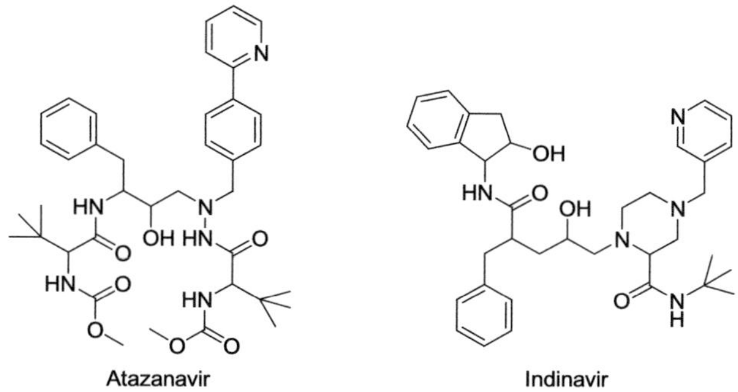

Figure 18.
Antiviral drugs containing pyridine moiety as part of their structure.

2.2.11 Antiseptics

These are antimicrobial agents that can be applied on living tissues or skin in order to reduce the possibility of infection or putrefaction. Cetylpyridinium chloride and Laurylpyridinium chloride are used as antiseptic in oral and dental care products. Both of these are simply the derivatives of pyridinium chloride salt (**Figure 19**).

Additionally, there are many other important pyridine-based drugs like Bisacodyl as laxative, Disopyramide as antiarrhythmic, Nikhetamide as respiratory stimulant, Pioglitazone as antidiabetic, and Torsemide as diuretic and so on (**Figure 20**) [20–22].

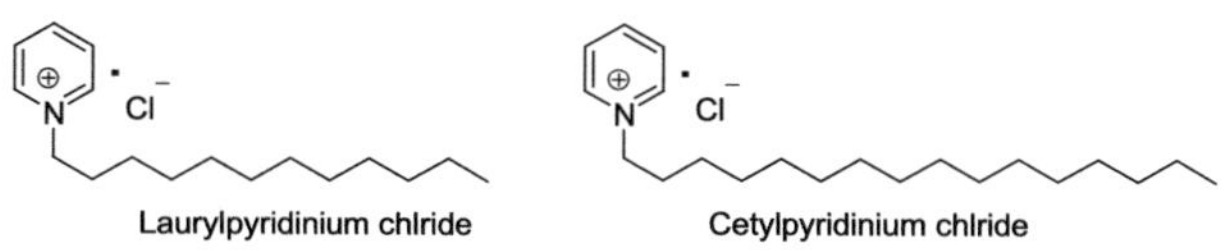

Figure 19.
Antiseptics based on pyridinium salts.

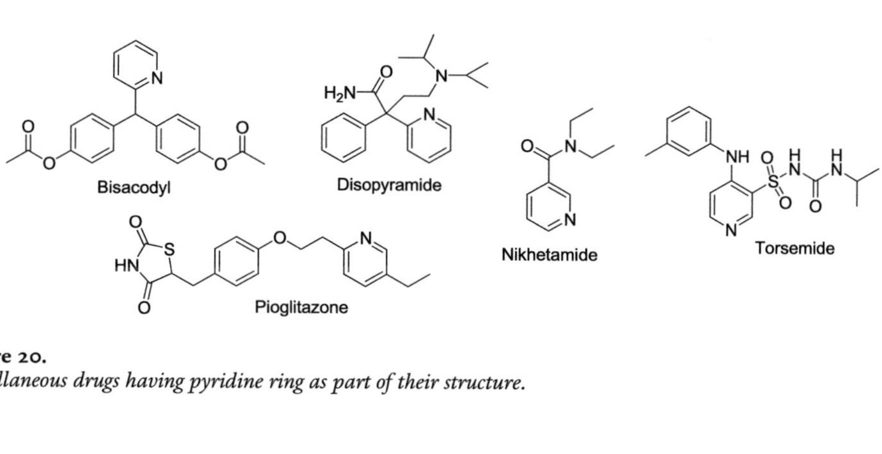

Figure 20.
Miscellaneous drugs having pyridine ring as part of their structure.

3. Schiff bases of pyridine: the excellent bioactive ligands and efficient chemosensors

3.1 Schiff bases and their metal complexes

3.1.1 Schiff base

Schiff bases are generally the condensation products of primary amines and carbonyl compounds. They are considered as a sub-class of imines which are the organic compounds containing carbon-nitrogen double bond. Structurally, Schiff base is an analogue of an aldehyde or ketone in which the carbonyl (C═O) group has been replaced by an imine or azomethine (>C═N—) group. Schiff bases are generally synthesized by the condensation reaction between primary amines and aldehydes or less commonly ketones (**Figure 21**). Schiff bases are more readily formed with aldehydes as compared to ketones. Schiff bases derived from aliphatic aldehydes are unstable in nature and readily get polymerized whereas those derived from aromatic aldehydes are more stable especially due to their effective conjugation systems.

Schiff bases have an interesting range of applications in various field of science ranging from synthesis to catalysis, analysis and medicine to modern technologies. For example, they are widely used in organic synthesis especially as the precursor of heterocyclic compounds and as the catalysts in many catalytic reactions. Several Schiff bases can be used for the qualitative and quantitative detection of metal ions. Some Schiff bases can be used as optical, fluorescent as well as electrochemical sensors. The most important application of Schiff bases is in the field of medicinal chemistry. Some important drugs consist azomethine group of Schiff base in their structure e.g., Thiocetazone, Nitrofurazone, Nitrofurantoin etc. (**Figure 22**).

In recent years, various Schiff base containing derivatives have been synthesized and evaluated for their biological activities including antimicrobial, antitubercular, antifungal, antioxidant, anti-inflammatory, anticonvulsants, antidepressant, antihypertensive and anticancer activity. As they possess a wide variety of biological activities, they are considered as a versatile pharmacophore and emerged as a potent class of pharmaceuticals. Several studies showed that the presence of a lone pair of electrons in sp^2 hybridized orbital of nitrogen atom of the azomethine group is of considerable chemical and biological importance as it interferes in normal cell processes by the formation of hydrogen bond between the active centers of cell constituents and

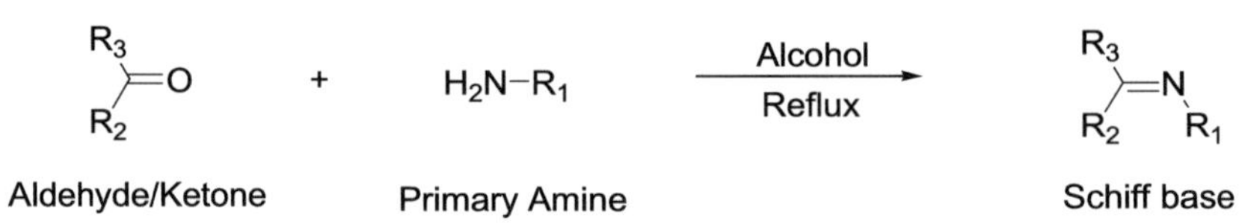

Figure 21.
Reaction scheme for Schiff base condensation.

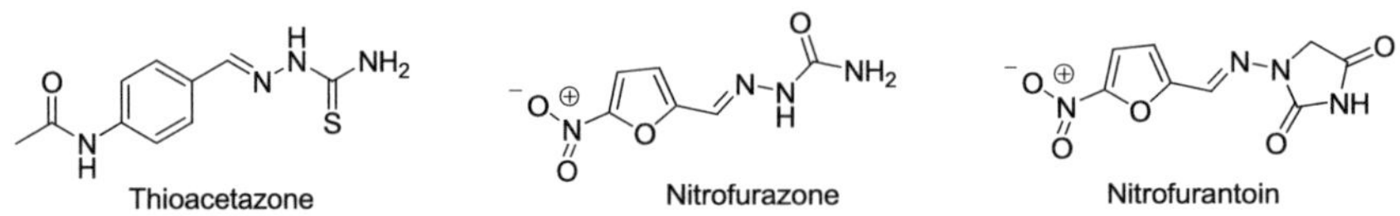

Figure 22.
Important drugs containing azomethine (—CH═N—) group.

sp^2 hybridized nitrogen atom. Thus, Schiff bases have key role in design and development of novel compounds which are more potent and have interested biological activities. Due to the vast pharmacological activities, they constitute a significant class of compounds for new drug development and continue to be an active area of research in medicinal chemistry [23–27].

3.1.2 Schiff base metal complexes

Schiff bases are widely used as ligands in coordination chemistry due to the presence of imine nitrogen which is basic in nature and exhibits π-acceptor properties. These act as Flexi-dentate ligands due to presence of nitrogen of azomethine group and other hetero atoms like nitrogen, oxygen or sulfur of specific functional group if present. The metal complexes of Schiff bases are also known as metallo-imines and they play a central role in coordination chemistry. Jacobsen's catalyst is a well-known example of Schiff base metal complex which is derived from chiral tetradentate Salen ligand (**Figure 23**).

Some metal complexes play a vital role in the bioactivity of life saving drugs especially anticancer drugs. Cisplatin, Carboplatin and Oxiplatin are anticancer drugs designed from binding of organic ligands with platinum metal ion (**Figure 24**).

In organic synthesis the Schiff base reactions are very useful in making carbon-nitrogen bonds. Schiff base are considered as a very important class of organic ligands which can be used as building blocks and find extensive applications in organic synthesis as well as in organocatalysis. Thus, Schiff base appears to be an important intermediate in a number of enzymatic reactions that involves interaction of an enzyme with an amino or a carbonyl group of the substrate. It is a well-known fact that the binding of bioorganic molecules or drugs to the metal ions drastically change their biomimetic properties, therapeutic effects and pharmacological activities. Thus, both the Schiff base ligands and their metal complexes have further extensive applications ranging from material sciences to biological sciences. Due to their biological activities and clinical usage, they are of worth attention. Their successful application

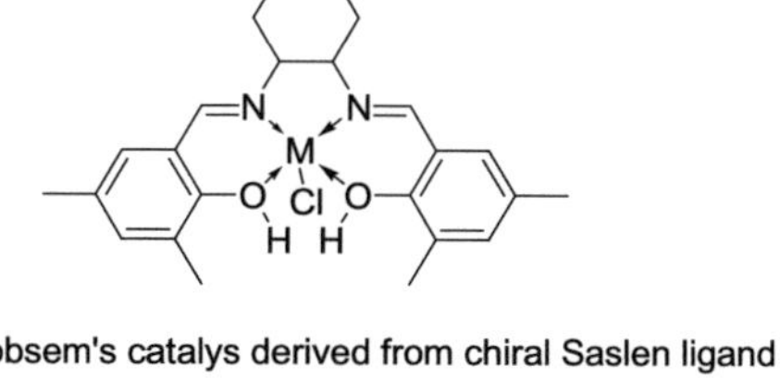

Figure 23.
Jacobsen's catalyst.

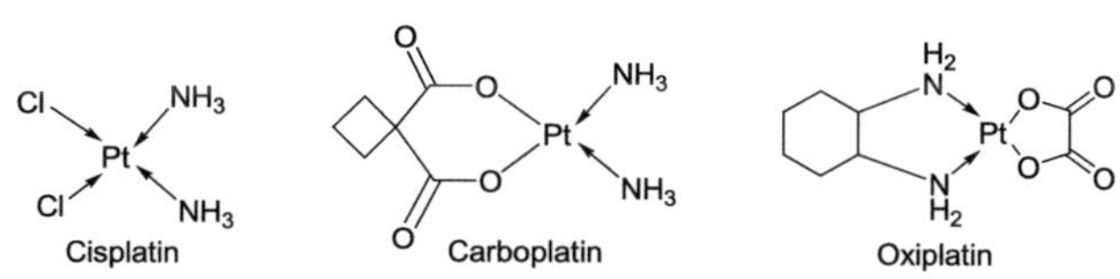

Figure 24.
Metal complexes of platinum used as anticancer drugs.

can lead to the formation of series of novel compounds with wide range of physical, chemical and biological activities [28–33].

3.2 Schiff bases of pyridine as bioactive ligands and versatile pharmacophore

3.2.1 Protein-ligand interactions

Protein-ligand interactions are essential for all processes happening in living organisms as proteins are the fundamental units of all living cells that play a vital role in various cellular functions. It is a reversible non-covalent interaction comprises biological recognition at molecular level in which the molecules i.e. protein and ligand recognize each other by stereo specificity. The evolution of the protein functions depends on the development of specific sites which are designed to bind ligand molecules. Ligand binding capacity is important for the regulation of biological functions which occur through the molecular mechanics involving the conformational changes in proteins. This change initiates a sequence of events leading to different cellular functions. A detailed understanding of the protein–ligand interactions is therefore central to understand biology at the molecular level. Moreover, knowledge of the mechanisms responsible for the protein-ligand recognition and binding helps to understand the drug-receptor interaction in detail and facilitate the discovery, design, and development of drug molecules. A modern computational technique based on protein-ligand interactions is Molecular docking which is now routinely used for drug designing and development processes [34, 35].

3.2.2 DNA-Metal complex interactions

Many transition metal complexes are known to bind with DNA via both covalent and non-covalent interactions. Formation of a protein-ligand complex is based on molecular recognition between biological macromolecules and ligands which depends on affinity and specificity. The interaction between transition metal complexes and DNA has aroused the widespread interest because it helps not only to understand the life processes at the molecular level but also to promote the development of chemistry discipline itself. The interest in preparation of new metal complexes gained the tendency of studying on the interaction of metal complexes with DNA for their applications in biotechnology and medicine. Cisplatin, Carboplatin, Oxiplatin and their derivatives are widely used as anticancer drugs which are based on DNA-Metal complex interactions but they create several side effects such as anemia, diarrhea, alopecia, petechia, nephrotoxicity, emetogenesis, ototoxicity, neurotoxicity etc. Efforts are continuously made to prepare the chemotherapeutic drugs without side effects or fewer side effects. In recent times, the treatment of cancer with a chemotherapeutic approach is based on DNA-Metal complex interactions [36–38].

3.2.3 Role of Schiff bases as bioactive ligand

The Schiff bases display significant biological activities due to presence of imine (>C=N—) functional group. Thus, Schiff base derived from aromatic aldehyde and aromatic amines have enormous applications in biological fields. Pyridine carboxaldehyde derivatives of Schiff bases are of great interest due to their role in natural and synthetic organic chemistry as these can exhibit physiological effects similar to pyridoxal-amino acid systems which are considered to be very important in numerous metabolic reactions (**Figure 25**).

They show diverse biological activities in terms of antibacterial, antiviral, antitubercular, antipyretic, anti-inflammatory, antiulcer, antihistaminic, antitumor etc. (**Figure 26**). The bonding interaction between aromatic ring of Schiff base ligand and aromatic amino acid side chains of receptor has also been revealed in most of the X-ray crystal structures of protein complexes. This protein ligand interaction involves some non-covalent interactions and the evaluation of the structure-activity relationship of Schiff bases also demonstrates their desired biological activity. This ensures the application of Schiff bases in drug designing process and they are widely used as prodrugs as well as the drug molecules itself [39–41].

A series of Schiff bases have been synthesized using 2-vinylaniline and various aldehydes including pyridine-2-aldehyde (**Figure 27**). These Schiff bases were then complexed to transition metal ions like Mn^{+2}, Co^{+2}, Ni^{+2} and Cu^{+2}. All of these compounds were evaluated for their antibacterial activity against bacterial species like *E. coli*, *Staphylococcus aureus* and *Pseudomonas aeruginosa* as well as for their antifungal activity against fungal species like *Candida albicans* and *Candida krusei*. It was concluded that different Schiff bases and their metal complexes had varying degree of antibacterial and antifungal activities. However, all the metal complexes had enhanced antimicrobial activity as compared to their ligand [42].

A combination of pyridine-2-aldehyde with S-methyl and S-benzyl dithiocarbazate resulted in synthesis of Schiff bases (**Figure 28**) which were allowed to form

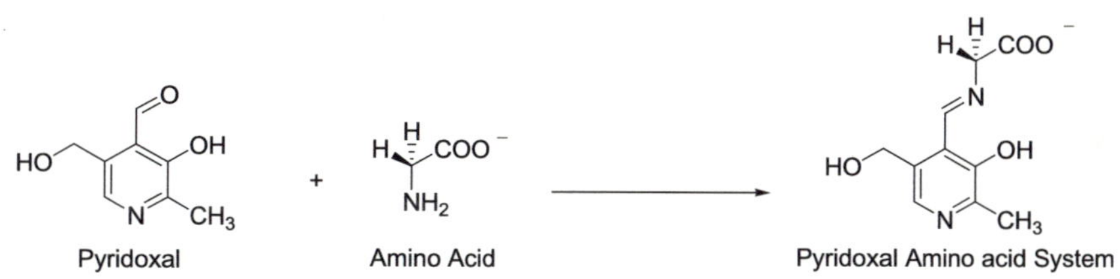

Figure 25.
Pyridoxal amino acid system.

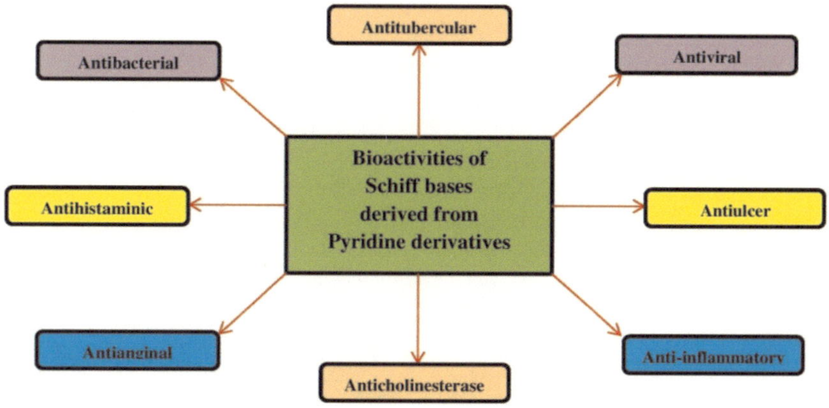

Figure 26.
Bioactivities of Schiff bases derived from pyridine derivatives.

complexes with Mn^{+2} and Zn^{+2} ions. These Schiff bases and metal complexes were evaluated for their biological activities against bacteria, fungi and K562 leukemia cell line. It was observed that Schiff base with S-methyl dithiocarbazate and its complex with Zn^{+2} had broad antimicrobial activity as compared to the Schiff base with S-benzyl dithiocarbazate and its complex with Mn^{+2}. Further only S-methyl dithiocarbazate and its complex with Mn^{+2} showed significant antitumor activity against K562 leukemia cell line [43].

Schiff bases have been derived from pyridine-4-carbaldehyde and various aromatic amino compounds such as 2, 3 and 4-aminobenzoic acids, 4-aminoantipyrene, 2-aminophenol, 2-aminothiophenol etc. (**Figure 29**). The synthesized compounds were evaluated for their antioxidant activities and DNA binding interaction studies. It was found that the Schiff base of pyridine-4-carboxaldehyde and aminophenol was an efficient antioxidant with 74% inhibition of free radicals generated by DPPH. Further most of the synthesized Schiff bases showed efficient binding with DNA which was in good agreement with molecular docking studies [44].

A Schiff base was derived from 2,6-diaminopyridine and salicylaldehyde by microwave irradiation (**Figure 30**) which form complexes with transition metal ions such as Co^{+2}, Ni^{+2}, Cu^{+2}, Zn^{+2} and Cd^{+2}. It was found that all the complexes were non electrolyte and possessed an octahedral geometry in which N donor sites of imine and O donor site of phenolic groups were coordinated to the metal ions [45].

A series of Schiff bases was derived from Isoniazid and various aromatic aldehydes like 2-benzyloxybenzaldehyde and its derivatives as well as with various ketones like n-hexanophenone, cyclohexanone etc. (**Figure 31**). All these novel Schiff bases were then evaluated for their antitubercular activities. It was found that these compounds

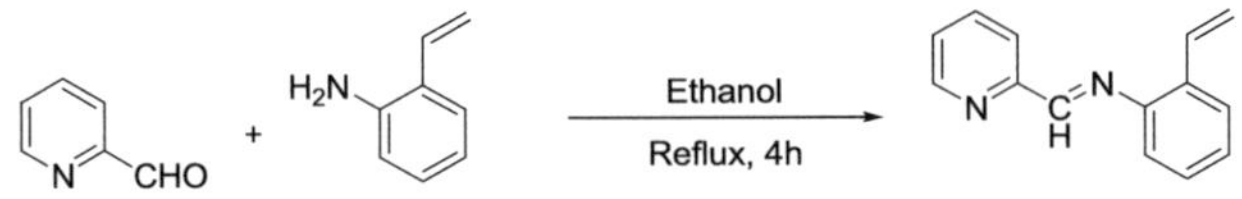

Figure 27.
Schiff base derived from pyridine-2-aldehyde and 2-vinylaniline.

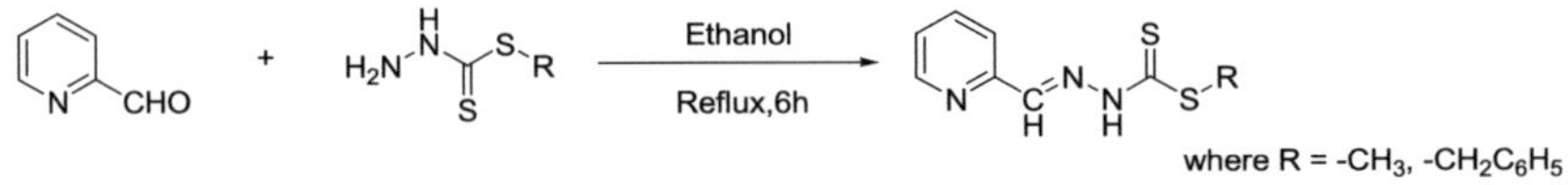

Figure 28.
Schiff bases derived from pyridine-2-aldehyde with dithiocarbazate derivatives.

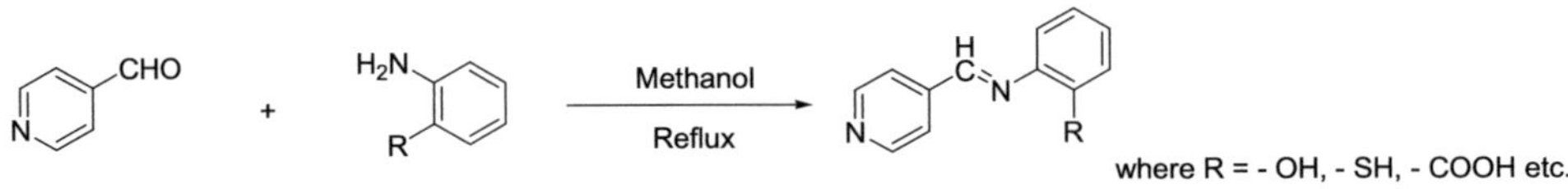

Figure 29.
Schiff bases derived from pyridine-4-aldehyde with different aromatic amino compounds.

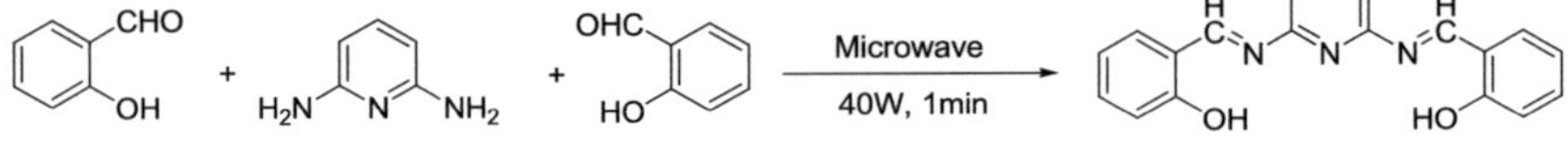

Figure 30.
Schiff base derived from salicylaldehyde and 2,6-diaminopyridine.

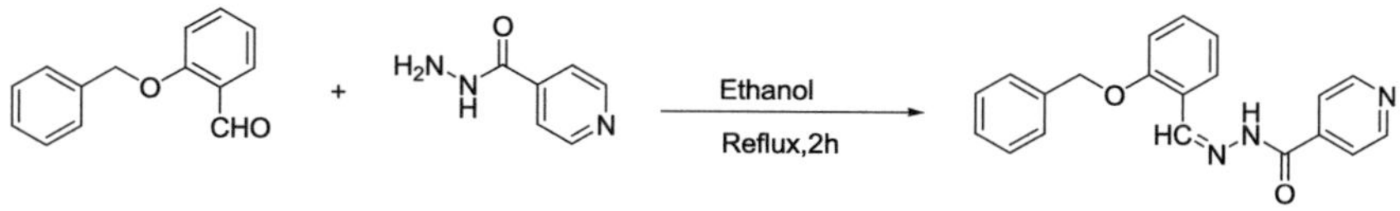

Figure 31.
Schiff base derived from 2-bezyloxybenzaldehyde and isoniazid.

showed high level of activity against Mycobacterium tuberculosis in vitro and in vivo and they had also low toxicity [46].

Schiff bases were derived by the reaction of Isoniazid with 2-acetylfuran and 2-acetyl-5-methylfuran (**Figure 32**). Antibacterial and antifungal activity of the Schiff bases and their complexes were evaluated. It was observed that all these compounds were active against all the microbial strains and their metal complexes with Pd^{+2} and Pt^{+2} were far more active as compared to their parent Schiff base [47].

A Schiff base was derived from Isoniazid and 2-hydroxy-5-methoxybenzaldehyde (**Figure 33**). The metal complexes of this Schiff base were prepared using transition metal ions Mn^{+2}, Ni^{+2}, Cu^{+2} and Zn^{+2}. It was observed that Mn^{+2}, Ni^{+2} and Cu^{+2} complexes had moderate activity against gram positive *Staphylococcus aureus* and gram-negative *E. coli*. It was found that Zn^{+2} complexes showed the highest antifungal activity against the fungal species *Aspergillusflavus* [48].

A Schiff base synthesized from Isoniazid and 2-hydroxynaphthaldehyde (**Figure 34**) was complexed with various transition metal ions like Co^{+2}, Ni^{+2}, Cu^{+2} and Zn^{+2}. The biological activity of Schiff base as ligand and its metal complexes were tested on gram-positive bacteria *E. coli* and gram-negative bacteria *Staphylococccus Aurous* as well as two fungi *Aspergillusflavus* and *Candida albicans*. It was observed that all the metal complexes possessed biological activity and some of them were more potent than their parent Schiff base [49].

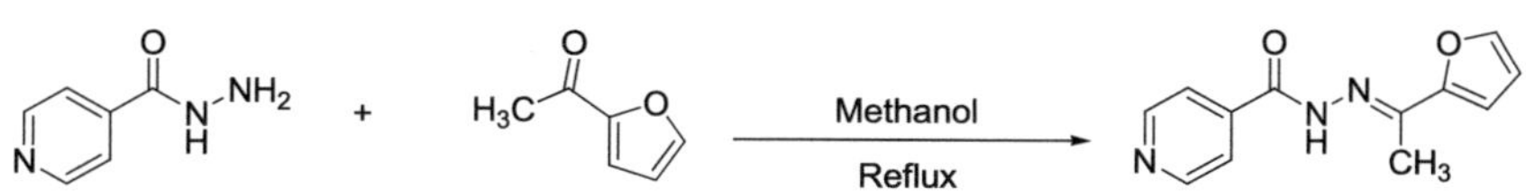

Figure 32.
Schiff base derived from isoniazid and 2-acetylfuran.

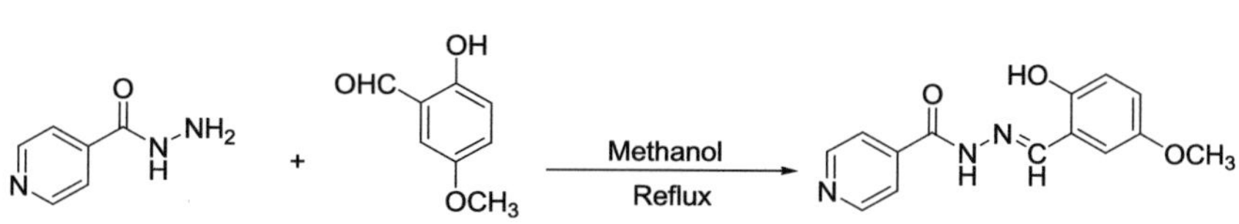

Figure 33.
Schiff base derived from isoniazid and 2-hydroxy-5-methoxybenzaldehyde.

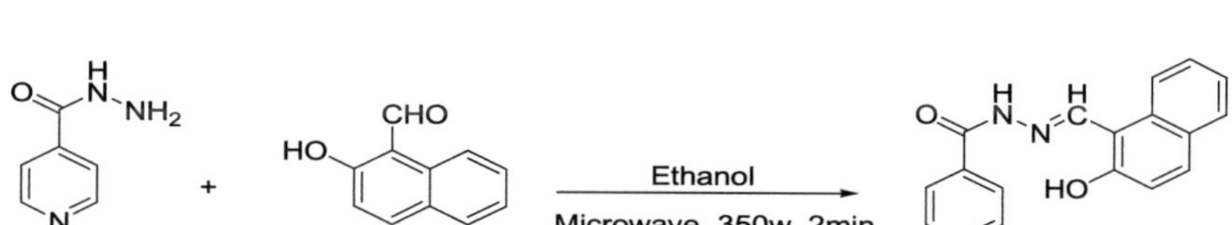

Figure 34.
Schiff base derived from isoniazid and 2-hydroxynaphthaldehyde.

Schiff base derived from Nicotinic acid hydrazide and 2,5-dimethoxybenzaldehyde (**Figure 35**) were complexed with various transition metal ions. In-vitro antimycobacterial activities of these complexes were evaluated against *Mycobacterium tuberculosis* and *H37Rv*. It was found that some of the metal complexes showed higher activity than the Isoniazid and the Schiff base whereas some others showed moderate activity. However, all these metal complexes were found to be more toxic as compared to Isoniazid [50].

Schiff base synthesized from the reaction of Isoniazid and Ketoprofen (**Figure 36**) was found to be a bioactive compound due to large energy gap between HOMO and LUMO as observed from Frontier orbital theory analysis. It was also found to be a more potent against *Mycobacterium tuberculosis* infection as compared to Isoniazid with the help of Molecular docking studies [51].

Two schiff bases were developed by the condensation of 3,4-diaminopyridine with 3,5-difluorine-2-hydroxybenzaldehyde and 5-fluorine-2-hydroxybenzaldehyde (**Figure 37**). The antifungal activity of both the schiff bases were assessed against yeast among which the schiff base obtained from 3,5-difluorine-2-hydroxybenzaldehyde was found to give good results [52].

A schiff base was synthesized by the reaction between 2-benzoylpyridine and 2-aminopyrimidine (**Figure 38**). The binuclear complexes of the schiff base with transition metal ions V(IV), Co (II) and Cu (II) were obtained and examined for their antibacterial properties against three bacterial strains *Escherichia coli*, *Klebseilla pneumonia* and *Staphylococcus aureus*. The antifungal activity was also determined against three fungal strains *Candida albicans*, *Candida glabrata* and *Candida parapsilosis*. It

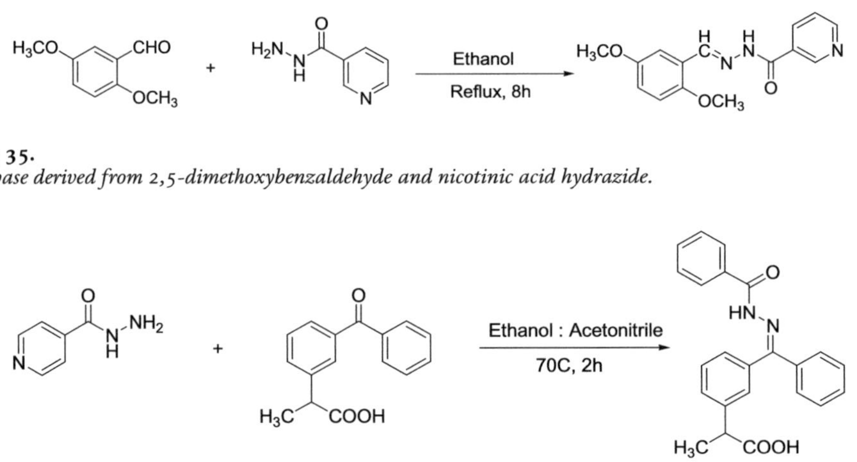

Figure 35.
Schiff base derived from 2,5-dimethoxybenzaldehyde and nicotinic acid hydrazide.

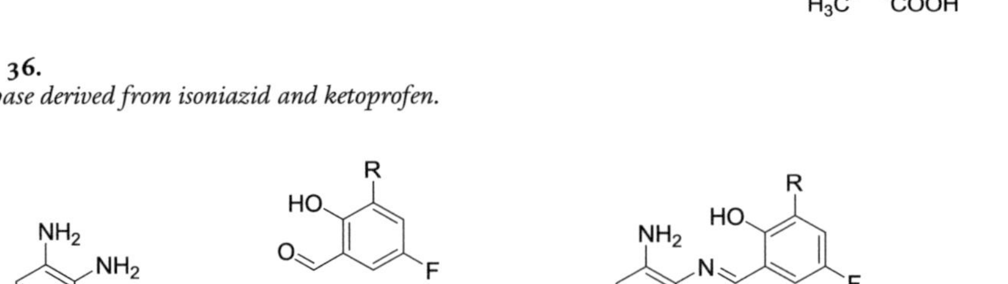

Figure 36.
Schiff base derived from isoniazid and ketoprofen.

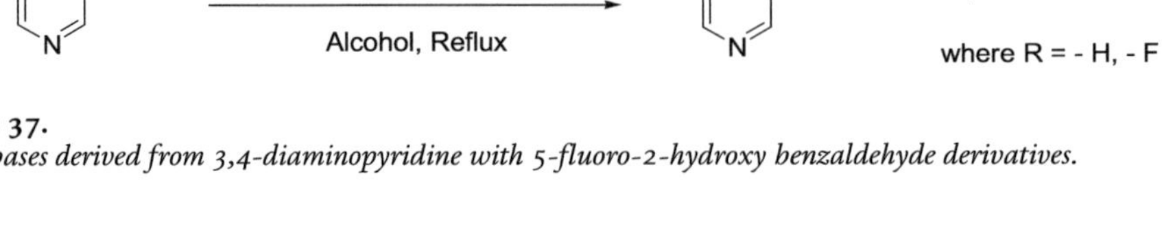

Figure 37.
Schiff bases derived from 3,4-diaminopyridine with 5-fluoro-2-hydroxy benzaldehyde derivatives.

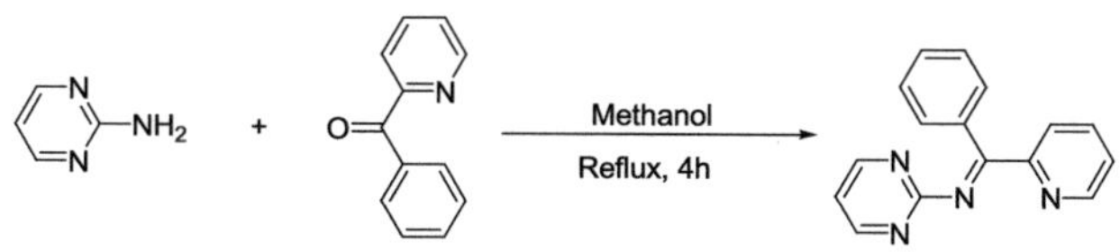

Figure 38.
Schiff base derived from 2-aminopyrimidine and 2-benzoylpyridine.

was revealed that the schiff base showed good to moderate antibacterial and antifungal activities [53].

A novel methyl substituted pyridine Schiff base was obtained by reacting 2,4-dihydroxybenzaldehyde and 2-amino-4-methylpyridine (**Figure 39**). Its metal complexes were also designed with transition metal ions Fe(III), Co(III), Cu(II) and Ni (II). The schiff base and all of its metal complexes were examined for their antimicrobial and antioxidant properties which were found to be moderate to good against reference standards [54].

A series of schiff bases were synthesized from syringaldehyde by reaction with different aminopyridines (**Figure 40**) and their antibacterial properties were evaluated for different gram-positive and gram-negative bacteria. It was observed that compound3 was more effective against gram negative bacteria *P. aeruginosa* in comparison to standard ampicillin drug. The antioxidant potential was also determined and predicted [55].

A pyridine-based Schiff base (S)-N-benzylidene-2-(benzyloxy)-1-(5-(pyridine-2-yl)-1,3,4-thiadiazol-2-yl) ethanamine was synthesized (**Figure 41**). Its antioxidant and antimitotic activities were correlated with standards Ascorbic acid and Methotrexate respectively and both of these activities were found in good agreements to standards [56].

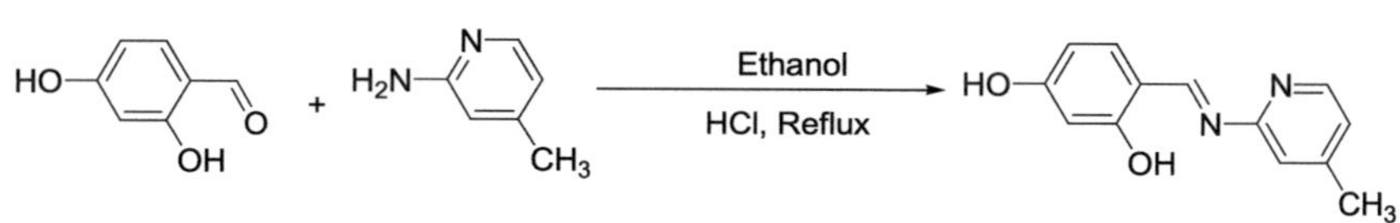

Figure 39.
Schiff base derived from 2,4-dihydroxybenzaldehyde and 2-amino-4-methylpyridine.

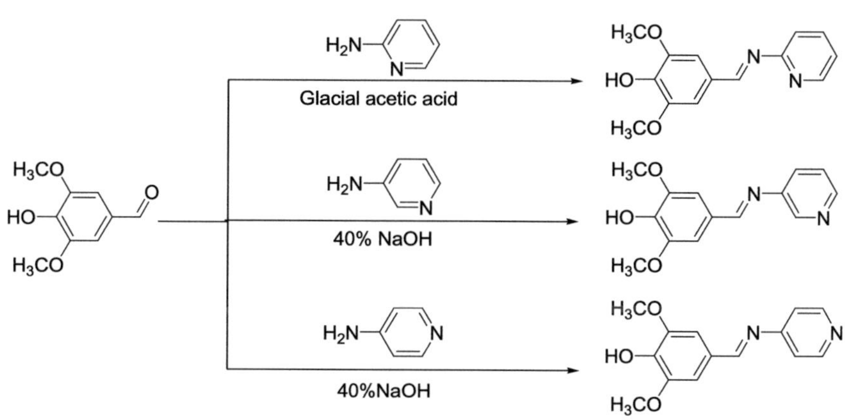

Figure 40.
Schiff bases derived from syringaldehyde with different aminopyridines.

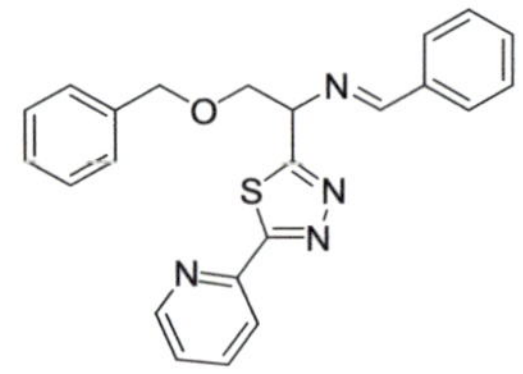

Figure 41.
Schiff base derived from pyridine-thiadiazol based compound and benzaldehyde.

3.3 Schiff bases of pyridine as chemosensors for ion recognition studies

3.3.1 Chemosensors

A chemosensor is a molecular structure i.e. an organic or inorganic complex that can be used for sensing of an analyte to produce a detectable change or a signal. In general, chemosensors are the chemical molecules that bind selectively with the guest moiety and produce a detectable or measurable change in physical, chemical or spectral properties of the system. As shown in **Figure 42**, the designing of a chemosensor is simply based on Host-Guest recognition.

These changes may be the color development or masking, modulation of emission intensity or redox potential which can be detected with the help of UV–visible absorption spectroscopy, fluorescence spectroscopy and voltammetry respectively. Thus, chemosensors are designed to contain a signaling moiety and a recognition moiety that gives rise to change in either UV–visible absorption or the emission properties. The color change or spectral change observed in either case is due to the formation of host-guest complex i.e., the complex formed between the receptor and ion. The visualization of color is based on the coordination between organic molecules having lone pair of electrons which act as donors and the metal ion or a specific anion which act as receptor [57–60].

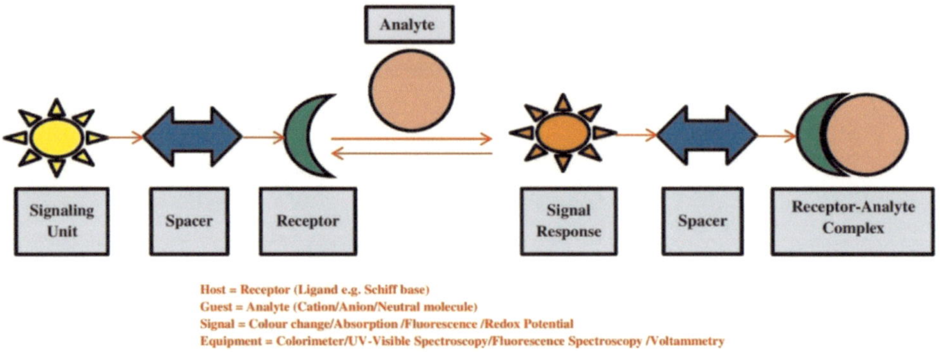

Figure 42.
Designing of chemosensor based on host-guest recognition.

3.3.2 Need for cation recognition

There are several transition metal ions which are very crucial for the life of living organisms. Some of them are required in trace quantity but if their concentration exceeds than the trace amount, they become toxic for the biological systems and may lead to various diseases and disorders. There are certain non-essential elements for living system which are widely used in industries and daily life. Their frequent and larger use can lead to overloading of such elements in the human body which may cause a large number of diseases like bone disorder, neurodegenerative diseases, sclerosis, dialysis encephalopathy etc. Their high concentration in water is harmful to growing plants and aquatic life. Transition metal ions as pollutants have some toxic impact on human health as well as on environment. The detection of these ions has gained extreme importance in recent years in the field of chemical, biological and environmental sciences. There is an urgent need to develop some efficient approaches to detect such metal ions with high selectivity and sensitivity so that to control the harmful effect on human health and environment [61, 62].

3.3.3 Need for anion recognition

Anion recognition plays a vital role in aqueous medium due to analysis of various anions in biological and environmental systems. Anion sensing continues to be a developing field in supramolecular chemistry because of its significance in industrial chemistry, environmental sciences as well as in biological fields. However, anion sensing in pure water is challenging job because they have large variation in size as compared to metal cations. Moreover, they have large solvation energy in aqueous medium and there is a strong competition occurs between solvent and anions for binding with the receptor. These problems can be overcome to certain extent by the use of chemosensors. A large number of chemosensors have also been reported for anion recognition and sensing with high selectivity as well as sensitivity. Literature review revealed that most of these sensors have complicated structure and hard synthetic routes. Moreover, some of them have poor yields and troublesome purification process. It can be expected that chemosensors derived from Schiff bases may solve these issues up to certain extent as they do not have much complex structure and can be synthesized easily with good yield and purity [63–65].

3.3.4 Role of Schiff bases in ion recognition

Schiff bases are organic molecules that contain azomethine group and are capable of donating lone pair of electrons, so that they can coordinate with large number of metal ions especially transition metal ions. Schiff bases of nitrogen-based heterocycles such as pyridine or their derivatives can act as excellent ligands due to presence of ring nitrogen atom with a localized pair of electrons leading to the formation of very stable complexes with transition metal ions. It has been demonstrated that the presence of nitrogen atom of azomethine group and oxygen atom of phenolic or carbonyl group in Schiff base has strong affinity towards metal ions which results in metal-oxygen-nitrogen cycle i.e. chelatogenic cycle. Due to this, the intramolecular charge transfer is improved between the π-conjugated rings which displays unique emission enhancement. Schiff bases have the strong binding abilities to the various ions and also have individual photophysical properties. This property of Schiff base can be used in ion recognition and their derivatives are extensively used in development of chemosensors for detection of metallic cations and anions in various kinds of environmental and biological media. **Figure 43** represents the different kind of chemosensors based on Schiff bases that can be derived from pyridine derivatives [66–70].

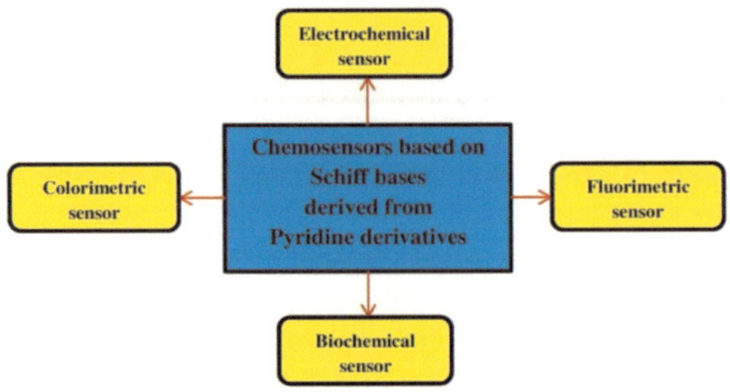

Figure 43.
Chemosensors based on Schiff bases derived from pyridine derivatives.

A pyridylazo compound (**Figure 44**) was designed which showed a very high affinity towards Al^{+3} ions. The turn on fluorescence behavior showed that the synthesized compound could be used for detection of Al^{+3} ions with high selectivity in qualitative as well as quantitative estimations [71].

A condensation reaction between 4′-amine-2,2′6′2″-terpyridine with benzaldehyde derivatives resulted in the synthesis of Schiff bases (**Figure 45**) which were studied for its cation recognition properties for various ions. It was observed that the synthesized Schiff bases selectively recognized Al^{+3} ions due to enhancement in fluorescence [72].

A reversible fluorescent colorimetric imino-pyridyl bis Schiff base receptor was developed (**Figure 46**) for the detection of $Al^{+3}$and HSO_3^- in aqueous medium.

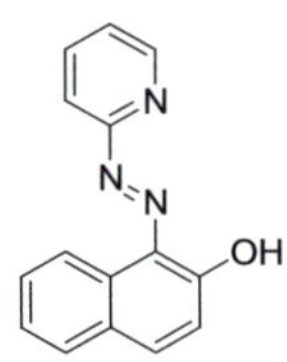

Figure 44.
Fluorescent chemosensor based on pyridylazo compound.

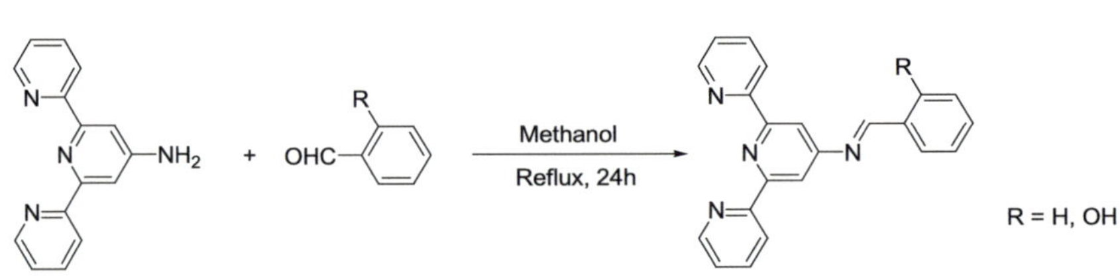

Figure 45.
Schiff base derived from 4′-amine-2,2′,6′,2 -terpyridine with benzaldehyde derivatives.

Figure 46.
Schiff base derived from pyridine-4-aldehyde and 4-aminoaniline.

The receptor exhibited excellent fluorescent colorimetric response towards Al^{+3} ions with high selectivity and also selective colorimetric response towards HSO_3^- ions [73].

A fluorescent chemosensor based on 2-(7,10-diphenylfluoranthen-8-yl)-pyridine (**Figure 47**) was designed and examined for its cation recognition ability. It was found to show excellent selectivity towards Fe^{+3} ions by exhibiting a great decrease in emission intensity [74].

A series of donor-acceptor systems was synthesized in which pyridine moiety acted as acceptor unit and carbazole moiety acted as donor unit (**Figure 48**). The synthesized compounds were then investigated for their sensing properties towards various metal cations. The compound showed a remarkable enhancement in fluorescence in presence of Cu^{+2} ions and could be used as sensor for Cu^{+2} ions with high selectivity over various other metal ions [75].

A chemosensor based on naphthalimide and pyridine moiety was designed (**Figure 49**) and found to show good response towards Cu^{+2} ions with high selectivity and sensitivity in the presence of wide range of metal ions in aqueous media [76].

A fluorescent chemosensor based on BODIPY with two pyridine ligands was synthesized (**Figure 50**) and examined for detection of various cations and anions. It was found to display very high selectivity and sensitivity towards Cu^{+2} ions by giving a visible color change from pink to blue and quenching of fluorescence emission. Further, it was noted that on addition of S^{-2} anions to the Cu^{+2} complex the color could be restored [77].

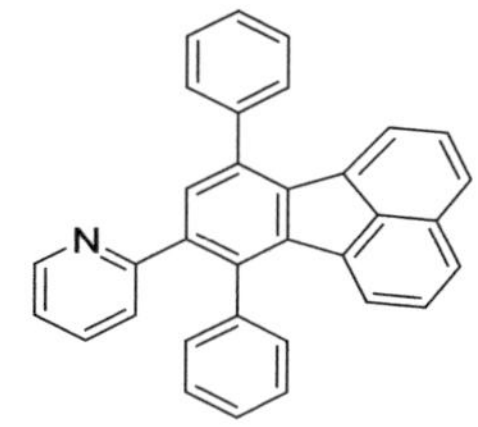

Figure 47.
Fluorescent chemosensor based on 2-(7,10-diphenylfluoranthen-8-yl)-pyridine.

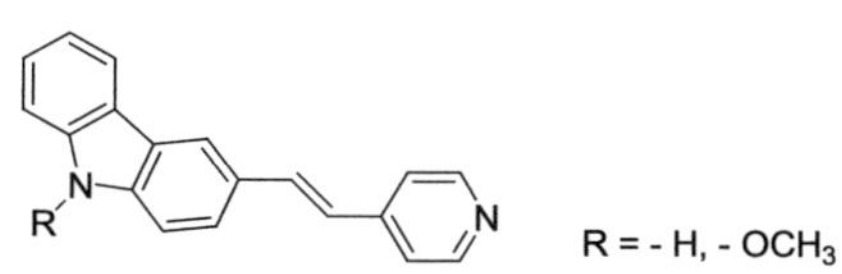

Figure 48.
Fluorescent chemosensor based on pyridine-carbazole based compound.

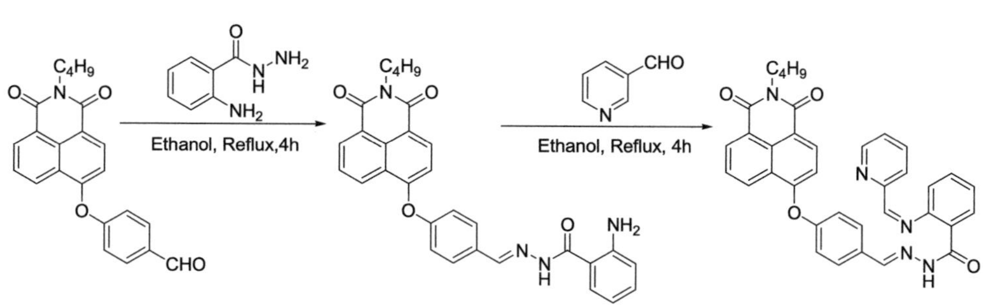

Figure 49.
Schiff base derived from naphthalimide based compound and pyridine-3-aldehyde.

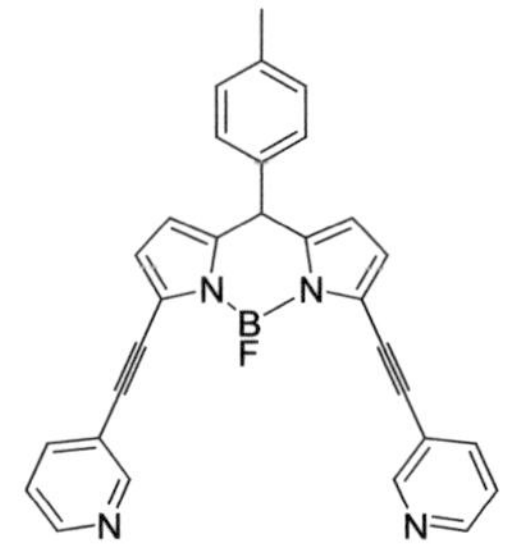

Figure 50.
Fluorescent chemosensor based on BODIPY with two pyridine ligands.

A schiff base ligand was synthesized from 4-hydroxy-3,5-dimethoxybenzaldehyde and pyridine dicarbohydrazide (**Figure 51**) which was then examined for its ion sensing ability and it was found to recognize Cu^{+2} ions over the other metal ions. Further the Schiff base complex with Cu^{+2} ions was able to detect CN^{-} ion over different anions [78].

A Schiff base was synthesized from 2,6-diaminopyridine and salicylaldehyde whereas another schiff base was synthesized from pyridine-3-carbohydrazide and 2,5-dimethoxybenzaldehyde (**Figure 52**). Both of them were evaluated for their cation sensing properties and were found to form complexes with transition metal ions such as Co^{+2}, Ni^{+2}, Cu^{+2}, Zn^{+2} and Cd^{+2}, thus had potential to act as chemosensors for detection of these ions over other competing ions in aqueous media [45].

A chemosensor derived from pyridine-dicarbohydrazide and benzothiazole aldehyde (**Figure 53**) for the detection of various cations and anions. The sensor allowed the naked eye recognition of toxic Cu^{+2} ions in presence of many other cations as well as the recognition of some biologically relevant anions like F^{-}, AcO^{-} and AMP^{-2} ions with great sensitivity [79].

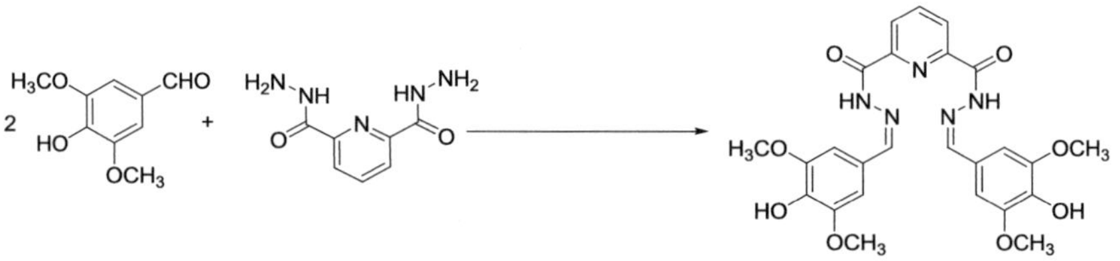

Figure 51.
Schiff base derived from 4-hydroxy-3,5-dimethoxybenzaldehyde and pyridine dicarbohydrazide.

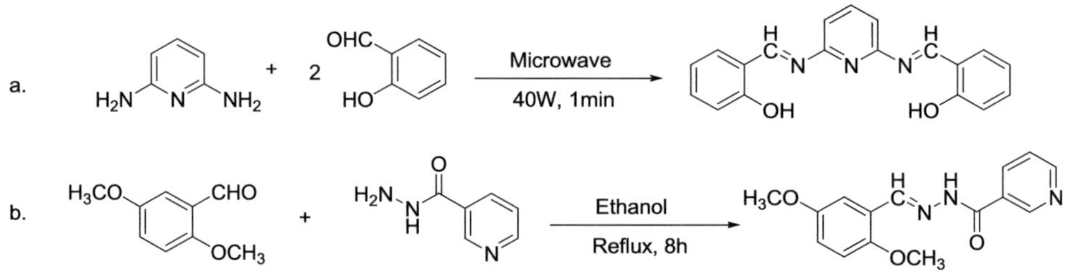

Figure 52.
a. Schiff base derived from 2,6-diaminopyridine and salicylaldehyde. b. Schiff base derived from 2,5-dimethoxybenzaldehyde and pyridine-3-carbohydrazide.

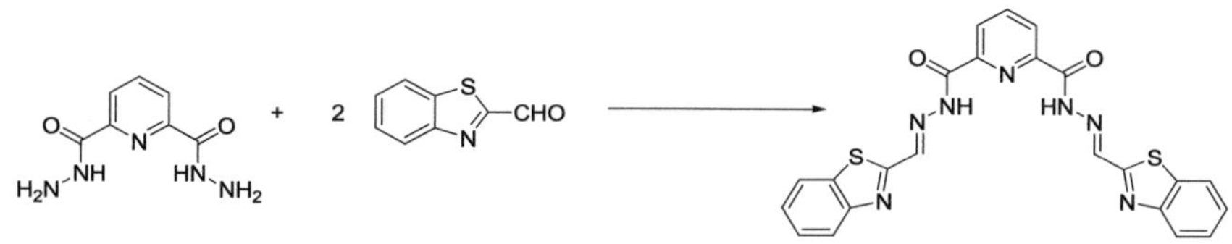

Figure 53.
Schiff base derived from pyridine dicarbohydrazide and benzothiazole aldehyde.

A chemosensor was designed from schiff base based on the condensation reaction between pyridoxal and 2-aminoethanol (**Figure 54**). The chemosensor produced a selective chromogenic behavior towards Ag^+ ions by changing the color of solution from light yellow to red observable by naked eye and also have excellent specificity and sensitivity towards Ag^+ ions over various other interfering cations in aqueous solution [80].

A Schiff base was derived from 4-E-2-phenyldiazenylaniline and pyridine-2-carboxaldehyde (**Figure 55**) and investigated for its cation recognition ability. The schiff base was found to be highly sensitive and selective for sensing of Ag^+ ions and Cd^{+2} ions and could act as chemosensor for the detection of Ag^+ and Cd^{+2} in presence of other interfering ions [81].

A porphyrin appended terpyridine compound was synthesized (**Figure 56**) and designed as chemosensor for its cation recognition ability. It was observed that the synthesized compound exhibited enhanced fluorescence in the presence of Cd^{+2} ions

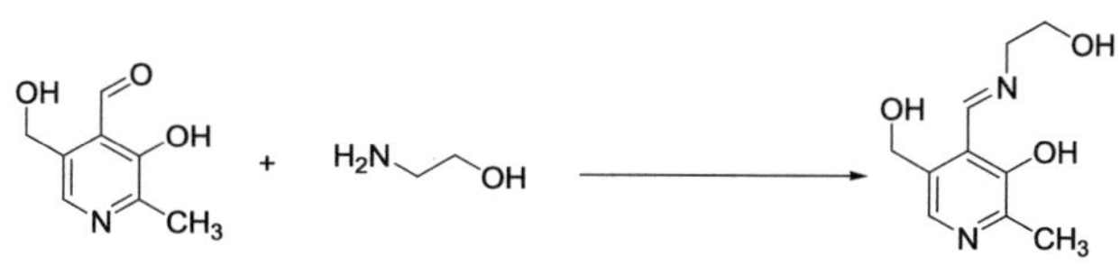

Figure 54.
Schiff base derived from pyridoxal and 2-aminoethanol.

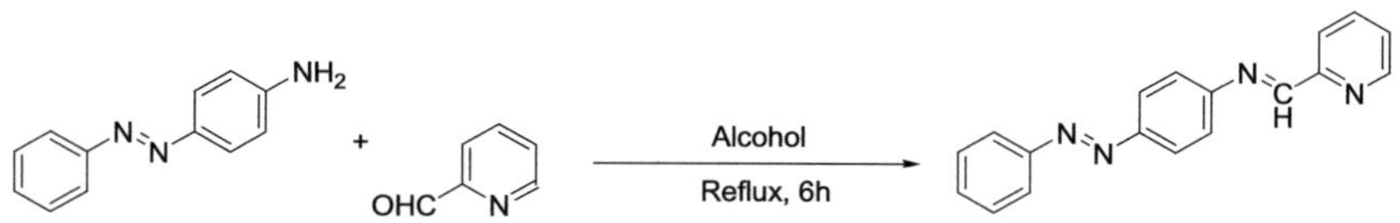

Figure 55.
Schiff base derived from 4-E-2-phenyldiazenylaniline and pyridine-2-aldehyde.

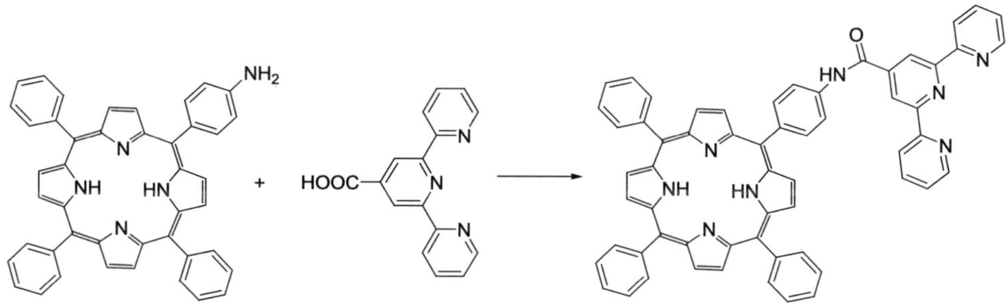

Figure 56.
Fluorescent chemosensor based on porphyrin appended terpyridine compound.

with high selectivity and sensitivity and could act as fluorescent chemosensor for Cd^{+2} ions in the presence of various other metal ions [82].

A schiff base based on 2,6-diaminopyridine was synthesized (**Figure 57**) and evaluated for its binding affinity with various metal ions. It was observed that the synthesized compound has prominent selectivity towards Pb^{+2}ions among various other metal ions and therefore could act as chemosensor for detection of Pb^{+2} ions [83].

A new bipyridine based ruthenium complex was synthesized (**Figure 58**) and investigated for its cation recognition ability. It was found that the synthesized compound was able to recognize Hg^{+2} ions in aqueous solution with high selectivity and could be used as chemosensor for the selective and sensitive detection of Hg^{+2} ions over various other cations [84].

A pyridine-based derivative of (Z)-2-(4-amino-phenyl)-3-(pyridine-4-yl) acrylonitrile was designed (**Figure 59**) and evaluated for its cation recognition properties. It was observed that the compound could selectively recognize Hg^{+2} ions by exhibiting a visible color change from light yellow to orange and could be used as a naked-eye sensor for detection of Hg^{+2} ions in presence of various other cations [85].

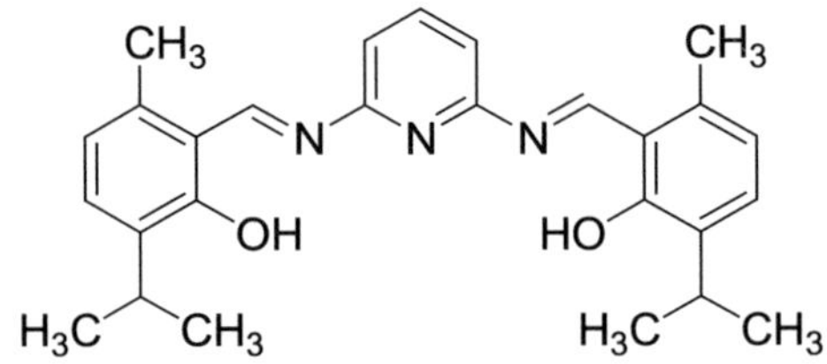

Figure 57.
Schiff base derived from 2,6-diaminopyridine and salicylaldehyde derivative.

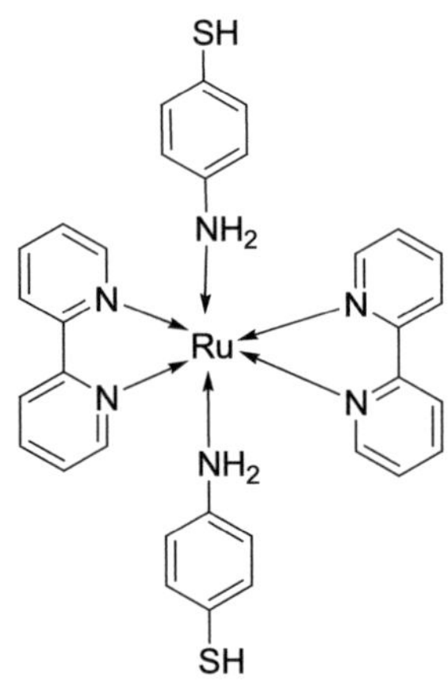

Figure 58.
Chemosensor based on bipyridine based ruthenium complex.

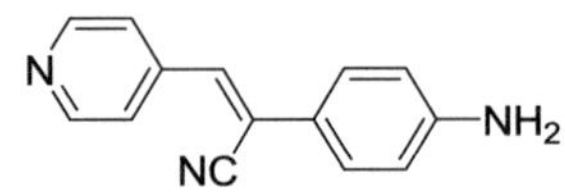

Figure 59.
Colorimetric sensor based on (Z)-2-(4-amino-phenyl)-3-(pyridine-4-yl) acrylonitrile.

Isoniazid functionalized silver nanoparticles were synthesized by wet chemical method (**Figure 60**) and it was observed to exhibit good absorbance and emission peaks with visible color change in the presence of Hg^{+2} ions. Therefore, these isoniazid capped silver nanoparticles could act as a selective chemosensor for the detection of Hg^{+2} ions in aqueous media [86].

Two schiff bases derived from fluorescein by condensation with 3-aminopyridine and 4-aminopyridine respectively (**Figure 61**) were evaluated for their ion recognition properties for various cations and anions. The compound 1 was able to detect Ce^{+3} cation in presence of various other metal ions and also F^- anion over other interfering anions and therefore could act as chemosensor for Ce^{+3} and F^- ions [87].

A simple, colorimetric and fluorimetric chemosensor was designed from an acylhydrazone based schiff base synthesized from Isoniazid and 2-hydroxynaphthaldehyde (**Figure 62**). The sensor was found to produce an immediate

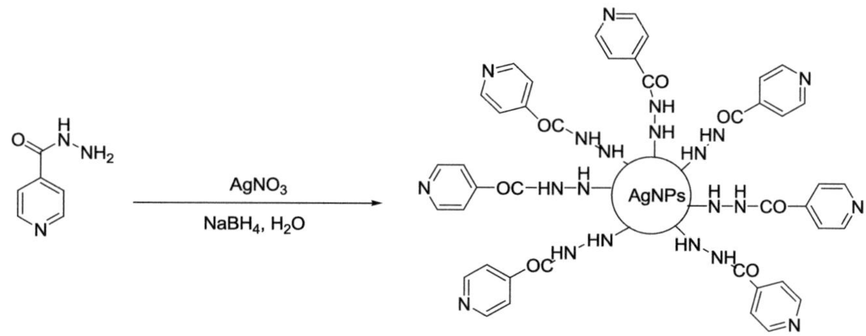

Figure 60.
Chemosensor based on isoniazid functionalized silver nanoparticles.

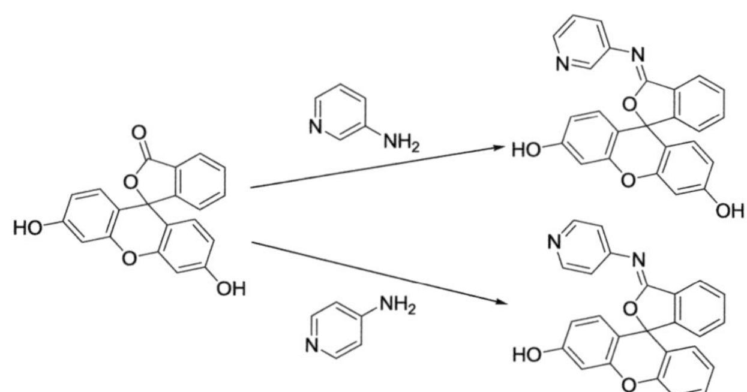

Figure 61.
Schiff bases derived from fluorescein with 3-aminopyridine and 4-aminopyridine.

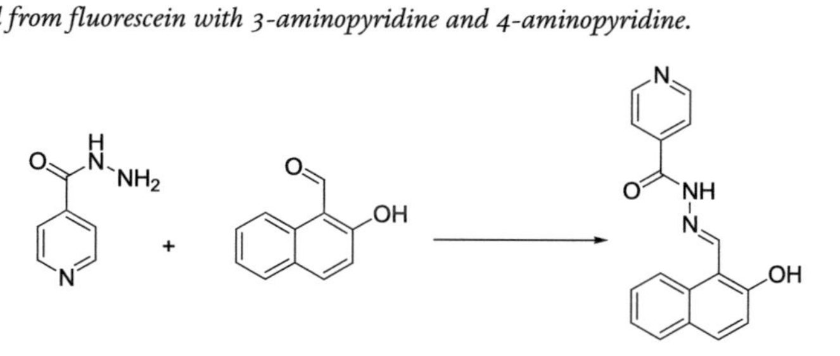

Figure 62.
Schiff base derived from Isoniazid and 2-hydroxynaphthaldehyde.

visible color change from colorless to yellow in the presence of CN^- ions in aqueous media with high selectivity and sensitivity [88].

Two schiff bases were prepared from pyridine-2-hydrazide with 5-nitrofuran-2-carboxaldehyde and 5-nitrothiophene-2-carboxaldehyde respectively (**Figure 63**) and tested for their anion sensing properties. The compound could selectively detect F^- and CO_3^{-2} ions over other interfering anions whereas compound could detect CO_3^{-2} ion with high selectivity and sensitivity. Finally, the compound was able to distinguish between F^- and CO_3^{-2} due to difference in their bathochromic shift [89].

A Hantzsch ester fluorescent probe based on thienyl-pyridine appended to dihydropyridine ring was synthesized (**Figure 64**) and applied for fluorescent sensing of nitric oxide in aqueous solution. The sensor showed extremely strong blue fluorescent which was switched off in the presence of NO and also possessed high selectivity and sensitivity towards NO [90].

A chemosensor based on 3,3′-(4-(2-amino-4,5-dimethoxyphenyl) pyridine-2,6-diyl) dianiline was synthesized (**Figure 65**) and found that it could detect formaldehyde through fluorescence enhancement and show the visible color change from yellow to blue. The compound could act as chemosensor for detection of formaldehyde qualitatively as well as quantitatively [91].

A simple Schiff base chemosensor was developed by the condensation reaction between 8-hydroxyjulolidine-9-carboxaldehyde and 2-hydrazinylpyridine (**Figure 66**). The ion recognition ability was determined for four transition metal ions

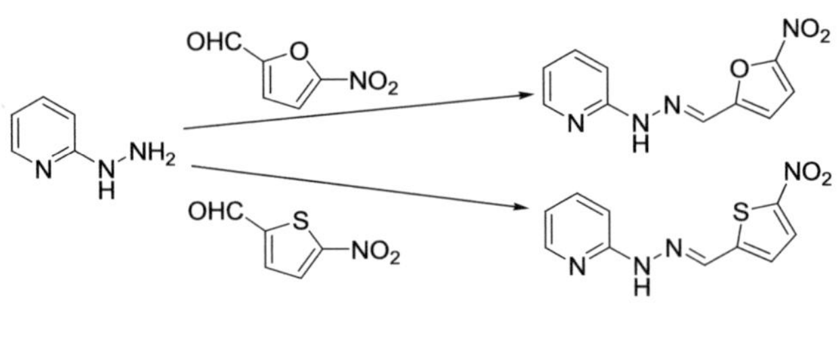

Figure 63.
Schiff bases derived from pyridine-2-carbohydrazide with 5-nitrofuran-2-aldehyde & 5-nitrothiophene-2-aldehyde.

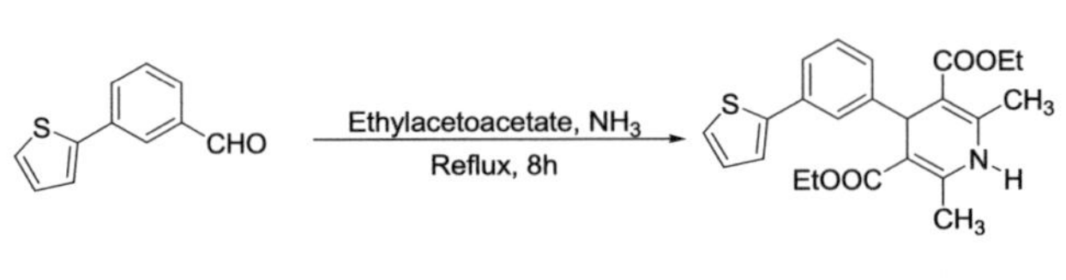

Figure 64.
Fluorescent probe based on thienyl-pyridine appended to dihydropyridine.

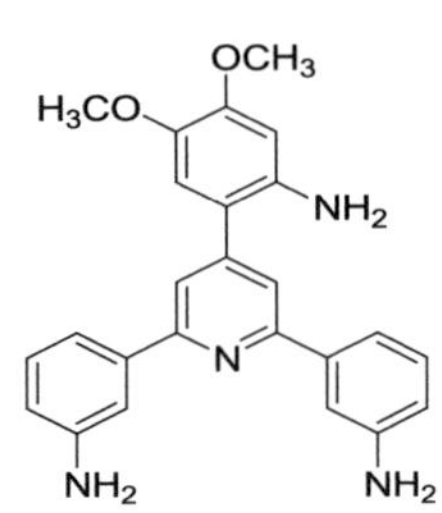

Figure 65.
Chemosensor based on 3,3′-(4-(2-amino-4,5-dimethoxyphenyl) pyridine-2,6-diyl) dianiline.

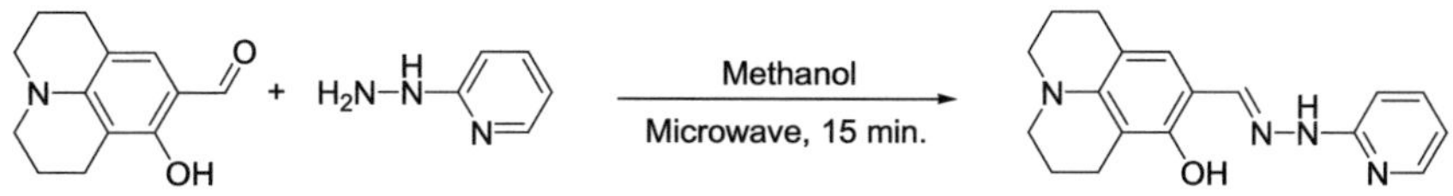

Figure 66.
Schiff base derived from 8-hydroxyjulolidine-9-carboxaldehyde and 2-hydrazinylpyridine.

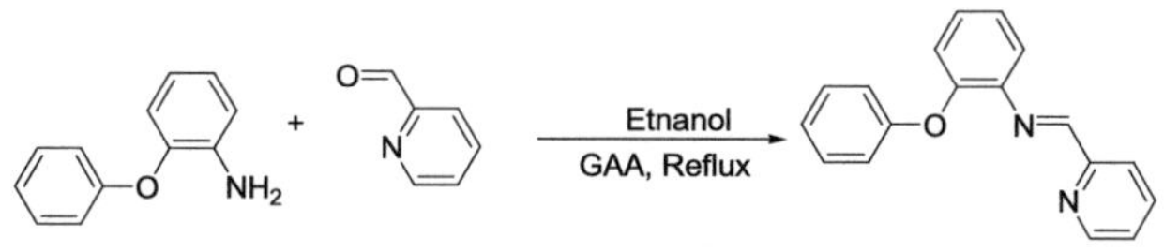

Figure 67.
Schiff base derived from 2-phenoxyaniline and pyridine-2-aldehyde.

Co^{+2}, Ni^{+2}, Cu^{+2} and Zn^{+2} using colorimetric and fluorescent analysis. It was revealed that the chemosensor can serve as an effective tool for the detection of all the four ions in environment as well as in biological applications [92].

A new fluorescent probe was designed from Schiff base 2-(pyridine-2-ylmethylene)-phenoxyaniline (**Figure 67**) and used for selective detection of Cd^{+2} ion. A significant fluorescence enhancement was observed and it gave satisfactory results for detection of Cd^{+2} ions in tap water and river water samples [93].

4. Conclusion

Pyridine is among the most valuable nitrogen-based heterocyclic compounds known for its important chemical and biological properties. The pyridine moieties are widely distributed in nature as in many naturally occurring compounds, vitamins, essential oils and metabolites which are required for various cellular functions. Additionally, pyridine derivative is used on large scale as precursor or intermediates in chemical and agrochemical products. Further, these derivatives possess therapeutic potentials due to their important bioactivities and with their proper structural modification or derivatization they can be led to important prodrugs or drugs. Literature review reveals that when pyridine-based nucleus is modified to some extent by introducing new functional group or even new molecule at appropriate positions, the bioactivity may be enhanced significantly. Thus, Schiff bases are continuously designed from amino or carboxaldehyde derivatives of pyridine since last few years and evaluated for their biological potential. As they possess a wide variety of biological activities, they are considered as a versatile pharmacophore and emerged as a potent class of pharmaceuticals for new drug development and continue to be an active area of research in medicinal chemistry. Development of novel drugs as a pharmacophore group is a constantly growing need that concerns researchers throughout the world as increasing number of diseases continue to be an emerging problem. The chemistry of pyridine-based Schiff bases is less extensive and not much work has been done in this field. In the view of the stated pharmacological properties of pyridine compounds, it is expected that they have high potential in the field of various biological activities that are still unexplored. Further, owing to their strong binding abilities towards various ions and unique photophysical properties, Schiff bases find applications in ion recognition and widely used as chemosensors for selective detection of ions. The ion recognition studies have gained extreme importance in recent years in the field of chemical, biological and environmental sciences. There is an urgent need to develop some efficient approaches to detect metal ions with high selectivity and sensitivity so that to control their harmful effect on human health and environment. It can be expected that the chemosensors derived from Schiff bases of pyridine derivatives do not have much complex structure and can be synthesized easily with good yield and purity as compared to most of other chemosensors. Thus, designing of specific chemosensor for the recognition of various ions is one of the most demanding areas of present chemical research due to their significant contribution in analytical, industrial, agricultural, environmental and biological fields. Keeping all these facts in the mind, it is of extreme importance to synthesize some Schiff bases derived from pyridine derivatives and to evaluate their potential as bioactive ligands and chemosensors. This chapter covers not solely the chemistry and biological significance of pyridine derivatives, but also reflects the light on Schiff bases derived from them with their pharmacological importance and ion recognition properties. It is worthwhile to have a full overview about pyridine, its derivatives and Schiff bases derived from them, all at one place with recent researches that will provide a single platform for potential researchers of these fields. Thus, the main objective of this chapter is to promote the research and development of some new pyridine-based Schiff bases and to evaluate their various biological activities for their effective use in drug designing process as well as their applications in ion recognition studies to develop more efficient chemosensors.

Acknowledgements

The author is greatly thankful to Dr. Gurpinder Singh for his valuable guidance with immense support and also the Department of Chemistry, Lovely Professional University for providing necessary facilities.

Conflict of interest

The author declares no conflict of interest.

Author details

Kaushal K. Joshi
Lovely Professional University, Phagwara, India

*Address all correspondence to: kaushalj28@gmail.com

References

[1] Arora P, Arora V, Lamba H. Importance of heterocyclic chemistry: A review. International Journal of Pharmaceutical Sciences and Research. 2012;**3**(9):2947-2955

[2] Abbas A. A review: Biological importance of heterocyclic compounds. Der Pharma Chemica. 2017;**9**(13):141-147

[3] Kerru N, Gummidi L, Maddila S. A review on recent advances in nitrogen-containing molecules and their biological applications. Molecules. 2020;**25**(8):1909

[4] Baumann M, Baxendale I. An overview of the synthetic routes to the best selling drugs containing 6-membered heterocycles. Beilstein Journal of Organic Chemistry. 2013;**9**: 2265-2319

[5] Kishbaugh T. Six-membered ring systems: Pyridine and benzo derivatives. Progress in Heterocyclic Chemistry. 2012;**24**:343-391

[6] Nevase M, Pawar R, Munjal P. Review on various molecule activity, biological activity and chemical activity of pyridine. European Journal of Pharmaceutical and Medical Research. 2018;**5**(11):184-192

[7] Patil P, Sethy S, Sameena T. Pyridine and its biological activity: A review. Asian Journal of Research in Chemistry. 2013;**6**(10)

[8] Chaubey A, Pandeya S. Pyridine: A versatile nucleus in pharmaceutical field. Asian Journal of Pharmaceutical and Clinical Research. 2011;**4**(4):1-4

[9] Altaf A, Shahzad A, Gul Z. A review on the medicinal importance of pyridine derivatives. Journal of Drug Design and Medicinal Chemistry. 2015;**1**(1):1-11

[10] Zakharychev V, Kuzenkov A, Martsynkevich A. Good pyridine hunting: A biomimic compound, a modifier and a unique pharmacophore in agrochemicals. Chemistry of Heterocyclic Compounds. 2020;**56**: 1491-1516

[11] Guan A, Liu C, Sun X. Discovery of pyridine-based agrochemicals by using intermediate derivatization methods. Bioorganic & Medicinal Chemistry. 2016;**24**(3):342-353

[12] Gujjarappa R, Nagaraju V, Malakar C. Recent advances in pyridine-based organocatalysis and its application towards valuable chemical transformations. ChemistrySelect. 2020; **5**(28):8745-8758

[13] Chelucci G. Metal complexes of optically active amino and imino based pyridine ligands in asymmetric catalysis. Coordination Chemistry Reviews. 2013; **257**(11–12):1887-1932

[14] Wang X, Qin C, Wang E. Synthese, structures and photoluminescence of a novel class of d10 metal complexes constructed from pyridine-3,4-dicarboxylicacid. Inorganic Chemistry. 2004;**43**(6):1850-1856

[15] Main group metal chalcogenidometalates with transition metal complexes of 1,10-phenanthroline and 2,2′-bipyridine. Coordination Chemistry Reviews. 2017; **330**:95-109

[16] Etaibi A, Apasery M. A comprehensive review on the synthesis and versatile applications of biologically active pyridine-based disperse dyes. International Journal of Environmental Research and Public Health. 2020;**17**(13): 4714

[17] Kashouti M, Molla M, Elsayad H. Synthesis of several new pyridine-2(1H) thiones containing an arylazo function and their applications in textile printing. Pigment & Resin Technology. 2008; **37**(2):80-86

[18] Ishihara M, Tsuneya T, Shiga M. New pyridine derivatives and basic components in spearmint oil (Menthagentilis) and peppermint oil (Menthapiperita). Journal of Agricultural and Food Chemistry. 1992; **40**(9):1647-1655

[19] Vernin G. Heterocyclic compounds in flavours and fragrances. Perfumer and Flavorist. 1982;7:23-35

[20] Ling Y, Hao Z, Liang D. The expanding role of pyridine and dihydropyridine scaffolds in drug design. Drug Design, Development and Theory. 2021;**15**:4289-4338

[21] Chiacchio M, Iannazzo D, Romeo R. Pyridine and pyrimidine derivatives as privileged scaffolds in biologically active agents. Current Medicinal Chemistry. 2019;**26**(40):7166-7195

[22] De S, Kumar A, Shah S. Pyridine: The scaffolds with significant clinical diversity. RSC Advances. 2022;**12**: 15385-15406

[23] Raczuk E, Dmochowska B, Fiertek J. Different Schiff bases—Structure, importance and classification. Molecules. 2022;**27**(3):787

[24] Qin W, Long S, Panunzio M. Schiff bases: A short survey on an evergreen chemistry tool. Molecules. 2013;**18**(10): 12264-12289

[25] Brodowska K, Chruscinska E. Schiff bases—Interesting range of applications in various fields of science. Chemik. 2014;**68**(2):129-134

[26] Kajal A, Bala S, Kamboj S. Schiff bases: A versatile pharmacophore. Journal of Catalysts. 2013;**893512**:1-14

[27] Chaudhary A, Singh A. Schiff bases: An emerging potent class of pharmaceuticals. International Journal of Current Research in Medical Sciences. 2017;**3**(5):60-74

[28] Nworie F, Nwabue F, Elom N. Schiff bases and Schiff base metal complexes: From synthesis to applications. Journal of Basic and Applied Research. 2016; **2**(3):295-305

[29] Dalia S, Afsan F, Hossain M. A sort review on chemistry of schiff base metal complexes and their catalytic application. International Journal of Chemical Studies. 2018;**6**(3):2859-2866

[30] Dief A, Mohamed I. A review on versatile applications of transition metal complexes incorporating Schiff bases. Beni-Suef University Journal of Basic and Applied Sciences. 2015;**4**(2):119-133

[31] Maher K, Mohammed S. Metal complexes of Schiff base derived from salicylaldehyde—A review. International Journal of Current Research and Review. 2015;7(2):6-16

[32] Usharani M, Akila E, Ashokan R. Pharmacological properties of Schiff base metal complexes derived from substituted pyridine and aromatic amine—A review. International Journal of Pharmaceutical Science and Health Care. 2013;**5**(3):1-11

[33] Induleka R, Anushyaveera P, Tamilselve M. Evaluation of anticancer activity of Schiff bases derived from pyridine and their metal complexes. Oriental Journal of Chemistry. 2022; **38**(3):1-8

[34] Du X, Li Y, Xia Y. Insights into protein-ligand interactions: Mechanisms,

models and methods. International Journal of Molecular Sciences. 2016; **17**(2):144

[35] Fu Y, Zhao J, Chen Z. Insights into the molecular mechanics of protein-ligand interactions by molecular docking and molecular dynamics simulation. Computational and Mathematical Methods in Medicine. 2018;**3502514**:1-12

[36] Jayaseelam P, Prasad S, Rajavel R. Synthesis, characterization, anti-microbial, DNA binding and cleavage studies of Schiff base metal complexes. Arabian Journal of Chemistry. 2011;**2011**: 1-10

[37] Sujarani S, Ramu A. Synthesis, characterization, anti-microbial and DNA interaction studies of benzophenone-ethanamine Schiff base with transition metal complexes. Journal of Chemical and Pharmaceutical Research. 2013;**5**(4):347-358

[38] Subbaraj P, Ramu A, Rama N. Synthesis, characterization, DNA interaction and pharmacological studies of substituted benzophenone derived Schiff base metal complexes. Journal of Saudi Chemical Society. 2015;**19**(2): 207-216

[39] Casella L, Gullotti M. Stereochemistry of pyridoxal amino acid model systems. Inorganica Chimica Acta. 1983;**79**:260-261

[40] Kumar J, Rai A, Raj V. A comprehensive review on the pharmacological activity of Schiff base containing derivatives. Organic and Medicinal Chemistry International Journal. 2017;**1**(3):88-102

[41] Chaturvedi D, Kamboj M. Role of Schiff Base in drug discovery research. Chemical Sciences Journal. 2016;**7**(2):1-2

[42] Mittal P, Joshi S, Panwar V. Biologically active Co^{+2}, Ni^{+2}, Cu^{+2} and Mn^{+2} complexes of Schiff bases derived from vinyl aniline and heterocyclic aldehydes. International Journal of ChemTech Research. 2009;**1**(2):225-232

[43] Zhang L, Ding T, Chen C. Biological activities of pyridine-2-carbaldehyde Schiff base derived from s-methyl and s-benzyl dithiocarbazate and their Zn^{+2} and Mn^{+2} complexes. Russian Journal of Coordination Chemistry. 2011;**37**(5): 356-361

[44] Shamim S, Murtaza S, Nazar M. Synthesis of Schiff bases of pyridine-4-carbaldehyde and their antioxidant and DNA binding studies. Journal of the Chemical Society of Pakistan. 2016; **38**(3):494-503

[45] Mohammed H, Taha N. Microwave preparation and spectroscopic investigation of binuclear Schiff base metal complexes derived from 2,6-diaminopyridine with salicylaldehyde. International Journal of Organic Chemistry. 2017;**7**(4):412-419

[46] Hearn M, Cynamon M, Chen M. Preparation and antitubercular activities in vitro and in vivo of novel Schiff bases of isoniazid. European Journal of Medicinal Chemistry. 2009;**44**(10): 4169-4178

[47] Sharma K, Singh R, Fahmi N. Synthesis, coordination behavior and investigations of pharmacological effects of some transition metal complexes with isoniazid Schiff bases. Journal of Coordination Chemistry. 2010;**63**(17): 3071-3082

[48] Prasanna M, Pradeep K. Synthesis, characterization and antimicrobial studies of transition metal complexes of hydroxymethoxybenzaldehyde isonicotinoylhydrazone. Research

Journal of Chemistry and Environment. 2013;**17**(6):61-67

[49] Alarabi H, Suayed W. Microwave assisted synthesis, characterization and antimicrobial studies of transition metal complexes of Schiff base ligand derived from isoniazid with 2-hydroxy naphthaldehyde. Journal of Chemical and Pharmaceutical Research. 2014;**6**(1): 595-602

[50] Kehinde O, Joseph A, Cyrila E. Synthesis, characterization, theoretical treatment and antitubercular activity evaluation of N-(2,5-dimethoxybenzylidene) nicotinohydrazide and some of its transition metal complexes against mycobacterium tuberculosis, H37Rv. Oriental Journal of Chemistry. 2016; **32**(1):413-427

[51] Rehman N, Khalid M, Bhatti M. Schiff base of isoniazid and ketoprofen: Synthesis, x-ray crystallographic, spectroscopic, antioxidant and computational studies. Turkish Journal of Chemistry. 2018;**42**: 639-651

[52] Carreno A, Rodriguez L, Hernandez D. Two new fluorinated phenol derivatives pyridine Schiff bases: Synthesis, spectral, theoretical characterization, inclusion in epichlorhydrin-β-cyclodextrin polymer, and antifungal effect. Frontiers in Chemistry. 2018;**6**:312

[53] Kamga F, Mainsah E, Kuate M. Synthesis, characterization and biological activities of binuclear metal complexes of 2-benzoylpyridine and phenyl(pyridine-2-yl) methanediol derived from 1-phenyl-1-(pyridine-2-yl)-N-(pyrimidin-2-yl) methaniminedihydrate Schiff base. Open Journal of Inorganic Chemistry. 2021;**11**: 20-42

[54] Borase J, Mahale R, Rajput S. Design, synthesis and biological evaluation of heterocyclic methyl substituted pyridine Schiff base transition metal complexes. SN Applied Sciences. 2021;**3**(197):1-13

[55] Sahni T, Sharma S, Verma D. Experimental validation of syringic Schiff bases with pyridine moiety as antibacterial and antioxidant agents along with in silico studies. The Pharma Innovation Journal. 2022;**SP-11(4)**: 417-426

[56] Pund A, Shaikh M, Chandak B. Pyridine −1,3,4-thiadiazole-Schiff base derivatives, as antioxidant and antimitotic agent: Synthesis and in silico ADME studies. Polycyclic Aromatic Compounds. 2022;**2022**:1-6

[57] Wu D, Sedgwick A, Gunnlaugsson T. Fluorescent chemosensors: The past, present and future. Chemical Society Reviews. 2017;**46**:7105-7123

[58] Silva A, Moody T, Wright G. Fluorescent PET (photoinduced electro transfer) sensors as potent analytical tools. The Analyst. 2009;**134**:2385-2393

[59] Silva A, Gunaratne H, Gunnlaugsson T. Signalling recognition events with fluorescent sensors and switches. Chemical Reviews. 1997;**97**(5): 1515-1566

[60] Czarnik A. Chemical communication in water using fluorescent chemosensors. Accounts of Chemical Research. 1994;**27**(10): 302-308

[61] Fabbrizzi L, Licchelli M, Pallavicini P. Transition metals as switches. Accounts of Chemical Research. 1999;**32**(10):846-853

[62] Hamilton G, Sahoo S, Kamila S. Optical probes for the detection of

protons, and alkali and alkaline earth metal cations. Chemical Society Reviews. 2015;**44**:4415-4432

[63] Kaur R, Kaur A, Singh G. Anion recognition properties of chromone based organic and organic-inorganic hybrid nanoparticles. Analytical Methods. 2014;**6**:5620-5626

[64] Duke R, Veale E, Pteffer F. Colorimetric and fluorescent anion sensors: An overview of recent developments in the use of 1,8-naphthalimide-based chemosensors. Chemical Society Reviews. 2010;**39**: 3936-3953

[65] Gunnlaugsson T, Glynn M, Tocci G. Anion recognition and sensing in organic and aqueous media using luminescent and colorimetric sensors. Coordination ChemistryReviews. 2006;**250**(23–24): 3094-3117

[66] Bader N. Applications of Schiff base chelates in quantitative analysis: A review. Rasayan Journal of Chemistry. 2010;**3**(4):660-670

[67] Berhanu A, Mohiuddin I, Malik A. A review of applications of Schiff bases as optical chemical sensors. TrAC, Trends in Analytical Chemistry. 2019;**116**:74-91

[68] Kolhe S, Patil D. Application of Schiff Base as a fluorescence sensor. Journal of Emerging Technologies and Innovative Research. 2019;**6**(3):175-181

[69] Dalapati S, Jana S, Guchhait N. Anion recognition by simple chromogenic and chromo-fluorogenic salicylidene Schiff base or reduced Schiff base receptors. Spectrochimica Acta Part A Molecular and Biomolecular Spectroscopy. 2014;**129C**(33):499-508

[70] Jimoh A, Helal A, Shaikh M. Schiff base ligand coated gold nanoparticles for the chemical sensing of Fe (III) ions. Journal of Nanomaterials. 2015;**101694**: 1-7

[71] Gupta V, Kumar S, Kumar R. A highly selective colorimetric and turn on fluorescent chemosensor based on 1-(2-pyridylazo)-2-naphthol for the detection of Al^{+3}ions. Sensors and Actuators B, Chemical. 2015;**209**:15-24

[72] Xu J, Li H, Li L. A highly selective fluorescent chemosensor for Al^{+3} based on 2,2′:6′,2-terpyridine with a salicylal Schiff base. Journal of Brazilian Chemical Society. 2020;**31**:1-14

[73] GhoraiA MJ, Chandra R. A reversible fluorescent-colorimetric imino-pyridylbis-schiff base sensor for expeditious detection of Al^{+3} and HSO^{3-} in aqueous media. Dalton Transactions. 2015;**44**:13261-13271

[74] Xian Z, Zhang L, Zhao W. Fluoranthene based pyridine as fluorescent chemosensor for Fe^{+3}. Inorganic Chemistry Communications. 2011;**14**:1656-1658

[75] Feng X, Tian P, Xu Z. Fluorescence-enhanced chemosensor for metal cation detection based on pyridine and carbazole. The Journal of Organic Chemistry. 2013;**78**(22):11318-11325

[76] Zhang J, Wu Q, Yu B. A pyridine containing Cu^{+2} selective probes based on naphthalimide derivative. Sensors (Basel). 2014;**14**(12):24146-24155

[77] Huang L, Zhang J, Yu X. A Cu^{+2} selective fluorescent chemosensor based on BODIPY with two pyridine ligands and logic gate. Spectrochimica Acta Part A Molecular and Biomolecular Spectroscopy. 2015;**145**:25-32

[78] Yadav N, Singh A. Dicarbohydrazide based chemosensors for copper and

cyanide ions via a displacement approach. New Journal ofChemistry. 2018;**42**:6023-6033

[79] Kumar R, Jain H, Gahlyan P. A highly sensitive pyridine-dicarbohydrazide based chemosensor for colorimetric recognition of Cu^{+2}, AMP^{-2}, F^{-} and AcO^{-} ions. New Journal of Chemistry. 2018;**42**:8567-8576

[80] Annaraj B, Neelakantan M. Water soluble pyridine based colorimetric chemosensor for naked eye detection of silver ion: Design, synthesis, spectral and theoretical investigation. Analytical Methods. 2014;**6**:9610-9615

[81] Chen Z, Tang Y, Liang H. Synthesis, crystal structure and spectroscopic characterization of Ag^{+1}, Cd^{+2} complexes with the Schiff base derived from pyridine-2-carboxaldehyde and 4-e-2-phenyldiazenylaniline. Journal of CoordinationChemistry. 2006;**59**(2): 207-214

[82] Luo H, Jiang J, Zhang X. Synthesis of porphyrin-appended terpyridine as a chemosensor for cadmium based on fluorescent enhancement. Talanta. 2007; **72**(2):575-581

[83] Tayade K, Kuwar A, Fegade U. Design and synthesis of a pyridine based chemosensor: Highly selective fluorescent probe for Pb^{+2}. Journal of Fluorescence. 2014;**24**:19-26

[84] Hamid A, Al-Khateeb M, Tahat Z. A selective chemosensor for mercuric ions based on 4-aminothiophenol-ruthenium (II)bis(pyridine) complex. International Journal of Inorganic Chemistry. 2011; **843051**:1-6

[85] Pan J, Zhu F, Kong L. A simple pyridine-based colorimetric chemosensor for highly sensitive and selective mercury (II) detection with the naked eye. Chemical Papers. 2015;**69**(4): 527-535

[86] Sakthivel P, Karuppannan S. A sensitive isoniazid capped silver nanoparticles-selective colorimetric fluorescent sensor for Hg^{+2} ions in aqueous media. Journal of Fluorescence. 2020;**30**:91-101

[87] Yan F, Jiang Y, Fan K. Novel fluorescein and pyridine conjugated schiff base probes for the recyclable real-time determination of Ce^{+3} and F. Methods and Applications in Fluorescence. 2020;**8**(1):015002

[88] Hu J, Li J, Qi J. Selective colorimetric and turn-on fluorimetric detection of cyanide using an acylhydrazone sensor in aqueous media. New Journal of Chemistry. 2015;**39**:4041-4046

[89] Singh A, Mohan M, Trivedi D. Chemosensor based on hydrazinyl pyridine for selective detection of F- ions in organic media and CO_3^{-2} ions in aqueous media: Design, synthesis, characterization and practical application. Chemistry Select. 2019; **4**(48):14120-14131

[90] Ali S, Pramanik A, Samanta S. A thienyl-pyridine-based Hantzsch ester probe for the selective detection of nitric oxide and its bio-imaging applications. Journal of the Indian Chemical Society. 2017;**94**(7):1-10

[91] Hidayah N, Purwono B, Nurohmah B. Synthesis of pyridine derivative-based chemosensor for formaldehyde detection. Indonesian Journal of Chemistry. 2019;**19**(4): 1074-1080

[92] Liu H, Ding S, Lu Q. A versatile Schiff base chemosensor for the determination of trace Co^{+2}, Ni^{+2}, Cu^{+2} and Zn^{+2} in the water and its bioimaging

applications. ACS Omega. 2022;7:
7585-7594

[93] Ma J, Dong Y, Yu Z. A pyridine-based Schiff base as a selective and sensitive fluorescent probe for cadmium ions with "turn-on" fluorescence responses. New Journal of Chemistry. 2022;**46**(7):3348-3357

Chapter 10

2(4)-Aminopyridines as Ligands in the Coordination and Extraction Chemistry of Platinum Metals

Liliya Sergeevna Ageeva, Nikolai Alekseevich Borsch and Nikolay Vladimirovich Kuvardin

Abstract

The specific behavior of aromatic amines in the coordination and extraction processes of isolation and separation of platinum and other metals is discussed using the example of 2(4)-aminopyridines (2(4)-AP). As intrasphere ligands, 2(4)-AP have a high electron-donor capacity due to the pumping of an easily polarizable π-electron density. The chemistry of the extraction of platinum metals, iridium in particular, is considered: depending on the conditions, ion associates, coordination-solvated compounds or compounds containing an amine in the inner and outer coordination sphere of the metal are extracted. In the extraction of simple singly charged anions, there is a violation of the exchange-extraction series established for a large set of aliphatic amines. Soft anions (according to Pearson), for example, SCN- and I-, are best extracted, while for aliphatic amines such an anion is hard ClO_4^-. In the coordination compounds of platinum metals, 2(4)-AP acts as an electron donor, is coordinated by heterocyclic nitrogen with a redistribution of electron density not only to the accepting metal-complexing agent, but also further along the N-Me-X chain (X is an acido ligand in the composition of the complex), which leads to even greater covalence of the molecule as a whole.

Keywords: 2(4)-aminopyridines, platinum metals, extraction, complex formation, coordination compounds

1. Introduction

In recent years, interest has increased in the study of the extraction properties of high-molecular-weight aromatic amines, primarily because 2(4)-octylaminopyridines turned out to be good extractants for the isolation and separation of platinum metals [1–3]. Particularly interesting is the question of the specificity with respect to platinum metals of aromatic amines, as ligands, which differ from aliphatic amines in that the lone pair of electrons of the nitrogen atom largely acquires an π-donor character. Compared to aliphatic amines, aromatic amines demonstrate a number of new properties in coordination and extraction chemistry [2, 4]. All this determined the interest in this class of extractants, typical representatives of which are 2(4)-octylamino-pyridines (2(4)-OAP).

Research carried out by the authors [5–9], allow us to get an idea of the specifics of the behavior of 2(4)-aminopyridines in the coordination and extraction processes of isolation and separation of metals.

2. Specificity 2(4)-aminopyridinesas ligands

The specificity of 2(4)-aminopyridines as ligands is due to the nature of the nitrogen atom in aromatic amines, which can be judged from the results of studies of halides [8] and coordination compounds of 2-OAP with nickel, palladium, and platinum [9]. In metal complexes, 2-OAP, as an intrasphere ligand, has a high electron-donating capacity due to the pumping of an easily polarizable π-electron density. The mobility of the electron density in the 2-OAP molecule depending on the requirements of the acceptor is evidenced by the delocalization of the positive charge of the proton in the cation (outer sphere ligand), which is the higher, the greater the polarizability of the anion [8].

An idea of the mobility of the electron density in a 2(4)-OAP molecule can be obtained within the framework of the theory of limiting structures (the theory of resonance) [10], giving an account of a certain formalism of this theory. The conclusions obtained in the framework of the theory of resonance and the theory of perturbations of molecular orbitales (PMO), which have a deep quantum-chemical substantiation [11], are quite adequate.

Conventionally, the 2-OAP molecule (**Figure 1**) [2], as well as the 4-OAP molecule (**Figures 2** and **3**) [5], can be represented as an average between the amine and pyridonimine limiting structures.

The contribution of the pyridoniminine structure increases the electron density on the heterocyclic nitrogen and decreases the electron density on the amine nitrogen. This contribution can be estimated if the energy of the N1s level of heterocyclic and amine nitrogen atoms is known. **Figure 4** shows, as an example, the experimental X-ray electron spectra of the N1s level of 4-OAP with band separation obtained on a

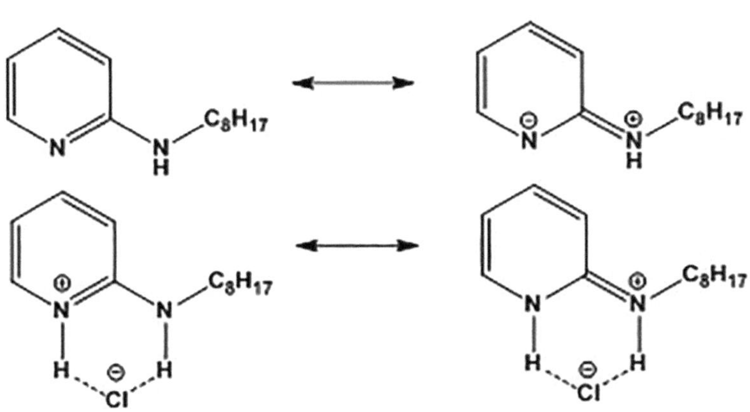

Figure 1.
Limiting (resonant) structures of the neutral and the protonated 2-OAP molecule.

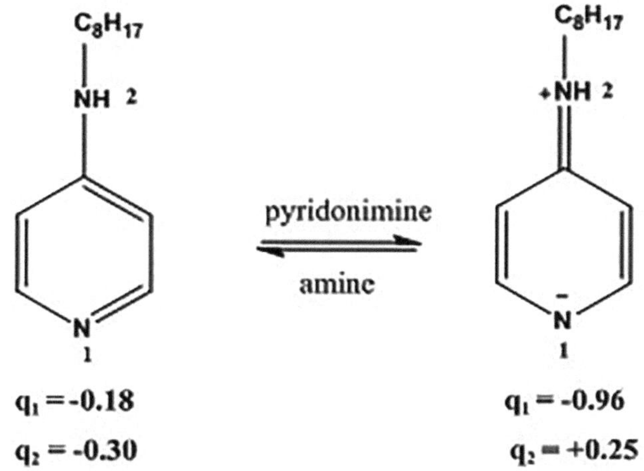

Figure 2.
Limiting (resonance) structures of 4-OAP molecules and the effective charges on nitrogen atoms calculated for them.

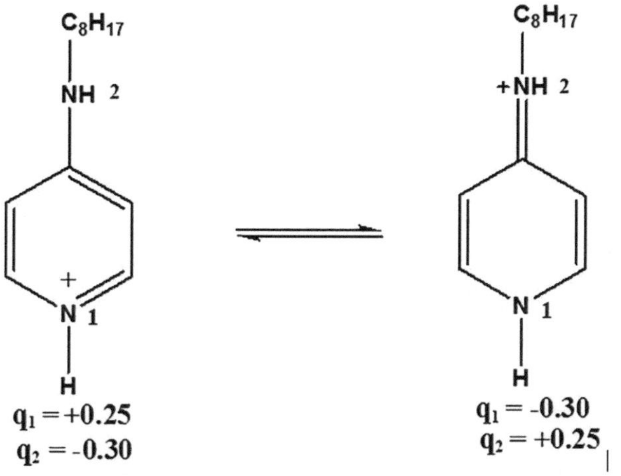

Figure 3.
Limiting (resonant) structures in protonated 4-OAP and effective charges on nitrogen atoms calculated for them.

Riber SIA-200 X-ray photoelectron spectrometer. It can be seen that the nitrogen atoms are not equivalent, the lower level refers to heterocyclic nitrogen.

The energy of the N1s level correlates with the effective charge on the nitrogen atom. Satisfactory correlation of these values for a large group of nitrogen-containing compounds of various structures was obtained in [12]. Effective charges on nitrogen atoms for limiting structures can be calculated using the concept of ionic nature (**Figures 2** and **4**) [10]. From the charge balance equations for nitrogen atoms, the contribution pyridoniminine structure in 2-OAP and 4-OAP molecules: 9 [2] and 48.6%, respectively.

Thus, in the first approximation, it can be assumed that 2(4)-OAP molecules represent a resonant structure with a contribution from the pyridoniminine component. This leads to an increased basicity of heterocyclic nitrogen compared to pyridine

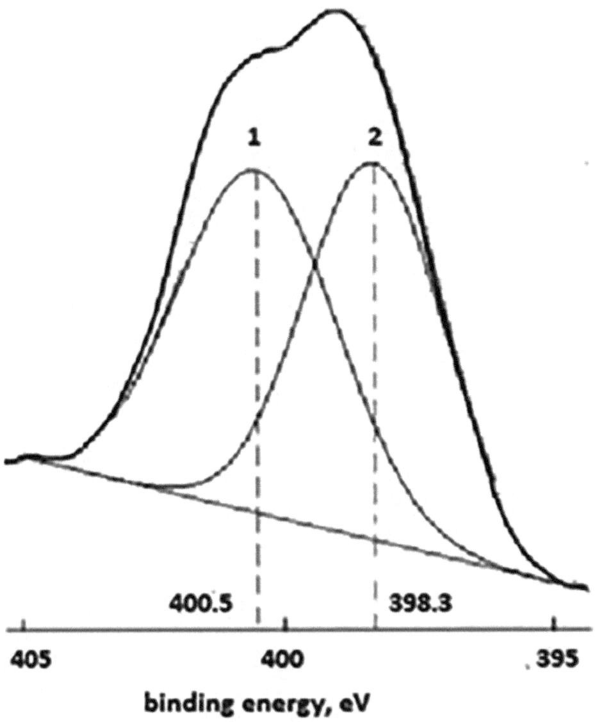

Figure 4.
X-ray spectrum of 4-OAP. Peak area: 1–50.63, 2–49.37%.

due to the pumping of electron density from the amino group in the ortho and para positions of the pyridine ring and partial delocalization of the charge in the cation. Since the "depth" of the resonance is higher in the case of 4-OAP, its basicity exceeds that of 2-OAP by two orders of magnitude.

All this points to the "soft" nature of 2(4)-AP as ligands. If we use Pearson's classification [13], then free 2(4)-AP should be attributed to "soft" bases (inner sphere ligand), protonated – to "soft" acids (outer sphere ligand). Soft and intermediate bases include other aromatic amines, while aliphatic amines are "hard" bases.

Factors such as the energetic and spatial arrangement of the top donor orbital of the nitrogen atom are thought to be responsible for the "soft" or "hard" behavior of the amine. From the standpoint of the quantum theory of perturbed molecular orbitals (PMO), one can consider the energy of the metal–ligand interaction and the resulting extraction chemistry depending on the nature of the amine, metal, and extraction conditions [14].

The formation of a coordination-solvated compound or associate, where the metal is present in the composition of the acid complex, depends on the result of the competitive process of complexation of the amine and proton, on the one hand, and the metal, on the other. In the first approximation, the quantitative side of this process is expressed by the main equation of the PMO theory [14]. An ionic associate or a coordination-solvated compound is formed depending on the relative contribution of the Coulomb or covalent component to the metal-nitrogen interaction energy.

If the contribution of the covalent component is much greater than that of the Coulomb component, then the coordination of the amine by the metal in the presence of a proton is possible. The higher the energy of the donor orbital and the lower the energy of the acceptor orbital of the amine and metal, the greater the contribution of the covalent component. These energy parameters of the interacting orbitals within the PMO are characterized by the orbital electronegativity of the donor and acceptor according to Klopman [13]. In addition, the covalent component increases with the length of the amine donor orbital.

If we talk about the nature of the amine, then in the presence of a proton, only amines with a low orbital electronegativity of the lone electron pair (OELEP), which depends on the valence state of the nitrogen atom in the amine molecule, can be coordinated by the metal. The OELEP of nitrogen decreases with an increase in the ρ- and π-character of an unshared pair of electrons, that is, with a decrease in the energy of the donor orbital and with an increase in its population [13]. Consequently, the OELEP of nitrogen decreases as one goes from aliphatic amines to anilines and further to heteroaromatic amines. In the same series, the softness of amines and their ability to extract platinum metals in the form of coordination-solvated compounds increase.

Of no less interest is the behavior of protonated 2(4)-AP, which acts as an outer-sphere ligand with respect to acid complexes of platinum and other rare metals.

3. Chemistry of metal extraction

2(4)-OAP extract metals from acidic and slightly acidic solutions. The extraction of iridium and other platinum metals has been studied most fully [2, 6].

Iridium (III) is extracted with a 0.1 M solution of 2-OAP in chloroform from dilute hydrochloric acid solutions with distribution coefficients D = 100–200, however, a nonequilibrium minimum appears on the curve D = f (pH) at pH 2 (the contact time of the phases is 30 min). During extraction from 1 to 6 M HCl equilibrium is established slowly: the value of D increases by almost an order of magnitude with an increase in the duration of phase contact up to 50 hours. Iridium (IV) is reduced to iridium (III) during extraction. 4-OAP extracts iridium (IV) with high distribution coefficients from more acidic solutions [3].

When 2-OAP is introduced directly into the aqueous phase (0.1 M solution in acetone) and the solution is heated to boiling in the presence of a tin (II) chloride catalyst for 30–40 min followed by extraction of the resulting compounds with chloroform ("heterogeneous" extraction), iridium is extracted with an unusually high partition coefficient for this element. The maximum extraction is observed from 1 to 2 M HC1 and reaches 99.9% for a single extraction.

Other platinum metals, as well as Au, under conditions optimal for the extraction of iridium, are extracted much worse (**Table 1**), and gold is quantitatively, and silver and palladium are partially concentrated at the phase boundary. The distribution coefficient of non-ferrous metals and iron, from which iridium usually needs to be separated, under these conditions by 3–5 orders of magnitude lower than iridium. Of these elements, only copper in the form of Cu (II) passes into the organic phase in a noticeable amount. At a high concentration of $SnCl_2$ in the aqueous phase, the organic phase contains tin.

Since alkylated 2(4)-AP – strong organic bases, they are able to extract halide and other metal acid complexes in the form of ion associates.

On the other hand, 2(4)-OAP can be considered as a potentially coordinating-active reagent due to the presence of heterocyclic aromatic nitrogen. In addition, during extraction, the formation of chelates due to the NH_2 group in α-position to the

Metal	HCl, M		Metal	HCl, M	
	one	3		one	3
Ir	800–1000	300–400	Fe	0.004	0.002
Rh	68	140	Ni	<0.001	<0.001
Pt	62	111	co	<0.004	<0.002
Pd	—	—	Zn	0.03	0.004
Ag	—	—	Sn (II)	0.03	0.014
Au	—	—	Sn (IV)	0.01	0.008
Cu	0.44	0.53			

Table 1.
Distribution coefficient of some metals in the extraction of 2-OAP under iridium extraction conditions: Ir, Pt, Au, Ag – $1{\cdot}10^{-4}$-$1{\cdot}10^{-5}$ M; Rh, Pd – $5{\cdot}10^{-4}$; 0.05 M 2-OAP, 0.1 M $SnCl_2$, heating for 40 min at 100°C, phase contact for 15 min, organic phase – Chloroform.

heterocyclic nitrogen. In the course of extraction, one or another mechanism is realized depending on the conditions [13].

Extraction of ion associates. In the form of ionic associates, platinum metals are extracted from HC1 solutions at a certain ratio of the concentration of components and the duration of phase contact [2]. Palladium (II) is extracted by 2-OAP predominantly in the form of an associate $(OAPH^{+})_2[PdC14]$ only from concentrated solutions of HC1 and when organic diluents are used solvents with a strong proton-donating ability. Platinum is extracted in the form of such a complex already from acidic solutions of HC1. Ir (III, IV) are predominantly extracted in the form of ionic associates of the composition $(OAPH^{+})_2[IrCl_6]$ and $(OAPH^{+})_3[IrCl_6]$ from 1 to 6 M HC1, especially with a short duration of phase contact.

According to this mechanism, Pd (II) is extracted from salicylate solutions with 4-heptylaminopyridine [15], and from oxalate solutions with 4-dodecylaminopyridine [16]. Ro (III) is extracted from citrate solutions with 2-dodecylaminopyridine [17], and from Ru (III) succinate solutions with 2-OAP [18]. From acetate solutions, 2-OAP extracts Ir (III) [19] from malonate – Au (III) [20]. 2-OAP and other metals are extracted in the form of ionic associates from chloride, malonate, succinate, salicylate, citrate media: To (IV) [21], Zr (IV) [22], V (V) [23], Mo (VI) [24], Cr (VI) [25], Bi (III) [26], Ga (III) [27], Tl (III) [28], Sm (III) [29], Hg (II) [30].

Anions of inorganic acids are also extracted as ionic associates [8].

Extraction of coordination-solvated compounds. Cu (II) is extracted with a solution of 2-OAP in chloroform from a neutral medium in the presence of at least 1 g-ion/l chloride ion, apparently in the form of a neutral coordination-solvated complex. In the case of Pd (II), a neutral diamine complex of the composition $Pd(OAP)_2C1_2$ is formed during extraction from solutions with a concentration of HC1 $\leq$ 3 M, Pt (II) – in the form of $Pt(OAP)_2Cl_2$ from weakly acid solutions of HC1 (pH >1.5); the phase contact duration is 30 min [2]. Most often, the organic phase contains compounds with 2(4)-OAP in the inner and outer coordination spheres of the metal ("mixed" extraction mechanism).

Mixed extraction mechanism. Iridium, under conditions optimal for its extraction, is extracted in the form of compounds containing 2-OAP in the inner and outer coordination spheres of the metal [7]. In addition to 2-OAP, the extractable compounds include $SnCl_2$, which is added to overcome the kinetic inertness of the initial complex iridium chlorides [6]. Complexation in the aqueous phase and subsequent extraction of the resulting compounds are described by the following Equations [7]:

$$[IrCl_6]^{3-}{}_w + m[SnCl_3]^-{}_w \rightarrow [Ir(SnCl_3)_m Cl_{6-m}]^{3-}{}_w + mCl_w \tag{1}$$

$$\begin{aligned}&[\mathbf{Ir(SnCl_3)_m Cl_{(6-m)}}]^{3-}{}_w + xOA\Pi H^+{}_w \rightarrow [\mathbf{Ir(SnCl_3)_{m-x}\,(OA\Pi)_x Cl_{(6-m)}}]^{x-3}{}_w \\ &+ xH_w + xC1_B + x[\mathbf{SnCl_3}]^-{}_w\end{aligned} \tag{2}$$

$$\begin{aligned}&[\mathbf{Ir(SnCl_3)_{m-x}\,(OA\Pi)_x Cl_{(6-m)}}]^{x-3}{}_w + (\mathbf{x}-\mathbf{3})OA\Pi \cdot HCl_o \\ &\rightarrow (OA\Pi H^+)_{x-3}[\mathbf{Ir(SnCl_3)_{m-x}\,(OA\Pi)_x Cl_{(6-m)}}]^{x-3}{}_o + (\mathbf{3}-\mathbf{x})\mathbf{Cl}^-{}_w\end{aligned} \tag{3}$$

Here x = (0–2), m = 1–6; component concentration interval: $1{\cdot}10^{-5}$-$1{\cdot}10^{-3}$ g-at/l Ir; 0.05–0.2 M; $\mathbf{SnCl_2} \leq 0.1$ M 2-OAP in acetone; 1–6 M HCl.

The ratio of ligands in the inner coordination sphere of iridium is determined by the concentration of the components in the specified range, as well as the temperature and duration of heating the solution before extraction. At a low concentration of 2-OAP, along with coordination-solvated compounds, anionic iridium chlorotin complexes are extracted that do not contain 2-OAP in the inner coordination sphere of the metal. The ratio between these two types of extractable compounds under these conditions can be estimated from the results of a physicochemical study of the extraction of iridium in the presence and absence of OAP in the aqueous phase upon heating [7].

The given chemistry of iridium extraction is confirmed by the study of extracts by high-voltage electrophoresis on paper and iridium compounds isolated from the extract using physicochemical and spectral methods of analysis [7]. These compounds are a dark brown pasty substance. The total content of the organic component (C, H, N), according to elemental analysis, is 41.95%, which indicates a high molecular weight of the anionic part of the associate; indirectly indicates the presence of tin. Direct evidence for the presence of tin in the complex is the Mossbauer spectrum of the compound on ^{119}Sn nuclei, which is characteristic of the $[SnCl_3]$ – ligand in the iridium coordination sphere (chemical shift 1.65 mm/s, quadrupole splitting 2.33 mm/s). Significant quadrupole splitting in the Mossbauer spectrum of the compound indicates the presence of 2-OAP in the inner coordination sphere of the metal.

This conclusion most convincingly follows from the data of PMR spectroscopy of substances before and after electrophoresis: in the spectrum of the substance after the separation of the cationic part, signals from the protons of the hetero ring and the octyl radical are clearly recorded; 2-OAP is indeed part of the anionic part of the associate and is coordinated by iridium. If we take into account the results of elemental analysis (32.38% C, 4.70% H, 4.87% N), then the probable composition of the compound is $(OAPH^+)[Ir(OAP)_2(SnCl_3)_3C1]^-$, possibly impurity of the complex $(OAPH^+)_2[Ir(OAP)(SnCl_3)_2Cl_3]^{2-}$.

Compounds containing OAP in the inner and outer coordination spheres of the metal can be extracted without preliminary heating of the metal solution with OAP in the aqueous phase, if its kinetic inertness is relatively low. In particular, the results of the study of platinum extracts using electron spectroscopy and thin layer chromatography [2] can be explained if the presence of the associate $(OAPH^+)[Pt(OAP)C1_3]$ – is assumed in the organic phase.

In principle, more than two molecules of 2(4)-AP can enter into the coordination sphere of a metal. In this case, the formation of complexes containing the metal in the

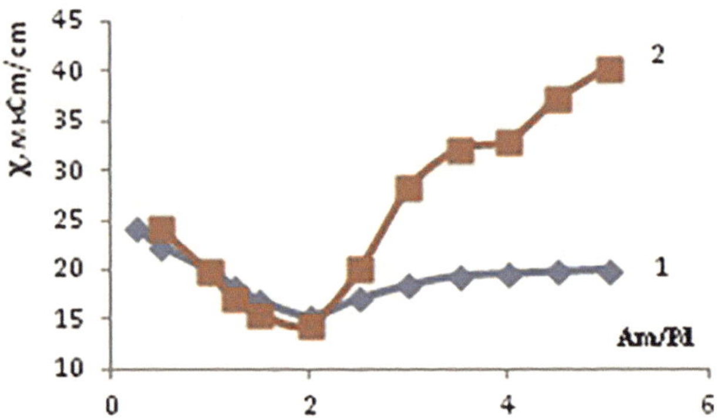

Figure 5.
Electrical conductivity of 1 10^{-4} M aqueous solution of $K_2[PdCl_4]$ depending on the metal/amine molar ratio with the addition of: 1–2-AP; 2–4-AP.

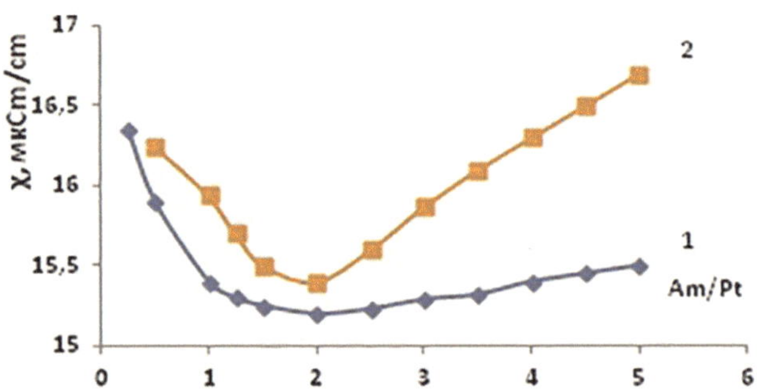

Figure 6.
Electrical conductivity of 1 10–4 M aqueous solution of $K_2[PtCl_4]$ depending on the metal/amine molar ratio with the addition of: 1–2-AP; 2–4-AP.

cationic form is not excluded. Under the conditions of extraction of iridium, such compounds should precipitate at the phase boundary, which is observed in the extraction of palladium, as well as gold and silver, and only if 2-OAP is present during heating in the aqueous phase [2].

Another proof of the possibility of the formation of cationic complexes are the results of conductometric and spectrophotometric studies of complex formation 2(4)-AP with Pd (II), Pt (II) in aqueous solutions at concentrations of reagents $1{\cdot}10^{-5}$ - $1{\cdot}10^{-4}$ M, simulating the extraction conditions (**Figures 5** and **6**). The conductometric curves χ = f (CAm/CMe) show breaks at $C_{Am/CMe}$ = 2 and 4.

Thus, the chemistry of 2(4)-AP metal extraction can be quite complex. Depending on the nature of the metal and extraction conditions, associates containing 2(4)-AP only in the cationic part, and the metal in the anionic part, associates with OAP in the inner and outer coordination spheres of the metal, neutral coordination-solvated compounds can pass into the organic phase; the formation of cationic complexes is also not excluded.

4. Interionic interactions in 2(4)-AP associates

Extraction of hydrochloric acid with a solution of 2(4)-OAP in chloroform according to the neutralization mechanism is described by the equation:

Anion, X^-	lgK_{AmH+X-}		-ΔHh, kcal/mol	-ΔSh, kcal/mol·grad	-ΔGh, kcal/mol	R, A°
	2AP	4-OAP				
I^-	−2.32 ± 0.05	3.84	67	8.05	64	—
SCN^-	−2.74 ± 0.06	3.63	74	20*	68	1.95
Br^-	−3.24 ± 0.05	3.12	76	13.42	72	—
ClO_4^-	−3.13 ± 0.10	3.53	54	13.30	50	2.36
NO_3^-	−3.27 ± 0.04	2.99	74	16.90	69	1.89
Cl^-	−3.31 ± 0.07	2.48	84	17.10	79	—
F^-	−3.35 ± 0.09	1.63	116	30.70	107	—

Table 2.
Distribution constants of 2-AP and 4-OAP salts between chloroform and water (25 ± 2° C, μ 1) and thermodynamic characteristics of anion hydration in infinitely dilute solutions at 298°K (– calculated by correlation dependence ΔSh = f (R, A°), where R– radius of ions in water.*

$K_{ex} = K_{AmH+Cl-}(K_aK_D)^{-1}$, where K_{ex} is the HCl extraction constant, $K_{AmH+Cl-}$ is the chloride distribution constant, K_a is the ionization constant of the protonated amine, K_D is the amine distribution constant.

2-OAP chloride with an intramolecular hydrogen bond passes into the organic phase and contains practically no water molecules, which apparently explains the low over stoichiometric extraction only from 12 M HCl [1]. On the contrary, for 4-OAP, a high over stoichiometric extraction is observed already from 6 M HCl, since in this case the chelate cycle based on the intramolecular hydrogen bond is not formed [5].

In the extraction of simple singly charged anions, there is a violation of the exchange-extraction series established for a large set of aliphatic amines. This conclusion follows from the data on the distribution constants of 2-aminopyridine [8] and 4-OAP salts between chloroform and water (**Table 2**), according to which, according to the extractability of 2(4)-OAP, singly charged anions are arranged in a row:

$$F^- < Cl^- < NO_3^- < Br^- < ClO_4^- < SCN^- < I^-$$

Soft anions (according to Pearson) are best extracted: SCN^- and I^-, while for aliphatic amines such an anion is hard ClO^{4-}. In addition, it is well known that for aliphatic amines there is a linear correlation between the exchange constants of singly charged anions and the extraction constants of monobasic acids with the heat of hydration of the anion or the free energy of hydration. In the case of 2(4)-OAP, such a correlation is observed separately in the series $Br^- < SCN^- < I^-$ and $F^- < Cl^- < NO_3^- < ClO_4^-$, but not for the entire series as a whole (**Figures** 7 and 8).

The study of 2-OAP halides by PMR, IR and X-ray electron spectroscopy showed [8] that they all have a structure similar to chloride (**Figure 9**):

The specificity of the interionic interaction in 2(4)-OAP associates manifests itself in a decrease in the polarization of the n-electron cloud of the aromatic cation, depending on the nature of the anion, on the one hand, and the formation of a chelate cycle based on hydrogen bonds in the case of 2-OAP – with another. The data of IR spectroscopy indicate that the strength of the chelate ring in the case of 2-OAP decreases on passing to an anion with better extractability [8]. Consequently, the selectivity of the extraction of soft anions is due to the redistribution of the electron density in the aromatic cation, depending on the nature of the anion. Degree of

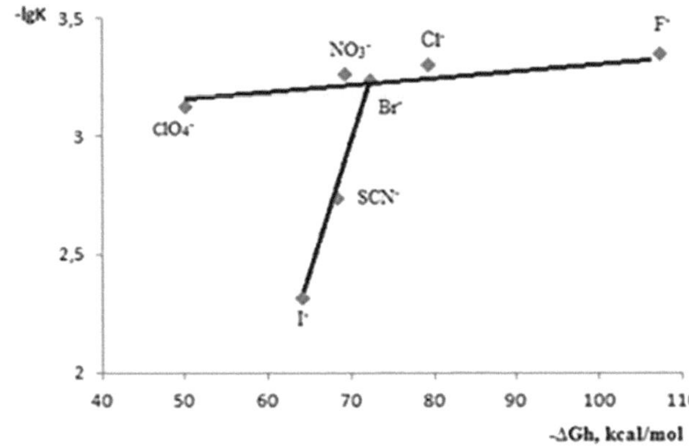

Figure 7.
Dependence of distribution constants of 2-aminopyridine salts on free energy anion hydration.

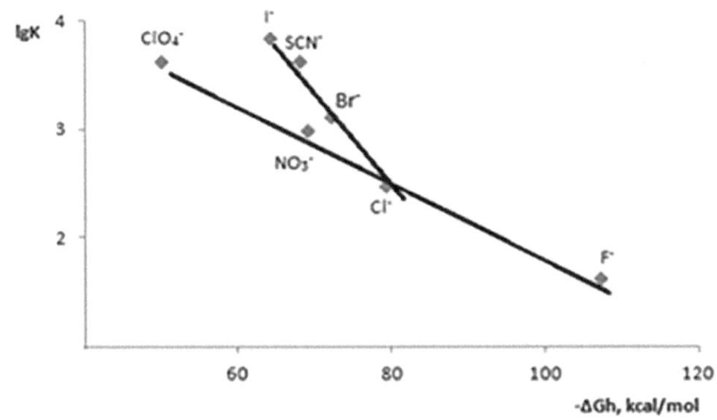

Figure 8.
Dependence of distribution constants of 4-OAP salts on free energy anion hydration.

indignation π-electron cloud of an aromatic cation can be quantified by the degree of charge delocalization (α) in the cation according to the data of X-ray electron or NMR spectroscopy [8]. For 2-OAP it is 90, 56, 54 and 48% in the series I^-, Br^-, Cl^-, $[GaCl_4]^-$, and for 4-OAP it is 90, 79, 65% in the series I^-, Br^-, Cl^-, respectively.

The distribution constants of 2-aminopyridine halides increase with increasing α.

Other processes involving aromatic cations show similar phenomena. In particular, on the surface of micelles RPy^+X^- (RPy^+ – long chain alkyl pyridinium ion; X^- – anions of different nature) in an aqueous solution, an interaction with charge transfer was found for soft anions, which increases in the series $Br^- < SO_3^{2-} < N_3^- < I^- < S_2O_3^{2-}$ according to an increase in the softness of the anion. Similarly, the interaction in the series $Cl^- < Br^- < I^-$ is observed for ion pairs in chloroform and is absent in the case of hard ClO_4^-. Charge transfer in ion pairs and on the surface of micelles is absent in the case of hard tetraalkyl- and tetraphenylammonium cations with soft Br^- and I^- [31]. The charge transfer is due to the mixing of the wave functions of nearby excited ones with the wave function of the ground state. Since in an ion pair the ground state is charged, and the excited – neutral, then it should be recognized that in the associates of a soft cation, for example, an $OAPH^+$ or RPy^+ cation with a soft anion, there is a covalent contribution (delocalization energy in terms of MO). This contribution is absent in associates with a hard cation, for example, the cation of an

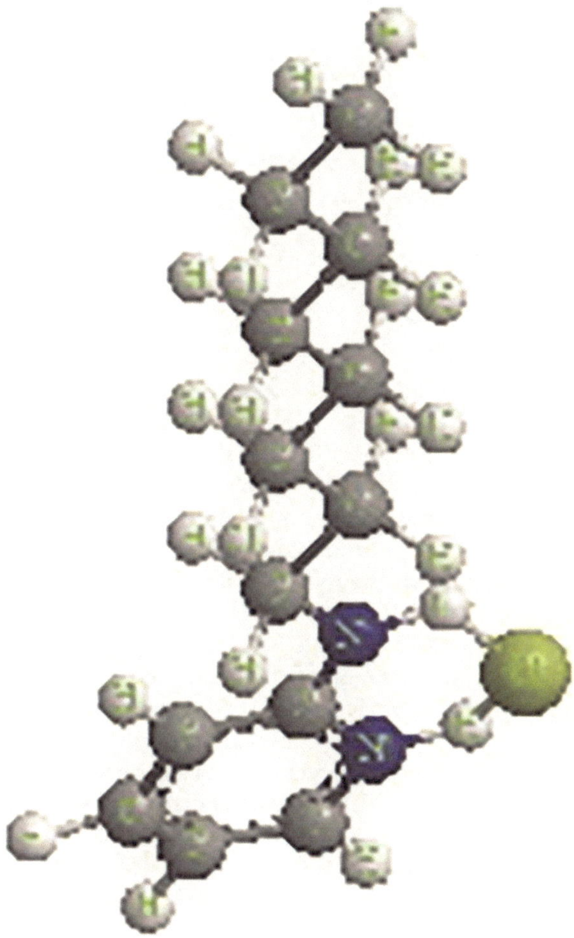

Figure 9.
3D structure of 2-OAP chloride.

aliphatic amine. This is also confirmed by the results of the study of 2(4)-OAP associates.

5. Structure of coordination compounds

The extraction of platinum metals by 2(4)-OAP in the form of coordination-solvated compounds is highly selective with respect to non-ferrous metals, in particular with respect to nickel. Therefore, it is of interest to study the structure of coordination compounds of the isovalent and isoelectronic series of metals with the composition $MeCl_2(OAP)_2$, where Me = Ni, Pd, Pt, i.e. complexes that pass into the organic phase during the coordination extraction of Pd and Pt with a solution of 2-OAP in chloroform.

The complexes were synthesized according to specially developed procedures [9]. Their composition was confirmed by the results of elemental analysis and the properties of the complexes. The formal oxidation state of the central atom of the complexes

Compound	Ni $2p_{3/2}$, Rd $3d_{5/2}$, Rt $4f_{7/2}$	Cl $2p_{3/2}$	N 1s
2-OAP			399.2
2-OAP·HCl	—	199.7	399.9; 401.0
$NiCl_2(OAP)_2$	856.0	198.4	400
$PdCl_2(OAP)_2$	338.4	198.4	399.7
$PtCl_2(OAP)_2$	73.2	198.5	399.6

Table 3.
Binding energy (eV ± 0.1) of internal electrons of metal and ligands in Ni, Pd, Pt complexes with 2-OAP.

is +2, which follows from X-ray electron spectroscopy data from the ionization energies of the Ni $2p_{3/2}$, Pd $3d_{5/2}$ and Pt $4f_{7/2}$ levels (**Table 3**).

The chlorine ion is a part of the coordination sphere of the central atom, which is due to the relatively high energy of the $2p_{3/2}$ level C1 in comparison with the corresponding energy values for ionically bound chlorine in the 2-OAP·HC1.

Formally, 2-OAP and 2-AP are ambidentate ligands with two donor nitrogen atoms, the heterocyclic nitrogen of the pyridine ring and the nitrogen of the amine group located in the α-position. The results of electron, IR, and NMR spectroscopy testify to the mode of coordination of 2-OAP by the metal [9]. 2-OAP and 2-AP are coordinated by Pd and Pt at the nitrogen of the heterocycle; The α-amino group does not interact directly with the metal. However, the spatial arrangement of the amino group, as well as the fact that coordinated chlorine has an excess negative charge, contribute to the formation of a chelate cycle due to the intramolecular H-bond, as in the case of associates (**Figures 10** and **11A**). The chelate cycle is absent in Pd and Pt complexes with 4-AP (**Figure 11B**).

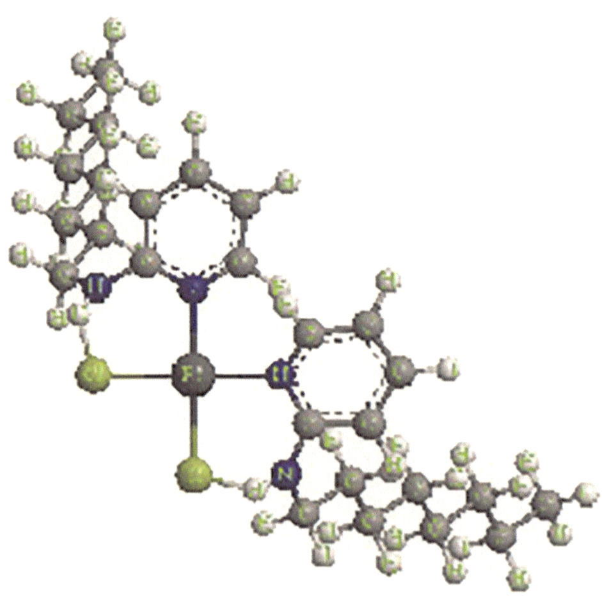

Figure 10.
3D structure of the Pd complex with 2-OAP.

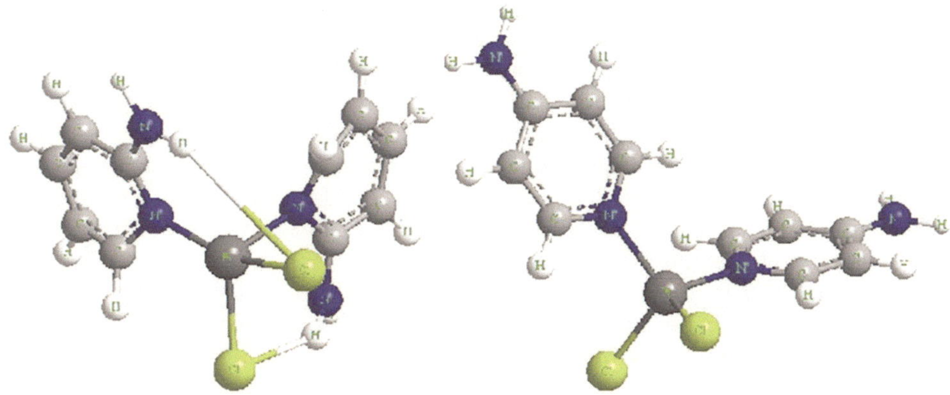

Figure 11.
3D structures of Pt(II) complexes: A – 2-AP, B – 4-AP.

The extraction of coordination-solvated complexes can be considered from the point of view of the formation of electron-donor-acceptor complexes by neutral halide complexes with the electron-donor OAP molecule. The results of X-ray electron spectroscopy indeed show that the binding energy of the N1s electrons of the nitrogen atom decreases upon passing from free 2-OAP to the complex for all the studied metals; 2-OAP is primarily an electron donor, and the energies of the N1s electrons of the aromatic and aliphatic nitrogen atoms are equalized during complexation. Based on the change in the energy of the N 1 s level of OAP during complex formation, the acceptor ability of Ni is significantly higher than the acceptor ability of Pd and Pt in the corresponding halides.

It is interesting to compare the electron ionization energy from the $2p_{3/2}$ level of chlorine in the compounds $MeCl_2(OAP)_2$ and $MeCl_2(NH_3)_2$, where Me = Pd, Pt [32], in compounds in which the central atom in one case forms a bond with a heterocyclic nitrogen 2-OAP, and in another – with ammonia nitrogen (the most rigid aliphatic amine). In complexes with 2-OAP, these values are much smaller; the electron density initially localized on the donor nitrogen atom is not only and not so much directly redistributed to the accepting complexing metal, but also further along the N—Me—C1 chain, which leads to an even greater covalence of the molecule as a whole. It is noteworthy that, in this respect, the complexes of palladium with OAP are similar to the complexes with triphenylphosphine and diphenylthiourea [33] – other soft ligands.

6. Conclusion

The specific behavior of aromatic amines is considered.in coordination and extraction processes for the isolation and separation of platinum and other metals on the example of 2(4)-aminopyridines (2(4)-AP). Toas intrasphere ligands2(4)-APhave a high electron-donating capacity due to the pumping of an easily polarizable π-electron density. In a protonated amine, electron density mobility is accompanied by delocalization of the positive proton charge over the ligand molecule, depending on the requirements of the acceptor. The degree of delocalization is the higher, the greater the polarizability of the anion. Chemistry of extraction of platinum metals2

(4)-AP, iridium in particular, can be quite complex. Depending on the nature of the metal and the extraction conditions, associates containing 2(4)-AP only in the cationic part, and the metal in the anionic part, associates with 2(4)-octylaminopyridine in the inner and outer coordination spheres of the metal, coordination neutral - solvated compounds; the formation of cationic complexes is also not excluded.

In the extraction of simple singly charged anions, the exchange-extraction series established for a large set of aliphatic amines is violated. Mild anions (according to Pearson), SCN- and I-, for example, are extracted best. For aliphatic amines, this anion is hard ClO4-. In coordination compounds of platinum metals, 2(4)-APact as an electron donor, coordinate on heterocyclic nitrogenwith the redistribution of the electron density not only to the accepting metal-complexing agent, but also further along the chain N—Me—X (X-acid ligand in the complex), which leads to an even greater covalence of the complex.

Author details

Liliya Sergeevna Ageeva*, Nikolai Alekseevich Borsch and
Nikolay Vladimirovich Kuvardin
Southwest State University, Kursk, Russian Federation

*Address all correspondence to: millfi@yandex.ru

References

[1] Borshch NA, Petrukhin OM. 2-Octyla minopyridine - a new extraction reagent. J. Analyt. Chem. 1978;**33**(9):1805-1812

[2] Borsch NA. Development of methods for the extraction isolation of iridium a nd other platinum metals. Moscow: Institute of Geochemistry and Analytical Chemistry, Academy of Sciences of the USSR; 1978

[3] Seregina NF, Petrukhin OM, Formanovsky AA, Zolotov YA. 4-Octyla minopyridine - extraction reagent for iridium. Reports of the Academy of Scie nces of the USSR. 1984;**275**(2):385-387

[4] Petrukhin OM, Borshch NA, Zolotov YA. International Solvent Extraction Conference. Vol. 3. Liege: ISEC; 1980. p. 200

[5] Borshch NA, Ageeva LS, Frolova AY. Effect of 4(2)-octylaminopyridine d istribution in a water (HCl)/chloroform system on the extraction of metal ions f rom chloride solutions. Russian Journal of Physical Chemistry A. 2019;**93**(5): 828-834. DOI: 10.1134/S00360244190 50066

[6] Borshch NA, Petrukhin OM. Extrac tive preconcentration of iridium and rh odium with 2-octylaminopyridine. Journal of Analytical Chemistry. 1978; **33**(11):2181-2190

[7] Borshch NA, Petrukhin OM, Sokolov AB, Marov II. Study of the chemistry of iridium extraction with 2-octylaminopyridine. J. Inorganic Chem. 1981;**26**(3):734-743

[8] Borsch NA, Maltseva NG. Specificity of interionic interaction in associates of 2-octylaminopyridine and some aspects of selective extraction of inorganic anions with aromatic amines. Journal of Inorganic Chemistry. 1982;**27**(9): 2355-2363

[9] Borshch NA, Petrukhin OM, Zolotov Yu A, Zhumadilov EG, Nefedov VI, Sokolov AB, et al. Coordination compounds of platinum (II), palladium (II) and nickel (II) with 2-octylaminopyridine. Journal of Coordination Chemistry. 1981;**7**(8): 1242-1249

[10] Pauling L. The Nature of the Chemical Bond. 3rd ed. New York; 1960. p. 97

[11] Dewar M, Dougherty R. Perturbation theory of molecular orbitals in organic chemistry. The Journal of Chemical Physics. 1977;**106**:695-711

[12] Nordberg R, Albridge RG, Bergmark T, Ericson U, Hedman J, Nordling G, et al. Ark. Kemi. Vol. 28. Almquist & Wiksells Boktryckeri; 1967. p. 257

[13] Klopman GJ. Chemical reactivity and the concept of charge- and frontier-controlled reactions. Journal of the American Chemical Society. 1968;**90**:223

[14] Klopman GN. Reactivity and Ways of Reactions. Moscow: Mir; 1977. p. 384

[15] Khogare BT, Anuse MA, Piste PB, Kokare BN. Development of a solvent extraction system with 4-heptylaminopyridine for the selective separation of palladium (II) from synthetic mixtures, catalysts and water samples. Desalination and Water Treatment. 2015;**57**:1-11. DOI: 10.1080/ 19443994.2015.1124054

[16] Umrao B. Shep, Rucha K. Pawar, Balasaheb R. Arbad, Solvent extraction

of palladium (II) from oxalate medium by 2-dodecylaminopyridine. Chemistry & Biology Interface. 2016;**6**(6):379-387

[17] Shep UB, Bagal MR, Arbad BR. 2-Dodecylaminopyridine assisted solvent extraction system for selective separation of rhodium (III) ion-pair complex from synthetic mixtures. J. Materials and Environmental Sciences. 2017;**8**(8):2894-2902

[18] Suryavanshi VJ, Patil MM, Zanje SB, Kokare AN, Gaikwad AP, Anuse MA, et al. Development of liquid-liquid extraction and separation method for ruthenium (III) with 2-Octylaminopyridine from succinate media: Analysis of catalysts. Russian Journal of Inorganic Chemistry. 2017;**62**(2):257-268. DOI: 10.1134/S003602361702019X

[19] Suryavanshi VJ, Patil MM, Zanje SB, Kokare AN, Kore GD, Mansing A, et al. Extraction of iridium(III) by ion-pair formation with 2-octylaminopyridine in weak organic acid media. Separation Science and Technology. 2016;**51**(10):1690-1699. DOI: 10.1080/01496395.2016.1177076

[20] Suryavanshi V, Kokare A, Zanje S, Mulik A, Pawar R, Patil M, et al. Ion-pair based liquid–liquid extraction of gold (III) from malonate media using 2-octylaminopyridine as an extractant: Analysis of alloys, minerals, and drug samples. Turkish Journal of Chemistry. 2018;**42**:1032-1044. DOI: 10.3906/kim-1712-34

[21] Kore GD, Patil SA, Anuse MA, Kolekar SS. An extractive studies on behavior of thorium (IV) from malonate media by 2-octylaminopyridine: A green approach. J. Radioanalyt. and Nucl. Chem. 2016;**310**(1):329. DOI: 10.1007/s10967-016-4857-7

[22] Noronha LE, Kamble GS, Kolekar SS, Anuse MA. Solvent extraction of zirconium (IV) with 2-octilaminopyridine from succinate media – analysis of real samples. Ind. J. Chem. Technology. 2013;**20**(7):252-258

[23] Noronha LE, Kamble GS, Kolekar SS, Anuse MA. Extractive separation of vanadium.(V) from succinate medium by solvent extraction using 2-n-octylaminopyridine. Int. J. of Analyt. And Bioanalyt. Chemistry. 2013;**3**(7):27-35

[24] Noronha LE, Kamble GS, Kolekar SS, Anuse MA. Recovery of molybdenum (VI) from hydrochloric acid medium by solvent extraction with 2-n-octylaminopyridine. Inter. J. Chem. Science Technology. 2013;**3**(1):15-24

[25] Mane CP, Mahamuni SV, Kolekar SS, Han SH, Anuse MA. Hexavalent chromium recovery by liquid–liquid extraction with 2-octylaminopyridine from acidic chloride media and its sequential separation from other heavy toxic metal ions. Arabian Journal of Chemistry. 2016;**9**(2):1420-1427. DOI: org/10.1016/j.arabjc.2012.03.021

[26] Mane CP, Anuse MA. Studies on liquid-liquid extraction and recovery of bismuth (III) from succinate media using 2-Octylaminopyridine in chloroform. J. Chinese Chem. Soc. 2008;**55**:807-817. DOI: 10.1002/jccs.200800121

[27] Sandip VM, Prakash PW, Anuse MA. Liquid–liquid extraction and recovery of gallium (III) from acid media with 2-octylaminopyridine in chloroform: Analysis of bauxite ore. Journal of the Serbian Chemical Society. 2010;**75**(8):1099-1113. DOI: 10.2298/JSC090630072M

[28] Sandip VM, Prakash PW, Anuse MA. Rapid liquid-liquid extraction of thallium (III) from succinatte media with 2-octylaminopyridine in chloroform as the extractant. Journal of the Serbian Chemical Society. 2008; **73**(4):435-451. DOI: 102298/ JSC0804435M

[29] Mandhare AM, Han SH, Anuse MA, Kolekar SS. Arab liquid–liquid anion exchange extraction studies of samarium (III) from salicylate media using high molecular weight amine. Journal of Chemistry. 2015;**8**(4):456-464. DOI: 10.1016/j.arabjc.2011.01.026

[30] Mane CP, Mahamuni SV, Gaikwad AP, Shejwal RV, Kolekar SS, Anuse AM. Journal of Saudi Chemical Society. 2015;**19**(1):46-51. DOI: 10.1016/ j.jscs.2011.12.016

[31] Rau A, Mukerjec RJ. Physical Chemistry. 1966;**70**:2144

[32] Nefedov VI. Structure of Molecules and Chemical Bond. Vol. 1. Moscow: VINITI; 1973. p. 148

[33] Petrukhin OM, Nefedov VI, Salyn OV, Shevchenko VNJ. Inorgan. Chem. 1974;**19**:1418-1420